INTRODUCTION TO MICROPROCESSORS

AF888600

INTRODUCTION
TO
MICROPROCESSORS

By

Abhishek Yadav
B.E. (E & C), M.E. (E & C)
Assistant Professor
IIMT Engineering College,
Meerut (U.P.)

Poonam Yadav
B. Tech and MBA

UNIVERSITY SCIENCE PRESS
(An Imprint of Laxmi Publications Pvt. Ltd.)
An ISO 9001:2008 Company
BENGALURU • CHENNAI • COCHIN • GUWAHATI • HYDERABAD
JALANDHAR • KOLKATA • LUCKNOW • MUMBAI • RANCHI • NEW DELHI
BOSTON (USA) • ACCRA (GHANA) • NAIROBI (KENYA)

INTRODUCTION TO MICROPROCESSOR

© by Laxmi Publications (P) Ltd.
All rights reserved including those of translation into other languages. In accordance with the Copyright (Amendment) Act, 2012, no part of this publication may be reproduced, stored in a retrieval system, or transmitted in any form or by any means, electronic, mechanical, photocopying, recording or otherwise. Any such act or scanning, uploading, and or electronic sharing of any part of this book without the permission of the publisher constitutes unlawful piracy and theft of the copyright holder's intellectual property. If you would like to use material from the book (other than for review purposes), prior written permission must be obtained from the publishers.

Printed and bound in India
Typeset at Shubham Composer, Delhi
First Edition : 2016
ISBN 978-93-80856-55-1

Limits of Liability/Disclaimer of Warranty: The publisher and the author make no representation or warranties with respect to the accuracy or completeness of the contents of this work and specifically disclaim all warranties. The advice, strategies, and activities contained herein may not be suitable for every situation. In performing activities adult supervision must be sought. Likewise, common sense and care are essential to the conduct of any and all activities, whether described in this book or otherwise. Neither the publisher nor the author shall be liable or assumes any responsibility for any injuries or damages arising here from. The fact that an organization or Website if referred to in this work as a citation and/or a potential source of further information does not mean that the author or the publisher endorses the information the organization or Website may provide or recommendations it may make. Further, readers must be aware that the Internet Websites listed in this work may have changed or disappeared between when this work was written and when it is read.

All trademarks, logos or any other mark such as Vibgyor, USP, Amanda, Golden Bells, Firewall Media, Mercury, Trinity, Laxmi appearing in this work are trademarks and intellectual property owned by or licensed to Laxmi Publications, its subsidiaries or affiliates. Notwithstanding this disclaimer, all other names and marks mentioned in this work are the trade names, trademarks or service marks of their respective owners.

Branches

✆	Bengaluru	080-26 75 69 30	
✆	Chennai	044-24 34 47 26,	24 35 95 07
✆	Cochin	0484-237 70 04,	405 13 03
✆	Guwahati	0361-254 36 69,	251 38 81
✆	Hyderabad	040-27 55 53 83,	27 55 53 93
✆	Jalandhar	0181-222 12 72	
✆	Kolkata	033-22 27 43 84	
✆	Lucknow	0522-220 99 16	
✆	Mumbai	022-24 91 54 15,	24 92 78 69
✆	Ranchi	0651-220 44 64	

PUBLISHED IN INDIA BY

UNIVERSITY SCIENCE PRESS
(An Imprint of Laxmi Publications Pvt. Ltd.)

An ISO 9001:2008 Company
113, GOLDEN HOUSE, DARYAGANJ,
NEW DELHI - 110002, INDIA
Telephone : 91-11-4353 2500, 4353 2501
Fax : 91-11-2325 2572, 4353 2528
www.laxmipublications.com info@laxmipublications.com

C—10109/015/04
Printed at:
Printed at Repro Knowledgecast Limited, Thane

CONTENTS

PREFACE

This book is intended to serve as a textbook for the subject of "Introduction to Microprocessor" for BE/B. Tech degree students of computer science, Electrical and Electronics and communication department.

Despite of publication of a number of textbooks on this subject, students are always very uncomfortable in this subject. The content of syllabus seems to be very wide. Purpose is very clear to give all the fundamentals in very easy language at a single place.

We are not saying that we are expoloring any new area in this field but it is our promise to the readers of this book that each chapter of this book starts from very basic principle and proceed systematically. The book is very useful for those students who are unable to attend regular classes (AMIETE, IETE etc.,)

We have tried our best to cover all the topics in detail and in simple way. Definitely like a general human beings, we have limitations. We will be highly grateful to all faculty members for their valuable suggestions and cooperation.

—Authors

ACKNOWLEDGEMENTS

We present our humble gratitude to Shri (Dr.) S.S. Pattnaik, professor and head of ETV Department, NITTTR Chandigarh without whom inspiration we can not image our today. We dedicate this small work to Mrs. Swapna Devi, Assistant Professor, ECE Department Chandigarh who not only always direct us as a teacher but also take care of us as a mother.

We are really thankful to your dear friend Mr. Manoj Kumar (Lecturer, ECE Department, G.B. Pant Engineering College, Pauri) and Mr. Subhash Chandra Arya (Lecturer, GIIT Greater Noida). They always encourage our efforts and rectify the mistakes from manuscript.

We are thankful to all faculty members who always supported me to write this book. We are thankful to Mr. Atul Sharma (Assistant Professor, Saraswati Institute of Technology, Ghaziabad), Mr. Amit Gupta (Assistant Professor, JP Institute of Technology, Meerut), Mr. Anurag Vijay Aggarwal (Sr. Lecturer , Vira College of Engineering, Bijnor), Mr. Manoj Bansal (Sr. Lecturer , Vira College of Engineering, Bijnor), Mr. Deepak Garg (Lecturer, IIMT Engineering College, Meerut), Miss Swati Singh (Lecturer, VIT, Meerut), Mr. Shailendra Pratap Singh (Lecturer, VIT, Meerut), Miss Khushboo Jain (Sr. Lecturer, VIT, Meerut), Mr. Sachin (Lecturer, VIT, Meerut) for their kind support.

A number of faculty members have given valuable suggestions for this edition. We are unable to give the names of individuals but we are grateful to them. We tried to follow all of their suggestions. We will be always in need of suggestions to improve ourselves.

At last but not least our great love to Krishna and Dimple who always filled our life with cheers.

With great regards to my parents and almighty GOD.

—Authors

1

INTRODUCTION

1.1 REVIEW OF COMPUTER FUNDAMENTALS

A **computer** is a programmable machine (or more precisely, a programmable sequential state machine). There are two basic kinds of computers: analog and digital.

Analog computers are analog devices. That is, they have continuous states rather than discrete numbered states. An analog computer can represent fractional or irrational values exactly, with no round off. Analog computers are almost never used outside of experimental settings.

A **digital computer** is a programmable-clocked sequential state machine. A digital computer uses discrete states. A binary digital computer uses two discrete states, such as positive/negative, high/low, on/off, used to represent the binary digits zero and one.

Computers are used for a wide variety of purposes.

Data processing is commercial and financial work. This includes such things as billing, shipping and receiving, inventory control, and similar business related functions, as well as the "electronic office".

Scientific processing is using a computer to support science. This can be as simple as gathering and analyzing raw data and as complex as modeling natural phenomenon (weather and climate models, thermodynamics, nuclear engineering, etc.).

Multimedia includes **content creation** (composing music, performing music, recording music, editing film and video, special effects, animation, illustration, laying out print materials, etc.) and multimedia playback (games, DVDs, instructional materials, etc.).

1.1.1 General Architecture and Brief Description of Elements

The classic crude over implication of a computer is that it contains three elements:

(*a*) Central processing unit (CPU),

(*b*) Memory, and

(*c*) I/O (input/output) devices.

The borders between those three terms are highly ambiguous, non-contiguous, and erratically shifting.

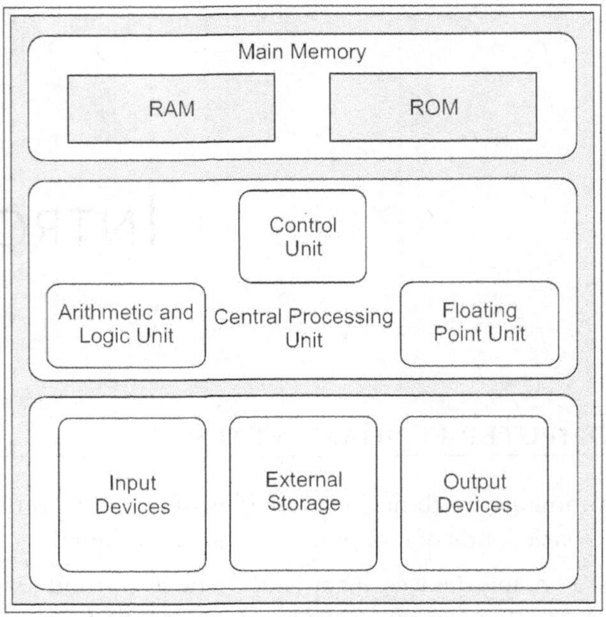

Fig. 1.1 Block Diagram of a Computer

(a) Central Processing Unit (CPU)

The **processor** is the part of the computer that actually does the computations. This is sometimes called an **MPU** (for main processor unit) or **CPU** (for central processing unit or central processor unit).

A processor typically contains an arithmetic/logic unit (**ALU**), control unit (including processor flags, flag register, or status register), internal buses, and sometimes-special function units (the most common special function unit being a floating point unit for floating point arithmetic).

Some computers have more than one processor. This is called **multi-processing**.

DSP stands for Digital Signal Processing. **DSP** is used primarily in dedicated devices, such as **MODEMs,** digital cameras, graphics cards, and other specialty devices.

Hybrid processors combine elements of two or three of the major classes of processors.

CPU also contains Arithmetic and logic unit (**ALU**), Floating-point unit (**FPU**) and Control unit. The brief description has been given as below:

Arithmetic and Logic Unit

An arithmetic/logic unit (**ALU**) performs integer arithmetic and logic operations. It also performs shift and rotate operations and other specialized operations.

Usually floating point arithmetic is performed by a dedicated floating-point unit (**FPU**), which may be implemented as a co-processor.

Control Unit

Control units are in charge of the computer. Control units fetch and decode machine instructions. Control units may also control some external devices.

(b) Main Storage (Memory)

The function of a memory is to store information. Main storage is also called **memory** or internal memory (to distinguish from external memory, such as hard drives).The ability to remember data after the power is switched off is the dividing line between the two main types of memory. If it loses its data when the power is switched off, then we call the memory **RAM** or volatile memory. If it can hold on to the data without power, we call it **ROM** or non-volatile memory (volatile means 'able to evaporate'). This is seen in Fig. 1.2.

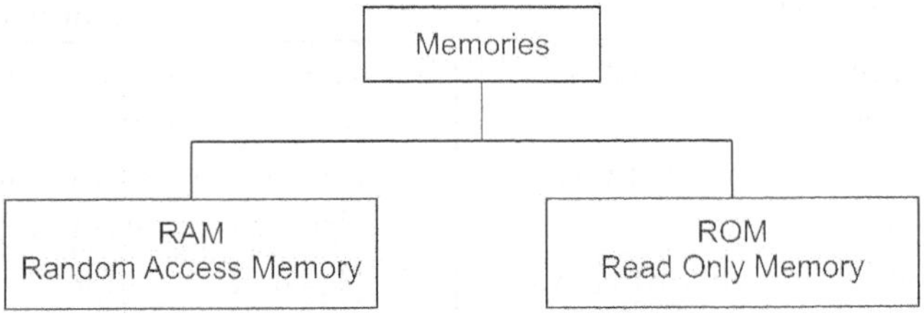

Fig. 1.2 The Two Classes of Memory

RAM

RAM is Random Access Memory, and is the basic kind of internal memory. RAM is called "random access" because the processor or computer can access any location in memory (as contrasted with sequential access devices, which must be accessed in order).

RAM has been made from reed relays, transistors, integrated circuits, magnetic core, or anything that can hold and store binary values (one/zero, plus/minus, open/close, positive/negative, high/low, etc.). Most modern RAM is made from integrated circuits. At one time the most common kind of memory in mainframes was magnetic core, so many older programmers will refer to main memory as **core memory** even when the RAM is made from more modern technology.

RAM chips can be designed in two different forms which we call static RAM (SRAM) and dynamic RAM (DRAM), as seen in Fig. 1.3.

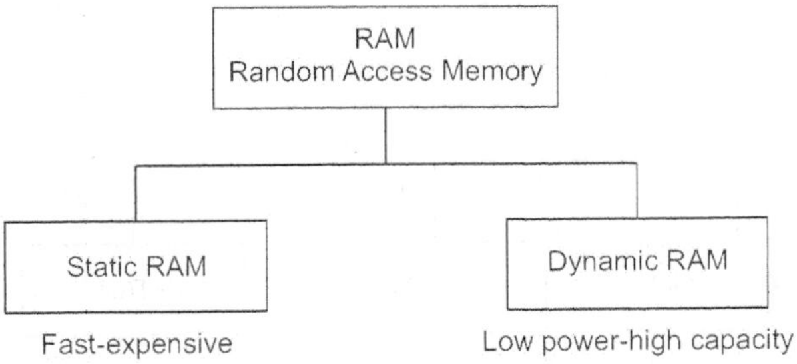

Fig. 1.3 Two Types of RAM

Static RAM is called static because it will continue to hold and store information even when power is removed. These are constructed of flip-flops. The problem with the flip-flop is that it draws current all the time. Therefore, it tends to get rather warm and, on a single chip, the components cannot be packed together very tightly. The benefit is that they are very fast and are used where speed of access is important. Static RAM is often called SRAM.

Dynamic RAM is called dynamic because it loses all data when power is removed. Transistors and integrated circuits are examples of dynamic memory. It is possible to have battery back up for devices that are normally dynamic to turn them into static memory.

Differences between static and dynamic RAMs are as follows:

Table 1.1 Comparison of SRAM and DRAM

Static RAM	Dynamic RAM
1. This semiconductor memory is constructed using bipolar transistors.	1. This semiconductor memory is constructed using MOS transistors.
2. Information is stored in the form of voltage levels in flip-flops.	2. Information is stored in the form of electrical charges in capacitors.
3. These voltage levels do not get drifted away	3. Has tendency of leakage.
4. No refresh logic is needed.	4. Refresh logic is necessary since leakage of electrical charges.
5. Power is required even when the chip is in standby mode.	5. Refresh login is inbuilt, so draws less power comparatively.
6. Four time larger in size compared to an equivalent dynamic cell.	6. Four times as many bits as a static RAM chip.

ROM

ROM is Read Only Memory (it is also random access, but only for reads). ROM is typically used to store things that will never change for the life of the computer, such as low level portions of an operating system. Some processors (or variations within processor families) might have **RAM** and/or **ROM** built into the same chip as the processor (normally used for processors used in standalone devices, such as arcade video games, ATMs, microwave ovens, car ignition systems, etc.).

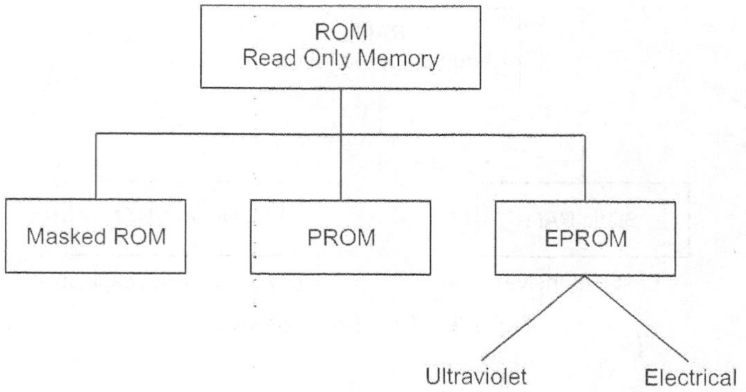

Fig. 1.4 Three Types of ROM

Masked ROM: A masked ROM is manufactured to our specification and cannot be changed. We must be very sure that the information is correct before it is made otherwise it all goes in the waste bin and the person responsible is looking for a new job. The initial cost is necessarily high due to the expense of the tooling required. It is only worthwhile if at least a few thousand identical chips are required (see Fig. 1.5).

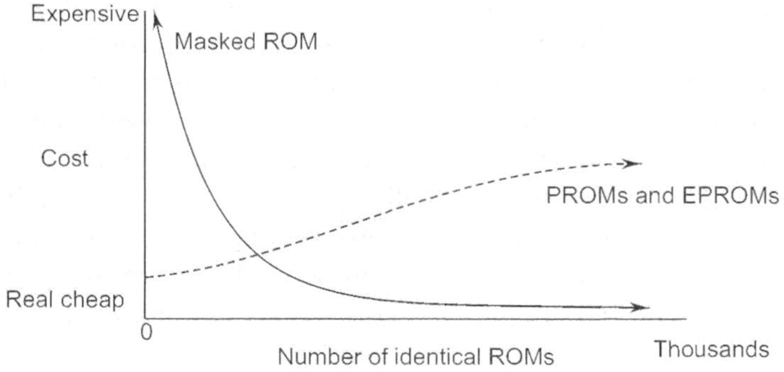

Fig. 1.5 The Economics of ROM Choice

Programmable ROM (PROM): This chip is supplied with all the data held at zero by means of small internal fuses. When one of the fuses is blown, the associated bit changes from 0 to 1. To blow the fuses, a piece of programming equipment is needed. This equipment can be purchased quite cheaply if only one **PROM** is to be programmed at a time. If a larger throughput is needed then this will inevitably increase the cost of the equipment.

Once the fuse is blown, it cannot be repaired so if you make a mistake, the chip is wasted. The ROM is useful for low volume production because the initial costs are much lower than the masked ROM but you do have to program them yourself.

Erasable Programmable ROM (EPROM): As the name would suggest, this chip allows us to program it, then change our mind and try again. To erase the data there are two methods—ultraviolet light or electrical voltage pulses. EPROMs are ideal for prototyping since it is so easy to change the data to make modifications.

The UVEPROM: The chip is bombarded with ultraviolet light via a transparent window on the chip. A specially constructed **EPROM** eraser provides the light. We pop the chip in, close the lid and switch on the timer. After a few minutes, the data is erased. When erased, all the data output is set to 1. We then put the chip into an **EPROM** programmer, usually the same piece of gear that was used to program the **PROM**. We can feed in the new data and within a couple of minutes, we have finished the process.

They can be erased and reprogrammed about 700 times before they become increasingly reluctant to erase and their life is over. Once programmed, the data is safe for about seven years. For long term storage, it is best to reload them or, better still, use a masked **ROM** if available.

Electrically Erasable Programmable ROM (EEPROM): This chip uses electrical voltage pulses as inputs to clear the previous data and is then reprogrammed in the same way as the UVEPROM. It has the added advantage that individual parts of the data can be reprogrammed without deleting everything first as is the case with the ultraviolet version. **EEPROM** can be found as serial access (**SAM**), as well as the more usual random access.

The reprogramming can be done while installed in the microprocessor- based system. It does not need a separate programmer. Their disadvantage is that they are slow to program and have a limited number of reprogramming cycles.

Registers and **flags** are a special kind of memory that exists inside a processor. Typically a processor will have several internal registers that are much faster than main memory. These registers usually have specialized capabilities for arithmetic, logic, and other operations. Registers are usually fairly small (8, 16, 32, or 64 bits for integer data, address, and control registers; 32, 64, 96, or 128 bits for floating point registers). Some processors separate integer data and address registers, while other processors have general-purpose registers that can be used for both data and address purposes. A processor will typically have one to 32 data or general purpose registers (processors with separate data and address registers typically split the register set in half). Many processors have special floating-point registers (and some processors have general purpose registers that can be used for either integer or floating point arithmetic). Flags are single bit memory used for testing, comparison, and conditional operations (especially conditional branching).

External Storage

External storage (also called **auxiliary storage**) is any storage other than main memory. In modern times, this is mostly hard drives and removable media (such as floppy disks, Zip disks, optical media, etc.). With the advent of USB and Fire Wire hard drives, the line between permanent hard drives and removable media is blurred. Other kinds of external storage include tape drives; drum drives, paper tape, and punched cards.

Random access or indexed access devices (such as hard drives, removable media, and drum drives) provide an extension of memory (although usually accessed through logical file systems). Sequential access devices (such as tape drives, paper tape punch/readers, or dumb terminals) provide for off-line storage of large amounts of information (or back ups of data) and are often called I/O devices (for input/output).

(c) Input/Output Devices

Most external devices are capable of both input and output (I/O). Some devices are inherently input-only (also called read-only) or inherently output-only (also called write-only). Regardless of whether a device is I/O, read-only, or write-only, external devices can be classified as block or character devices.

A **character** device is one that inputs or outputs data in a stream of characters, bytes, or bits. Character devices can further be classified as serial or parallel. Examples of character devices include printers, keyboards, and mice.

A **serial** device streams data as a series of bits, moving data one bit at a time. Examples of serial devices include printers and **MODEMs**.

A **parallel** device streams data in a small group of bits simultaneously. Usually the group is a single eight-bit byte (or possibly seven or nine bits, with the possibility of various control or parity bits included in the data stream). Each group usually corresponds to a single character of data. Rarely there will be a larger group of bits (word, long word, double word, etc.). The most common parallel device is a printer (although most modern printers have both a serial and a parallel connection, allowing greater connection flexibility).

A **block** device moves large blocks of data at once. This may be physically implemented as a serial or parallel stream of data, but the entire block gets transferred as single packet of data. Most block devices are random access (that is, information can be read or written from blocks anywhere on the device). Examples of random access block devices include hard disks, floppy disks, and drum drives. Examples of sequential access block devices include magnetic tape drives and high-speed paper tape readers.

Input Devices

Input devices are devices that bring information into a computer. Pure input devices include such things as punched card readers, paper tape readers, keyboards, and mice, drawing tablets, touch pads, trackballs, and game controllers. Devices that have an input component include magnetic tape drives touch screens, and dumb terminals.

Output Devices

Output devices are devices that bring information out of a computer. Pure output devices include such things as cardpunches, paper tape punches, **LED** displays (for light emitting diodes), monitors, printers, and pen plotters. Devices that have an output component include magnetic tape drives, combination paper tape reader/punches, teletypes, and dumb terminals.

1.2 CLASSIFICATION OF COMPUTERS

Computers are available in different shapes, sizes and weights; due to these different shapes and sizes they perform different sorts of jobs from one another.

They can also be classified in different ways.

1. **Supercomputers :** The biggest in size, the most expensive in price than any other is classified and known as supercomputer. It can process trillions of instructions in seconds. Governments specially use this type of computer for their different calculations and heavy jobs. Different industries also use this huge computer for designing their products. In most of the Hollywood's movies it is used for animation purposes. This kind of computer is also helpful for forecasting weather reports worldwide.
2. **Mainframe Computers :** Another giant in computers after the supercomputer is Mainframe, which can also process millions of instruction per second and capable of accessing billions of data. This computer is commonly used in big hospitals, airline reservations companies, and many other huge companies prefer mainframe because of its capability of retrieving data on a huge basis.
3. **Mini Computers :** This computer is next in the line and offers less than mainframe in work and performance. These are the computers, which are mostly preferred by the small type of business personals, colleges, etc.
4. **Personal Computers :** Almost all the computer users are familiar with the personal computers. They normally know what the personal computer is and what are its functions.

This is the computer mostly preferred by the home users. These computers are lesser in cost than the computers given above and also, small in size; they are also called **PCs** in short for Personal computers.

5. **Notebook Computers :** Having a small size and low weight the notebook is easy to carry anywhere. This is easy to carry around and preferred by students and business people to meet their assignments and other necessary tasks.

   It can store the same amount of data and having a memory of the same size as that of a personal computer. One can say that it is the replacement of personal desktop computer.

But technically computers are classified as given below:

1.2.1 The Von Neumann Model of Computer

The modern microcomputer has roots going back to USA in the 1940's. Of the many researchers, the Hungarian-born mathematician, John Von Neumann (1903–57), is worthy of special mention. He developed a very basic model for computers which we are still using today.

Von Neumann divided a computer's hardware into five primary groups:

1. CPU
2. Input
3. Output
4. Working storage
5. Permanent storage

This division provided the actual foundation for the modern PC, as Von Neumann was the first person to construct a computer which had working storage (what we today call **RAM**). And the amazing thing is, his model is still completely applicable today. If we apply the Von Neumann model to today's **PC**, it looks like this:

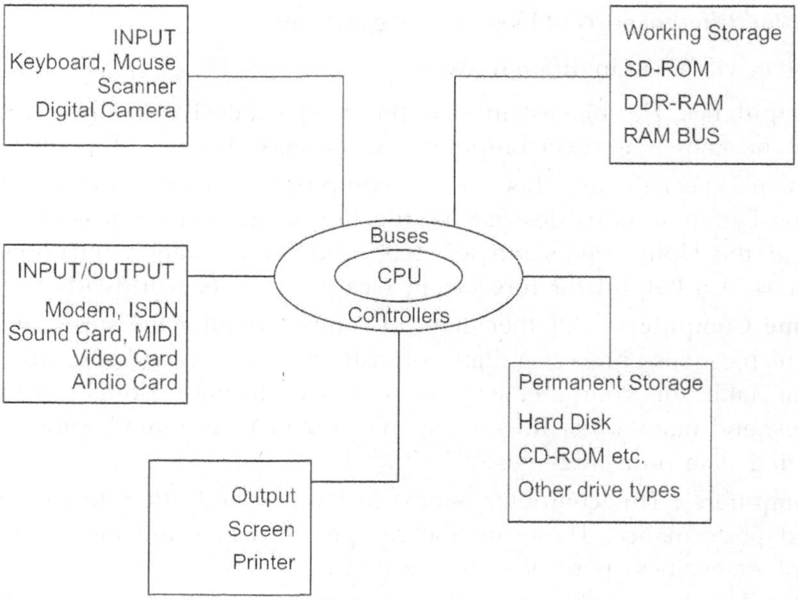

Fig. 1.6 The Von Neumann Model of Computer

1.2.2 Non-Von Neumann Models

Until recently, almost all general-purpose computers followed the Von Neumann design. However, the Von Neumann bottleneck continues to baffle engineers looking for ways to build fast systems that are inexpensive and compatible with the vast body of commercially available software. Engineers who are not constrained by the need to maintain compatibility with Von Neumann systems are free to use many different models of computing. A number of different subfields fall into the **non-Von Neumann category**, including neural networks (using ideas from models of the brain as a computing paradigm), genetic algorithms (exploiting ideas from biology and DNA evolution), quantum computation and parallel computers. Of these, parallel computing is currently the most popular.

The first parallel-processing systems were built in the late 1960s and had only two processors. The 1970s saw the introduction of supercomputers with as many as 32 processors, and the 1980s brought the first systems with over 1,000 processors. Finally, in 1999, **IBM** announced the construction of a supercomputer called the Blue Gene. This massively parallel computer contains over 1 million processors, each with its own dedicated memory. Its first task is to analyze the behavior of protein molecules.

Even parallel computing has its limits, however. As the number of processors increases, so does the overhead of managing how tasks are distributed to those processors. Some parallel-processing systems require extra processors just to manage the rest of the processors and the resources assigned to them.

1.2.3 Flynn's Classification of Computer

Flynn's classification is based on two concepts — Instruction Streams and Data Streams. Instruction Stream corresponds to a program counter. A system with *n* CPUs has *n* program counters, hence *n* instruction streams. Data Streams consists of a set of operands. A program computing the average temperature for one location has one data stream. A program computing the average temperature for each of 100 locations has 100 data streams. The biggest advantage of the Flynn model, compared to other models, is its simplicity. Disadvantage is **MISD**, **MIMD**, etc., are too big chunks, and even not used generally. In this classification, parallel processing is used. So it is better to understand the parallel processing.

Parallel processing works by distributing portions of a program's instructions to several different processors to be executed at the same time. In contrast, a uniprocessor executes the entire set of instructions itself and therefore takes longer to complete execution. Assuming the instructions can be divided evenly without causing any data hazards due to dependences or otherwise, n processors working in parallel (ideally) have the potential to speed up the program's execution time by a factor n. This ideal is difficult to approach, but a means of speeding up execution time by any amount is helpful nonetheless. There are two different parallel processing models that have been implemented to date, **SIMD** and **MIMD**, So now we discuss the different models of this classification.

1. **SISD-Single Instruction Stream Single Data Stream :** It includes all Von Neumann computers. Ex. IBM 370, DEC VAX, SUN, IBM PC, MacIntosh.

   SISD represents the organization of a single computer containing a control unit, a processor unit, and a memory unit. Instructions are executed sequentially and the system may or may not have internal parallel processing capabilities. Parallel processing in this case may be achieved by means of multiple functional units or by pipeline processing.

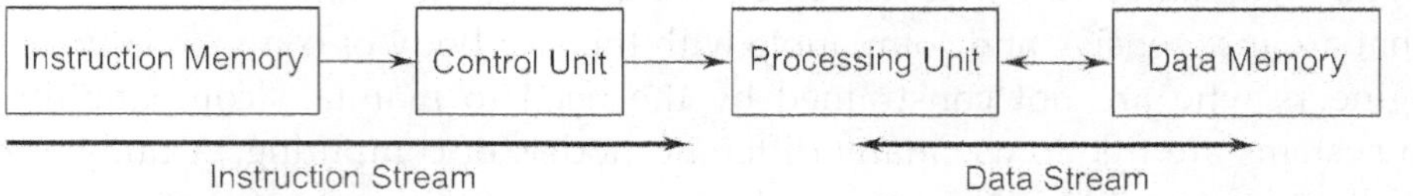

Fig. 1.7 Single Instruction Stream, Single Data Stream Model of Computer

2. **SIMD-Single Instruction Stream Multiple Data Stream :** Simple synchronization and processors can be smaller and more numerous than **MIMD** processors (they do not need to store their programs). Ex. CRAY-1, DAP CM-1, WARP, CM-2, ILLIAC IV.

   SIMD represents an organization that includes many processing units under the supervision of a common control unit. All processors receive the same instruction from the control unit but operate on different items of data. The shared memory unit must contain multiple modules, so that it can communicate with all the processors simultaneously.

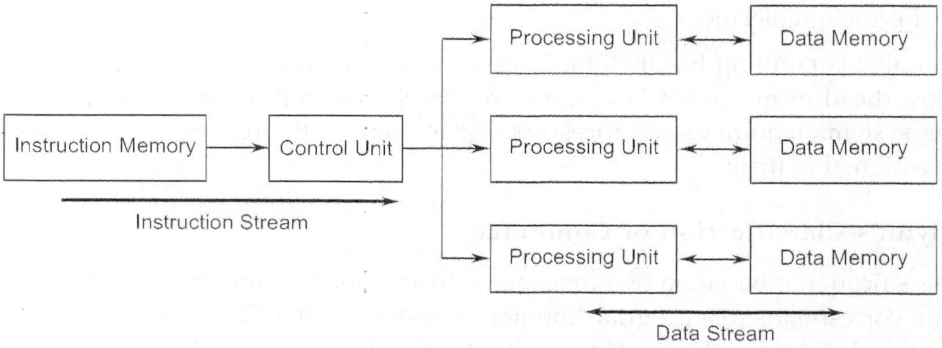

Fig. 1.8 Single Instruction Stream Multiple Data Stream Model of Computer

In **SIMD**, there may be two possible schemes of control unit along with instruction memory.

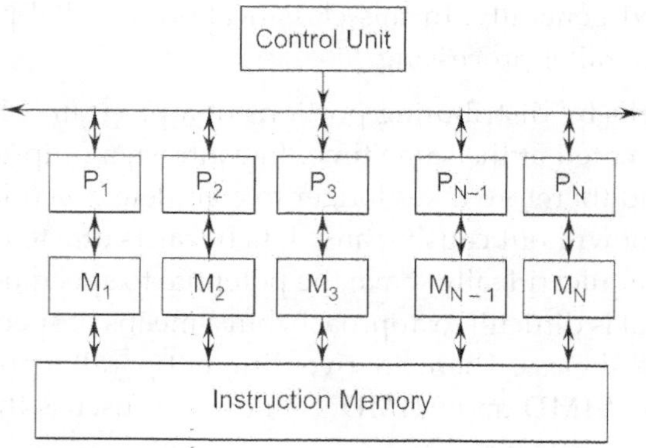

Fig. 1.9 Single Instruction Stream Multiple Data Stream Model of Computer with Separate Processor with Separate Memory

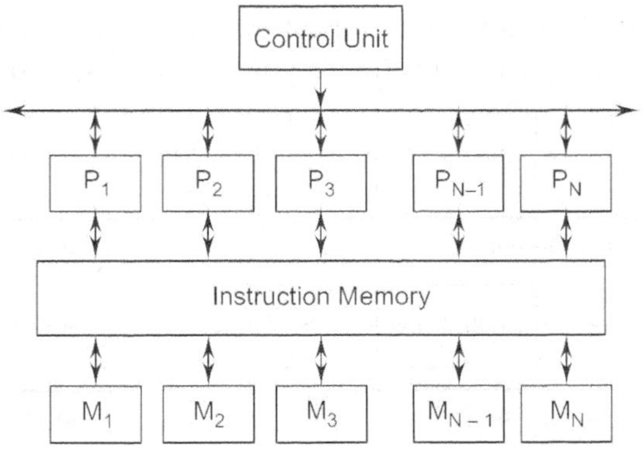

Fig. 1.10 Single Instruction Stream Multiple Data Stream Model of Computer with Separate Processor with Separate Memory but Accessible with a Common Instruction Memory

3. **MISD-Multiple Instruction Stream Single Data Stream :** Empty category with multiple instructions operating on same piece of data. It is not clear if any such machines exist, although some people regard pipelined machines as **MISD** and not **SIMD**.

   MISD structure is only of theoretical interest since no practical system has been constructed using this organization

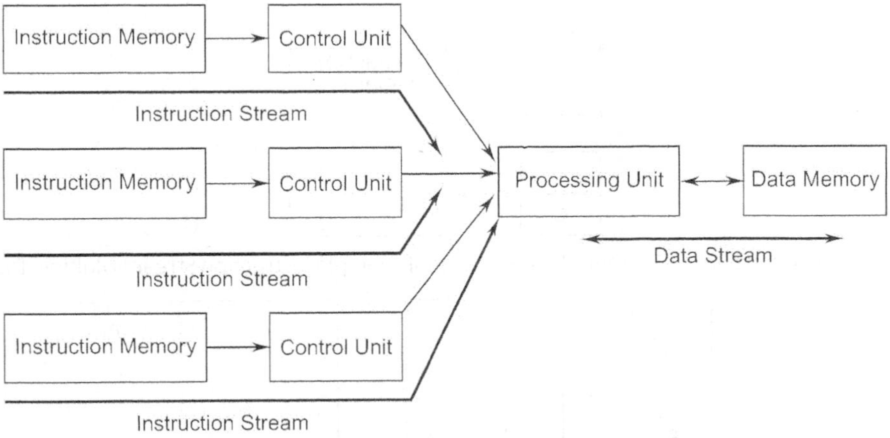

Fig. 1.11 Multiple Instruction Stream Single Data Stream

4. **MIMD-Multiple Instruction Stream Multiple Data Stream :** MIMD are multiple independent CPUs operating as part of a larger system. Most parallel processors fall into this category. Ex. Transputers, Supernode, DADO, N-cube, Ultracomputer, Butterfly, Alliant, Sequent Balance, CRAY X-MP.

   MIMD organization refers to a computer system capable of processing several programs at the same time. Most multiprocessor and multi-computer systems can be classified in this category.

In **MIMD** machines, memory can be arranged in a number of ways. There are two types of **MIMDs**: centralized shared-memory architectures, and distributed memory architectures. The first type is much more popular than the second type, but for a large number of processors (more than 12 or so) the second type is really the only option. The centralized architecture involves one main memory, and several processors with separate caches that all have access to it. The distibuted architecture involves an interconnection network that provides communication among the separate processors, each with their own cache and slice of main memory.

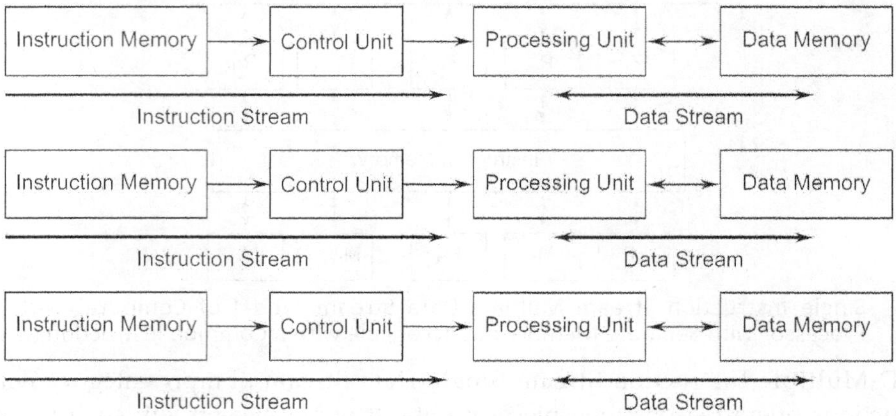

Fig. 1.12 Multiple Instruction Stream Multiple Data Stream

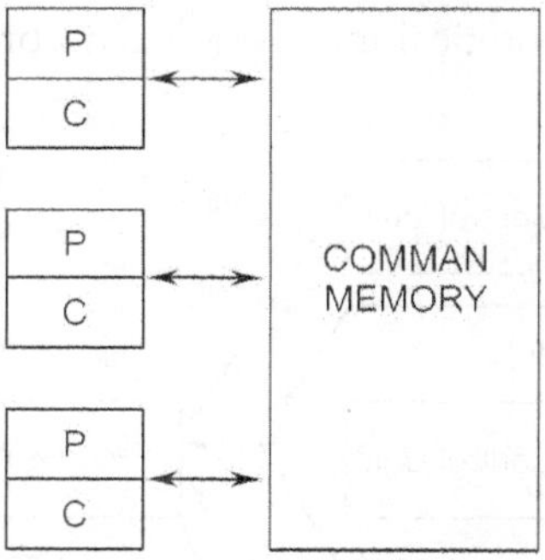

Fig. 1.13 Centralized Shared-Memory Architectures of Multiple Instruction Stream Multiple Data Stream

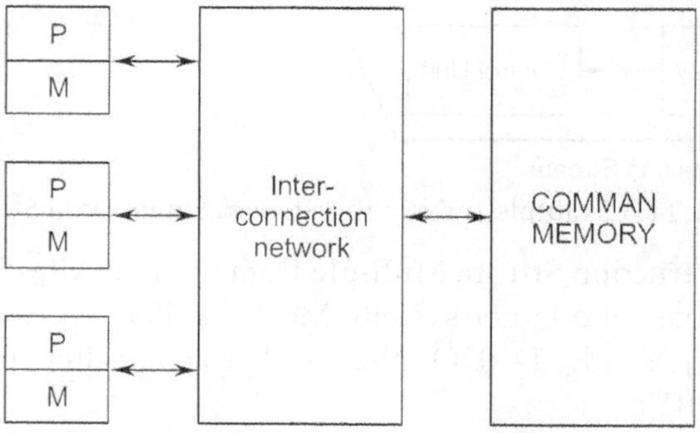

Fig. 1.14 Distributed Memory Architectures of Multiple Instruction Stream Multiple Data Stream

Flynn's classification depends on the distinction between the performance of the control unit and the data-processing unit. It emphasizes the behavioral characteristics of the computer system rather than its operational and structural interconnections. One type of parallel processing that does not Flynn's classification is pipelining. The only two categories used from this classification are SIMD array processors. Pipeline processing is an implementation technique where arithmetic sub operations or the phases of a computer instruction cycle overlap in execution.

So the different configurations along with there applications are given below:

Table 1.2 Different Configurations Along with their Applications

Instruction execution in one time cycle	Data stream input in one time cycle	Name	Example
1	1	**SISD**	Von Neumann Machine
1	Multiple	**SIMD**	Vector supercomputer, Array processor
Multiple	1	**MISD**	Not used
Multiple	Multiple	**MIMD**	Multiprocessor, Multicomputer

1.2.4 Comparison of Flynn's Classification Models

The idea of using parallelism in computer architectures to increase performance and improve availability has been around since the early days of computing. All computers can be placed into one of four categories if you look at the parallelism in the instruction and the data streams called for by the instructions at the most constrained component of the machine.

1. Single instruction stream, single data stream (SISD) computers are the typical uniprocessor machines.
2. Single instruction, multiple data stream (SIMD) computers are multiprocessor machines that execute the same instruction with multiple processors using different data streams. Each processor has its own data memory, but there is a single instruction memory and control processor that fetches and dispatches instructions. SIMD processors are typically special purpose, since full generality is not required. Multiple instructions, single data stream (MISD) computers are yet to be built commercially, but may be in the future.
3. Multiple instructions, multiple data stream (MIMD) computers often use off the shelf microprocessors. Each processor fetches its own instructions and operates on its own data. Not all machines fall into these categories, as there are several that combine features of more than one type, but in general these are fairly good guidelines. Many of the early model multiprocessors were SIMD, and there are still companies making SIMD computers today. However, the focus has shifted towards MIMD architectures because they make good general purpose microprocessors, and these tend to be more inexpensive and cost efficient than those that are custom made. In fact, almost all MIMDs built today use the same microprocessors found in workstations and small servers. MIMDs also offer flexibility, and can function as single-user machines running a single high-performance application, multiprogrammed machines running several tasks simultaneously, or some combination of both.

1.3 PROGRAMMING

A program is a list of instructions, together with any fixed information required to carry out those instructions. This applies to computers, of course, but also to any other subject that involves fixed instructions.

1.3.1 Programming Basics

Since a computer program consists of a series of instructions for the computer to carry out, plus any necessary fixed data required to perform those instructions, the process of programming consists of defining those instructions and that data. Thus, in order to design a program for a computer, you must determine three basic elements:

1. The instructions that must be performed.
2. The order in which those instructions are to be performed.
3. The fixed data required to perform the instructions.

Before your new program ever gets anywhere close to a computer, there are several steps you must take. These steps are:

1. **Define the problem :** Before you can tell a computer or even another person what you want to accomplish, you must know it yourself. You must be able to clearly state what the computer is to accomplish or produce as the end result of the activities it will carry out.
2. **Define the solution :** Now that you know what information the computer is to produce as its final output, you need to look at what information you have available and what information you still need, that will apply to that output. You also need to define the equations, logical procedures, or other methods you will need to use to manipulate the raw input data into becoming the final desired output.
3. **Map the solution :** The third paper step in programming is to layout the solution in its proper sequence. Remember that the order in which actions are taken is just as important as the actions themselves. You need to organize the solution procedure into its proper sequence, taking choices and alternatives into account.

1.3.2 Programming Languages

Early programming languages were designed for specific kinds of tasks. Modern languages are more general-purpose. In any case, each language has its own characteristics, vocabulary, and syntax. While this chapter will not by any means cover all of the available programming languages, we will take a look at a number of the better-known languages.

1. **FORTRAN :** One of the earliest computer languages, FORTRAN (an acronym for FORmula TRANslator) was designed to handle mathematical operations, originally on mainframe computers. FORTRAN was unable to handle text manipulations of any sort, and could just barely place quoted text in its printed output.
2. **COBOL :** The COmmon Business Oriented Language, or COBOL, is almost the exact opposite of **FORTRAN. COBOL** was devised to permit programs to be written for business data processing applications, using English-like statements. It is intended to handle business data records, so its mathematical capabilities are limited pretty much to dollars and cents, and percentages.
3. **ALGOL :** Another mathematically-oriented language, **ALGOL** (**ALGO**rithmic Language) does its work primarily in terms of numerical procedures called algorithms.
4. **Pascal :** Named after Blaise Pascal, a French philosopher, mathemetician, and physicist, Pascal was specifically designed as a teaching language. Its object was to force the student to correctly learn the techniques and requirements of structured programming.

Pascal was designed originally to be platform-independent. That is, a Pascal program could be compiled on any computer, and the result would run correctly on any other computer, even with a different and incompatible type of processor. The result was relatively slow operation, but it did work after its own fashion.

5. **BASIC** : The Beginner's All-purpose Symbolic Instruction Code, **BASIC** was the first interpreted language made available for general use. It is now in such widespread use that most people see and use this language before they deal with others. It has changed over time, and is now most commonly seen as Visual Basic in a Windows environment.
6. **FORTH** : Both a compiler and an interpreter, **FORTH** was originally developed to handle real-time operations and still allow direct user control and rapid program modifications. The name FORTH stems from its conception as a fourth-generation language, but it was developed on a computer system that permitted only five character filenames.
7. **Assembly Language** : Assembly language is a symbolic representation of the absolute machine code of a particular processor. Therefore, each processor has its own specific assembly language, although a family of processors, such as the Intel 80×86 series, may share some or all of its assembly code.
8. **C** : First came an experimental language called A, which was improved, corrected, and expanded until it was called B. This language in turn was improved, upgraded, and debugged and was finally called C. The C language has turned out to be quite versatile and amazingly powerful. The C language is amazingly simple, and is nevertheless capable of great things. The Unix operating system, which has been adapted to a wide range of platforms and is gaining in popularity, is written in C.
9. **C++** : When the concepts of *objects* and *object-oriented programming* were being developed, the standard C language didn't have the built-in structures to handle them. However, C was (and is) still highly useful and well worth keeping around, so a sort of extended C language was developed. This language was essentially "C and then some", or C-plus (C+). As the concepts of object-oriented programming continued to develop, C+ had to be upgraded, and became C++.
10. **Java** : The search for a platform-independent language is always in progress. Java is the latest language to be designed to meet this goal. Any computer with a Java Runtime Environment can run a Java program.

    One of the more useful aspects of Java is that Web browsers are now designed to be able to embed small Java applications, or "applets," into Web pages. Other Java programs, called "servlets," will run on Web servers. This allows extra communications between the page and the server, and permits a high degree of interactivity and dynamic page generation. The downside of using applets this way is that they inherently run much more slowly than native programs on your computer.
11. **Javascript** : Javascript is an interpreted version of Java, in most respects. Its most common application is within Web pages, where it can be used to provide interactivity and dynamic responses.
12. **Perl** : The Practical Extraction and Report Language (Perl) is very similar to C in many respects. However, it has a number of features which make it very useful in a wide range of applications. The most visible use of Perl is in CGI programming for the World Wide Web. Very often when you submit a form to a server, that form will be processed by a program written in Perl.

However, this is not the only use for Perl by any means. Perl is an excellent general-purpose programming that allows fast program development on a wide range of platforms.

1.3.3 Levels of Programming Languages

There is only one programming language that any computer can actually understand and execute: its own native binary machine code. This is the lowest possible level of language in which it is possible to write a computer program. All other languages are said to be *high-level* or *low-level* according to how closely they can be said to resemble machine code.

1. **Low-level Language :** In this context, a low-level language corresponds closely to machine code, so that a single low-level language instruction translates to a single machine-language instruction. Low-level languages have the advantage that they can be written to take advantage of any peculiarities in the architecture of the central processing unit (CPU) which is the "brain" of any computer. Thus, a program written in a low-level language can be extremely efficient, making optimum use of both computer memory and processing time. However, to write a low-level program takes a substantial amount of time, as well as a clear understanding of the inner workings of the processor itself. Therefore, low-level programming is typically used only for very small programs, or for segments of code that are highly critical and must run as efficiently as possible.
2. **High-level Language :** A high-level language instruction typically translates into a series of machine-language instructions. High-level languages permit faster development of large programs. The final program as executed by the computer is not as efficient, but the savings in programmer time generally far outweigh the inefficiencies of the finished product. This is because the cost of writing a program is nearly constant for each line of code, regardless of the language. Thus, a high-level language where each line of code translates to 10 machine instructions costs only one tenth as much in program development as a low-level language where each line of code represents only a single machine instruction.

   In addition to the distinction between high-level and low-level languages, there is a further distinction between *compiler languages* and *interpreter languages*.
3. **Absolute Machine Code :** The very lowest possible level at which you can program a computer is in its own native machine code, consisting of strings of 1's and 0's and stored as binary numbers. The main problems with using machine code directly are that it is very easy to make a mistake, and very hard to find it once you realize the mistake has been made.
4. **Assembly Language and Assembler :** Assembly language is nothing more than a symbolic representation of machine code, which also allows symbolic designation of memory locations. Thus, an instruction to add the contents of a memory location to an internal CPU register called the *accumulator* might be **add a number** instead of a string of binary digits (*bits*).

   No matter how close assembly language is to machine code, the computer still cannot understand it. The assembly-language program must be translated into machine code by a separate program called an *assembler*. The assembler program recognizes the character strings that make up the symbolic names of the various machine operations, and substitutes the required machine code for each instruction. At the same time, it also calculates the required address in memory for each symbolic name of a memory location, and substitutes those addresses for the names. The final result is a machine-language program that can run on its own at any time; the assembler and the assembly-language program are no longer needed. To help distinguish between the "before" and "after" versions of the program, the original assembly-language program is also known as the *source code*, while the final machine-language program is designated the *object code*.

If an assembly-language program needs to be changed or corrected, it is necessary to make the changes to the source code and then re-assemble it to create a new object program.

5. **Compiler Language :** Compiler languages are the high-level equivalent of assembly language. Each instruction in the compiler language can correspond too many machine instructions. Once the program has been written, it is translated to the equivalent machine code by a program called a *compiler*. Once the program has been compiled, the resulting machine code is saved separately, and can be run on its own at any time.

   As with assembly-language programs, updating or correcting a compiled program requires that the original (source) program be modified appropriately and then recompiled to form a new machine-language (object) program.

   Typically, the compiled machine code is less efficient than the code produced when using assembly language. This means that it runs a bit more slowly and uses a bit more memory than the equivalent assembled program. To offset this drawback, however, we also have the fact that it takes much less time to develop a compiler-language program, so it can be ready to go sooner than the assembly-language program.

6. **Interpreter Language :** An interpreter language, like a compiler language, is considered to be high level. However, it operates in a totally different manner from a compiler language. Rather, the interpreter program resides in memory, and directly executes the high-level program without preliminary translation to machine code.

   This use of an interpreter program to directly execute the user's program has both advantages and disadvantages. The primary advantage is that you can run the program to test its operation, make a few changes, and run it again directly. There is no need to recompile because no new machine code is ever produced. This can enormously speed up the development and testing process.

   On the down side, this arrangement requires that both the interpreter and the user's program reside in memory at the same time. In addition, because the interpreter has to scan the user's program one line at a time and execute internal portions of itself in response, execution of an interpreted program is much slower than for a compiled program.

1.3.4 Fourth-Generation Languages

Computers are designed to process, retrieve and store programmed information. Typically, they can only respond to electronic signals that correspond to binary numbers. Programming languages are designed to translate human language into commands that the computer can understand. There are currently five "generations" or levels of languages available.

1. Machine language, or First-Generation language, consists solely of binary numbers. Machine languages are computer dependent and therefore every type of computer has its own machine language.

2. Assembly language, or Second-Generation language, which came into use in the mid-1950s, uses a shorthand notation of letters and numbers to communicate with the computer in place of binary groupings. Assemblers are then utilized to translate the assembly language into machine language. Assembly language is still computer-dependent and therefore results in a unique assembly language for every type of machine.
3. Higher Order language (HOL), or Third-Generation language (3GL), which came into use in the 1960s, consists of statements which more closely resemble the spoken language, and therefore are easier to read and write since they require fewer statements per function. Special programs, or compilers, must then translate the HOL statements to assembly or machine language.
4. Expanding upon this, Fourth-Generation languages (4GLs), or Very High Level Languages (VHLL), also use instructions which resemble the spoken language, but they allow the programmers to define "what" they want the computer to do without necessarily telling the computer "how" to do it. Typically, the compilers, or interpreters, for 4GLs are not as efficient as HOL compilers in using available memory and processing speed.
5. Finally, the highest level of programming, Fifth-Generation languages (5GLs), would involve a computer which responds directly to spoken or written instructions, or English language commands. There exist only a few 5GLs or "natural languages", and they are typically used in artificial intelligence applications.

3GLs require a special program, or compiler, to translate the HOL statements into assembly or machine instructions. Compilers are not as efficient in terms of memory utilization or processing speed, so a tradeoff between performance and easiness is often required when employing 3GLs. Additionally, 3GLs require a vast amount of code to provide the same functionality and are time consuming to debug. Hence, the necessity for the creation of 4GLs, which capitalize on advanced techniques of programming while simplifying the man-machine interfaces. Currently, there are many 4GLs available, such as Oracle, VisiCalc, FOCUS, RAMIS II, and DBase IV, and new ones are still appearing daily. The languages tend to fall into four functional areas:

1. Query and Report Generators,
2. Graphic Languages,
3. Database management tools and
4. Spreadsheets.

But, they typically have a very limited range of application.

Advantages of 4GLs

(*a*) **Easy of Use :** Because the syntax of 4GLs closely relates to the human-language syntax, it is easier to learn. Additionally, due to the non-procedural nature of many of the languages, the techniques for accomplishing things are also simple, while the results are fast and impressive. Fourth-Generation languages also aim to remove the use of unnecessary acronyms, thereby allowing users to expend their effort on the purpose of the application, vice being bogged down with extraneous requirements.

(*b*) **Limited Range of Functions :** 4GLs are typically designed for a limited set of functions or specific applications. This is because the product then becomes easier to use than a full programming language.

(*c*) **Restrictions of Options :** Higher level languages often restrict the options available to users of lower level languages, such as the capability to modify themselves at execution time. To compensate for this, 4GLs permit automatic verification before testing.

(*d*) **Default Options :** A user of a 4GL is not required to specify everything. Instead, a compiler or interpreter is capable of making intelligent assumptions about what it thinks the user needs. Therefore, while 4GLs often require that many parameters be specified, they also provide a default option if the user does not make a selection. This saves time and debugging efforts.

(*e*) **Monologue and Dialogue :** With 4GLs, a dialogue occurs between the user and the computer. This allows for more opportunities to catch errors as they are being made. The software may ask the user questions, signal errors, and flag inconsistencies as the application is being created.

There are currently no standards for 4GLs, primarily because new ideas in language syntax, dialogue and semantics are appearing daily. Some 4GLs have already disappeared, and many of the best languages now come from relatively small, new vendors. As such, their application to Mission Critical Computer Resources (MCCR) systems, where obsolescence and supportability are key issues, is limited. Due to their limited range of application and their slow processing speeds, they will tend to be more prevalently utilized in Management Information Systems (MIS) developments.

REVIEW QUESTIONS

1. What do you understand by computer? What are the various applications of computer?
2. Differentiate between hardware and software of a computer. For a general purpose PC, what are the different components of hardware?
3. Explain Von Neumann model of computer.
4. Explain Non-Von Neumann model of computer and compare it with Von Neumann model of computer.
5. Explain Flynn's Classification of computer and compare different Flynn's Classification models.
6. What do you understand by computer program? What are the basic steps of computer programming?
7. Explain:
   (*a*) Low level programming language
   (*b*) High level programming language
   (*c*) Forth Generation level programming language.

2

FUNDAMENTALS OF MICROPROCESSOR

2.1 MICROPROCESSOR EVOLUTION AND TYPES

Microprocessors are simply a computer processor that has been configured into the design and function of a microchip. As part of the process, a microprocessor initiates the activation of all the components necessary to allow the computer to be used, such as waking (start operation) the operating system.

The basic functionality of a microprocessor is all based on the inherent ability to respond to and generate mathematical and logical operations. This function of small computer processors is made possible with the use of registers. Within the registers resides all the data required by the microprocessor to execute such basic functions as addition and subtraction. The configuration of the registers also allow the microprocessor to handle tasks such as comparing two different numbers, and retrieving numbers from various areas and redistributing them.

At the core of the design for the microprocessor is a series of logical instructions that regulate the order that tasks are received and executed. The presence of the instructions assures that tasks are carried out in a sequence that will flow from the completion of one task and into the logical initiation of the next task in the sequence. For example, when an end user boots up a computer, the first instruction that the microprocessor will execute is the activation of the BIOS, or basic input/output system.

From that point, the BIOS will load the operating system into the active memory and move on to the next task in the starting process. Each successive task will be achieved once the previous task is complete.

Generally, a microprocessor is a component that remains efficient and productive for the life of the computer. However, it is possible for the functionality of the chip to become damaged in some manner. When this happens, it is usually an easy process for a computer technician to replace the corrupted microprocessor with a fully functioning logic chip.

We divide the years of development of microprocessors as five generations:

(*a*) **First Generation (1971–73) :** Intel Corporation introduced 4004, the first microprocessor in 1971. It is evolved from the development effort while designing a calculator chip.

There were three other microprocessors in the market during the same period:

- Rockwell International's PPS-4 (4-bits)
- Intel's 8008 (8 bits)
- National Semiconductor's IMP-16 (16-bits)

They were fabricated using PMOS technology which provided low cost, slow speed and low output currents. They were not compatible with TTL.

(*b*) **Second Generation (1974–1978) :** Marked the beginning of very efficient 8-bit microprocessors. Some of the popular processors were:

- Motorola's 6800 and 6809
- Intel's 8085
- Zilog's Z80

They were manufactured using NMOS technology. This technology offered faster speed and higher density than PMOS. It is TTL compatible

(*c*) **Third Generation Microprocessors (1979–80) :** This age is dominated by 16-bits microprocessors. Some of them were:

- Intel's 8086/80186/80286
- Motorolla's 68000/68010

They were designed using HMOS technology. HMOS provides some advantages over NMOS as Speed-power-product of HMOS is four times better than that of NMOS. HMOS can accommodate twice the circuit density compared to NMOS. Intel used HMOS technology to recreate 8085A and named it as 8085AH with a higher price tag.

(*d*) **Fourth Generation (1981–1995) :**

- This era marked the beginning of 32-bits microprocessors
- Intel introduced 432, which was bit problematic
- Then a clean 80386 in launched.
- Motorola introduced 68020/68030.

They were fabricated using low-power version of the HMOS technology called HCMOS. Motorola introduced 32-bit RISC processors called MC88100.

(*e*) **Fifth Generation (1995 – till date) :** This age the emphasis is on introducing chips that carry on-chip functionalities and improvements in the speed of memory and I/O devices along with introduction of 64-bit microprocessors.

Intel leads the show here with Pentium, Celeron and very recently dual and quad core processors working with up to 3.5GHz speed.

2.1.1 Types of Microprocessors

There are different ways in which microprocessors are categorized. They are

(*a*) **CISC (Complex Instruction Set Computers) :** Earlier developments were based around the idea that making the CPU more complex and supporting a larger number of potential instructions would lead to increased performance. This idea is at the root of CISC processors, such as the Intel x86 range, which have very large instruction sets reaching up to and above three hundred separate instructions. They also have increased complexity in other areas, with many more specialized addressing modes and registers also being implemented, and variable length of the instruction codes themselves.

(*b*) **RISC (Reduced Instruction Set Computers):** RISC processors, such as the IBM PowerPC processor, have a greatly simplified and reduced instruction set, numbering in the region of one hundred instructions or less. Addressing modes are simplified back to four or less, and the length of the codes is fixed in order to allow standardization across the instruction set. Changing the architecture to this extent means that less transistors are used to produce the processors. This means that RISC chips are much cheaper to produce than their CISC counterparts.

(*c*) **VLIW(Very Long Instruction Word Computers) :** VLIW instructions are like RISC instructions except that they are longer to allow them to specify multiple, independent simple operations. A VLIW instruction can be thought of as several RISC instructions joined together. VLIW architectures tend to be RISC-like in most attributes.

The differences between these are given in table given below:

Table 2.1 Comparison of Different Types of Processors

ARCHITECTURE CHARACTERISTIC	CISC	RISC	VLIW
INSTRUCTION SIZE	Varies	One size, usually 32-bits	One size
INSTRUCTION FORMAT	Field placement varies	Regular, consistent placement of fields	Regular, consistent placement of fields
INSTRUCTION SEMANTICS	Varies from simple to complex; possibly many dependent operations per instruction	Almost always one simple operation	Many simple, independent operations
REGISTERS	Few, sometimes special	Many, general-purpose	Many, general-purpose
MEMORY REFERENCES	Bundled with operations in many different types of instructions	Not bundled with operations, *i.e.*, load/store architecture	Not bundled with operations, *i.e.*, load/store architecture
HARDWARE DESIGN FOCUS	Exploit microcoded implementations	Exploit implementations with one pipeline and no microcode	Exploit implementations with multiple pipelines, no microcode and no complex dispatch logic

Super Scalar Processors

A superscalar CPU architecture implements a form of parallelism called instruction-level parallelism within a single processor. It thereby allows faster CPU throughput than would otherwise be possible at the same clock rate. A superscalar processor executes more than one instruction during a clock cycle by simultaneously dispatching multiple instructions to redundant functional units on the processor. Each functional unit is not a separate CPU core but an execution resource within a single CPU such as an arithmetic logic unit, a bit shifter, or a multiplier.

Other types of specialized processors are

- General Purpose Processor (GPP)
- Special Purpose Processor (SPP)
- Application-Specific Integrated Circuit (ASIC)
- Digital Signal Processor (DSP)

2.2 MICROPROCESSOR ARCHITECTURE AND OPERATION OF ITS COMPONENTS

The Microprocessor is fabricated on a single chip using MOS technology. It comprises of

(*a*) Register section

(*b*) One or more ALU, and

(*c*) A control unit

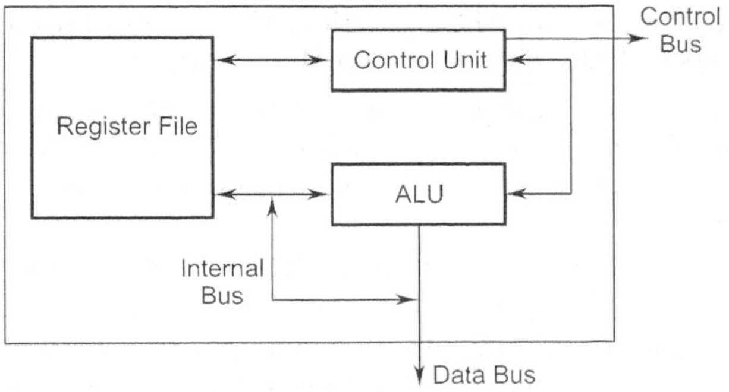

Fig. 2.1 Typical Schematic Arrangement of a Microprocessor

(a) Register Section

As already discussed that registers are a special kind of memory that exist inside a processor. Typically a processor will have several internal registers that are much faster than main memory. These registers usually have specialized capabilities for arithmetic, logic, and other operations. On the basis of registers; microprocessor can be classified as given below:

Accumulator based microprocessors – Intel's 8085, Motorola's 6809

- Here, one of the operands is assumed to be held a special register called "accumulator".
- All arithmetic and logic operations are performed using this register as one of the data source and result is stored back in the accumulator.
- One-operand instructions are predominant in this organization.

The general-purpose register based microprocessors – Intel's 8086/386, Motorola's 68000/020

- These processors have a set of registers which can be used to hold data, memory addresses or the results of an arithmetic or logic operations for indefinite amount of time.
- The number and size of these registers vary from processor to processor.
- Some registers are general purpose while others are earmarked with some functions.
- General purpose registers are used to store addresses or data for an indefinite time and are capable of manipulating data by shift or rotate operations.

- Typical dedicated registers include:
  (*i*) Program Counter (PC)
  (*ii*) Instruction Register (IR)
  (*iii*) Status Register or Flag Register
  (*iv*) Stack Pointer (SP)
  (*v*) Barrel Shifter

The **Program Counter (PC)** always contains the address of the next instruction to be executed. Its contents are automatically updated by ALU. The microcomputer executes a program sequentially unless it encounters a jump/branch/call instruction. At that time, PC will be loaded with the address present in the instruction. The size of the PC itself varies from one processor to another. For example, the 8085 has a 16-bit PC, while 68029 has 32-bits PC.

The **Instruction register** contains the instruction to be executed. After fetching the instruction from memory, microprocessor places it in IR for translation.

The **Status Register** contains individual bits each having a special meaning. The bits are termed as **flags**. Each flag is set or reset by an ALU operation. These flags or used by Conditional branch instructions. Typical flags include carry, sign, zero and overflow.

- The carry (C) flag is used to reflect whether or not an arithmetic operation such as ADD generates a carry. If carry is generated then CF = 1 else CF = 0. The carry is generated out of 8th bit for byte operations, 16th bit for word operations etc. Carry is used as Borrow flag for subtraction.
- The Zero (Z) flag is used to indicate whether the result of an arithmetic or logic operation is zero. ZF = 1 for zero result and ZF = 0 for a non-zero result.
- The Sign(S) flag indicates whether the result is positive or negative. SF = 1 indicates negative result means the most significant bit of the result is 1. If SF = 0, the result is a positive number. It is observed only for signed operations. This flag can be ignored for the result on an unsigned arithmetic or logic operations.
- The Overflow (O) flag is set if the result of an arithmetic and logical operation on signed numbers is too large for the microprocessor's maximum word size. OF can be shown as $OF = C7 \oplus C8$ where C7 is the final carry and C6 is the previous carry. Once again, this applies to signed numbers only.

The **Stack Pointer (SP)** register addresses the stack.

- A stack is a Last-In-First-Out read/write memory. The items that go in last will come out first. This is because all read (POP) and write (PUSH) operations will take place from one end called top of the stack(TOS).
- Stack is implemented using hardware or software.
- The hardware stack is designed by using a set of high speed registers to provide a fast response. The disadvantage is that stack size is limited. But PUSH and POP operations are very fast. Intel's 4040, an 8-bit processor used hardware stack.
- The software stack on the other hand, is implemented using a portion of memory. Some RAM locations are earmarked as stack. The advantage is that they provide unlimited space for stack, depends on the amount of memory we interface to microprocessor, though. But it is slower than hardware stack.
- The SP always contains the memory address of the last byte of the currently pushed item on TOS *i.e.*, it always points to the TOS. Stack is normally used by subroutines or interrupts for saving certain registers such as the program counter and status register.

PUSH and POP Operations

- If the stack is accessed from the top, the stack pointer is decremented after a PUSH operation and incremented before POP.
- On the other hand, if the stack is accessed from the bottom, SP is incremented after a PUSH and decremented after a POP.
- Typical microprocessors access stack from top.
- Depending upon the microprocessor, 8-, 16- or 32-bits can be pushed onto or popped from the stack.
- The value by which the SP is incremented or decremented after PUSH or POP operations depends on the register size.
- For example, in 8086 microprocessor, PUSH and POP operations can be done only on 16-bit data. Hence, SP is incremented or decremented by a value of 2 always.

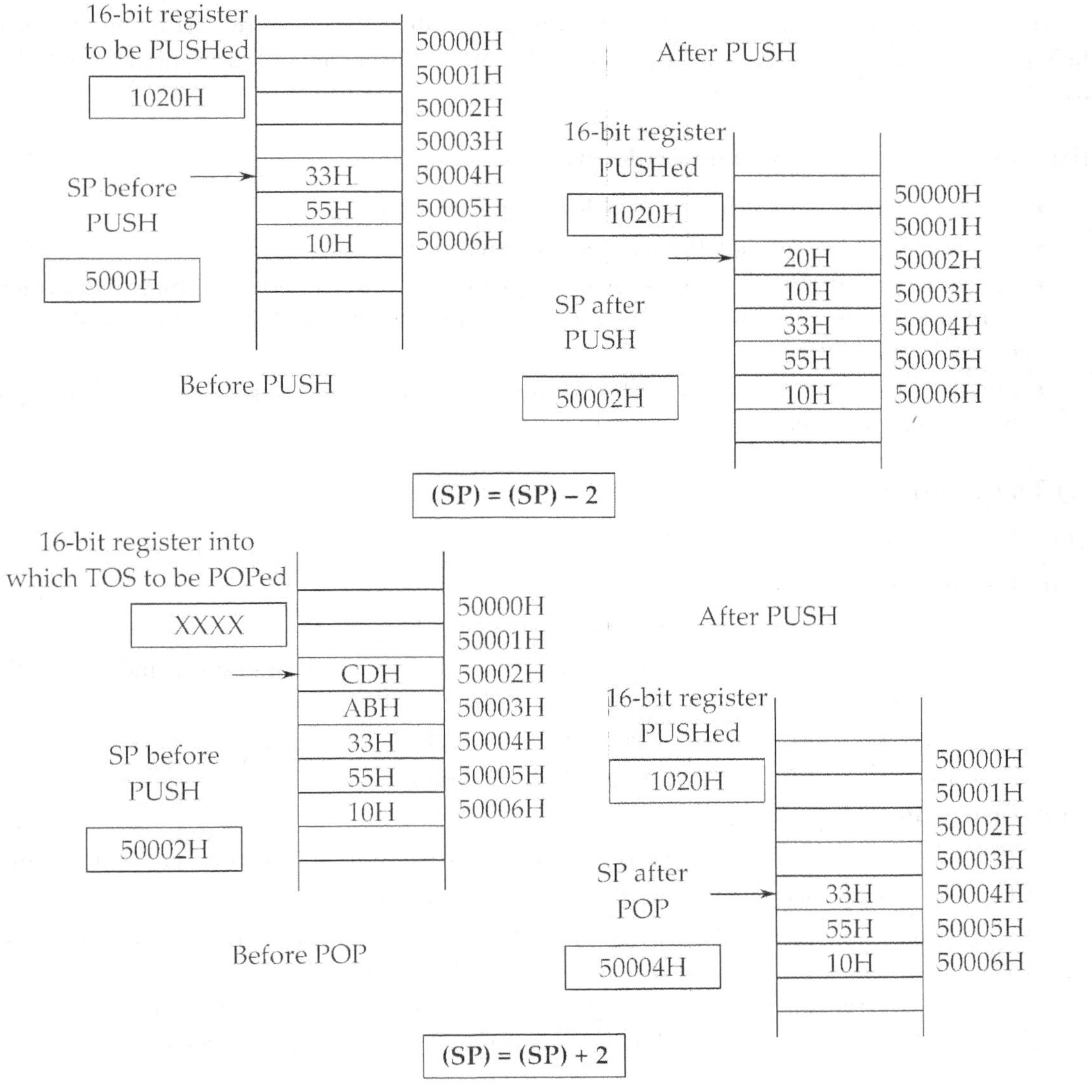

Fig. 2.2 PUSH and POP Operation

Consider the PUSH operation as shown in the Fig. 2.2, when the stack is accessed from the top. SP is decremented by 2 after PUSH.

Similarly, after POP operation SP is incremented by 2, since we are accessing stack from top.

Index Register :

Index register is useful with instructions where tables or arrays of date are accessed. Here, Index Register can be used to manipulate the address portion of the instruction. Thus appropriate data in the table can be accessed. The actual address called physical address of the data is calculated by adding address portion in the instruction with contents of the index register.

In 8086: MOV AL, 200[SI] means one byte present at an address (DS) + 200 + (SI) will be moved into AL register. (DS—Data Segment register).

Barrel Shifter :

32-bit processors include a special type of register called Barrel Shifter. This register provides faster shift operation. For example, Intel's 80386 barrel shifter can shift a number from 0 through 64 positions in one clock period.

(b) ALU (Arithmetic and Logical Unit)

- ALU performs all arithmetic and logic operations on data.
- The size of ALU defines the size of the microprocessor.
- For example Intel 8086 is a 16-bit microprocessor since its ALU 16-bits wide. Intel 8088 is also a 16-bit microprocessor even though its data bus is 8-bits wide. That is because of its 16-bit ALU
- Some 32-bit microprocessors like Motorola 68030 include multiple ALUs for parallel operations to achieve faster speed.

(c) The Control Unit

The CU performs basically two tasks:

(*i*) **Instruction Interpretation:**

- CU reads instruction from memory using PC
- It then recognizes the instruction type, gets necessary operands, and routes then to appropriate functional units of execution unit
- Necessary signals are issued to perform desired operation
- Results are routed to the specified destination.

(*ii*) **Instruction Sequencing:**

The CU determines the address of the next instruction to be executed and loads it into PC.

The CU is designed using one of the three techniques:

(*i*) **Hardwired Control :** Designed by physically connecting typical components such as gated and flip-flops. For example, Zilog's 16-bit Z8000

(*ii*) **Microprogramming :** This type of CUs includes a control ROM for translating the instructions. Intel's 8086 is a microprogrammed microprocessor.

(*iii*) **Nanoprogramming** : It includes two ROMs inside CU. The first ROM, which is called microROM stores all the addresses of the second ROM, which is called nanoROM. If the microinstructions repeat many times in a microprogram, use of two level ROMs provides tremendous memory savings.

Motorola's 68000, 68020 and 68030 are nanoprogrammed.

Fig. 2.3 shows a more descriptive block diagram of a microprocessor. Multiplexers and demultiplexers do not represent primary functions and are there to facilitate the flow of data between different blocks and also between different blocks and the outside world.

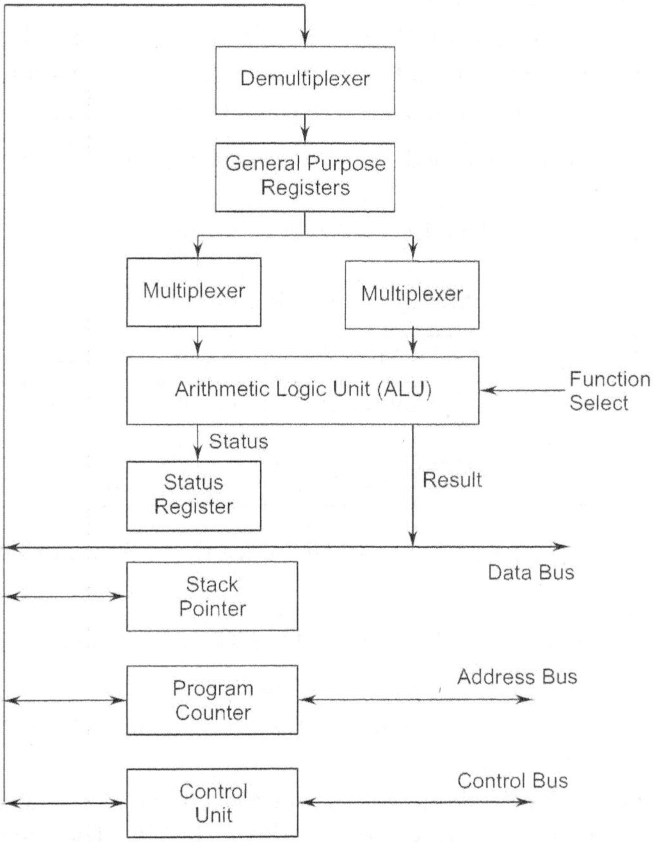

Fig. 2.3 Descriptive Block Diagram of a Microprocessor

2.3 INTERFACING AND DATA TRANSFER SCHEMES

Interfacing of Memory

In many applications, a memory of large size capacity is often realized by interconnecting several small size memory blocks. There are two kinds of techniques used for designing the main memory in such cases. They are

(*a*) Linear decoding

(*b*) Fully decoding

First, let us consider the block diagram of a typical RAM IC.

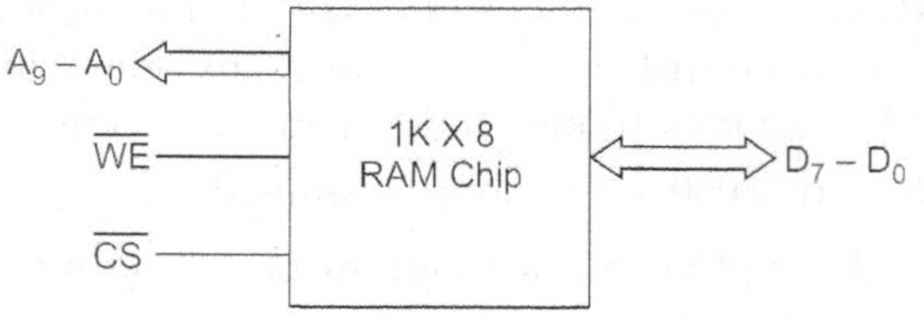

Fig. 2.4 RAM Chip

- The capacity of this chip is 1Kbytes.
- They are organized in the form of 1024 words with 8-bits/word.
- Each word has a unique address and is specified on 10-bit address lines $A_9 - A_0$ ($1024 = 2^{10}$).
- The inputs and outputs are routed through the 8-bit bidirectional data bus ($D_7 - D_0$). The operation of this chip is governed by two control inputs: $\overline{WE}$ (Write Enable) and $\overline{CS}$ (Chip Select).

The following truth table describes the operation of this chip:

Table 2.2 Operation of Chip

$\overline{CS}$	$\overline{WE}$	MODE	Status ($D_7 - D_0$)	Power
H	X	Not selected	High Impedance	Standby
L	L	Write	Input Bus	Active
L	H	Read	Output Bus	Active

- When $\overline{CS}$ is high, chip is not selected at all, hence D_7 to D_0 are driven to high impedance state
- When $\overline{CS} = 0$ and $\overline{WE} = 0$, data on lines $D_7 - D_0$ are written into the word addressed by A_0 through A_9.
- When $\overline{CS} = 0$ and $\overline{WE} = 1$, the contents of memory word whose address is on $A_9 - A_0$ will appear on lines $D_7 - D_0$

(*a*) **Linear Decoding :** Consider the problem where we have to connect 6 Kb memory to an 8-bit microprocessor whose address bus width is 16-bits. The memory chips are available as 1K X 8.

In linear decoding,

- Address lines A_9 through A_0 of the microprocessor used as common input to address lines of all memory chips
- The data lines of microprocessor are connected to data lines of all memory chips.
- The remaining address lines are used to select one of the chips ($\overline{CS}$) at a time. For example, 000001 selects chip1, 100000 selects chip 6 etc.
- R/W from microprocessor is connected to $\overline{WE}$ all RAM Chips

Primary Advantage is – This technique does not need any decoding circuit.

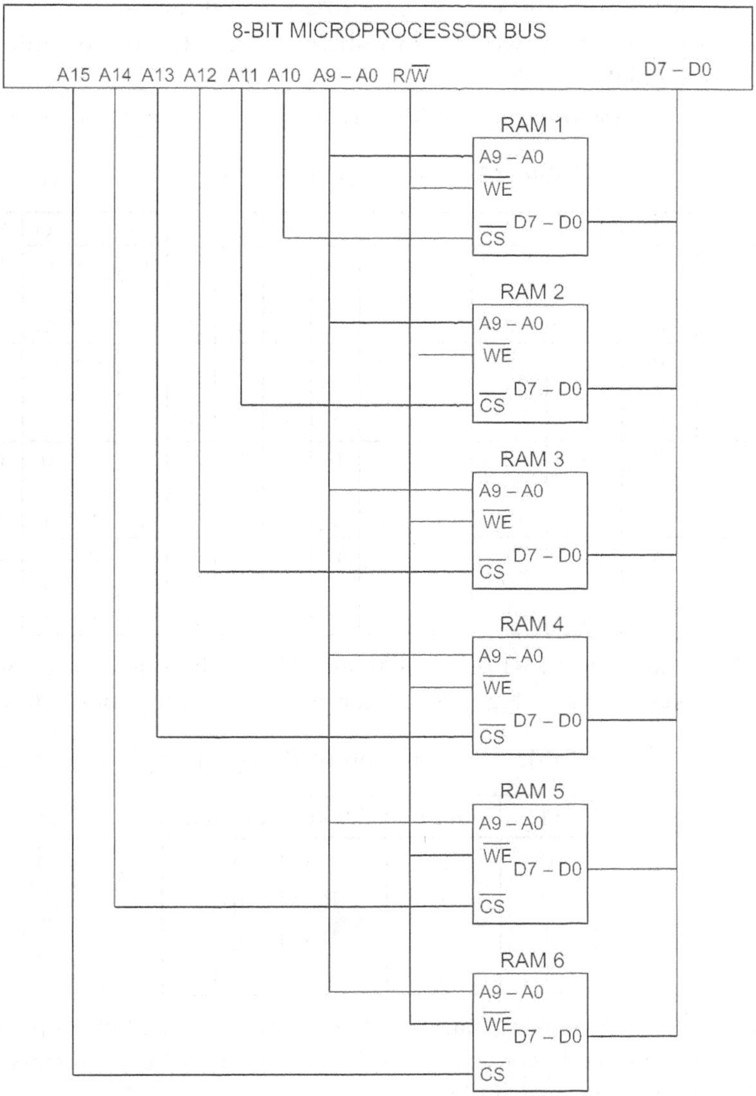

Fig. 2.5 Microprocessor and RAM Chip Connections

Some of the disadvantages of this approach are easily visible:

(*i*) Although there is an address bus of 16-bits wide, we could connect only 6KB of RAM. This idea clearly wasted address space

(*ii*) Address map is not contiguous. It is sparsely distributed.

(*iii*) Conflicts occur if two of the select lines become active at the same time.

(*iv*) If all unused address lines are not used as chip selectors, then these unused lines become don't cares. This result in foldback, meaning a memory location will have its image in memory map. For example, if A_{15} is don't care, then address 0000_{16} is same as address 8000_{16}. It wastes memory space.

(*b*) **Fully Decoding** : The problems of bus conflict and sparse address distribution are eliminated by the use of fully decoding address technique.

Consider an example where we interface 4Kb of RAM to an 8-bit microprocessor. The RAM chips are available in the form of 1K X 8.

First we have to write memory map to identify the address lines to be given to decoder logic:

Table 2.3 Decoding of RAM Chip

A15	A14	A13	A12	A11	A10	A9	A8	A7	A6	A5	A4	A3	A2	A1	A0	Address
0	0	0	0	0	0	0	0	0	0	0	0	0	0	0	0	0000
.	.	.	.	.	.	.	.	.	.	.	.	.	.	.	.	To
0	0	0	0	0	0	1	1	1	1	1	1	1	1	1	1	03FF
0	0	0	0	0	1	0	0	0	0	0	0	0	0	0	0	0400
.	.	.	.	.	.	.	.	.	.	.	.	.	.	.	.	To
0	0	0	0	0	1	1	1	1	1	1	1	1	1	1	1	07FF
0	0	0	0	1	0	0	0	0	0	0	0	0	0	0	0	0800
.	.	.	.	.	.	.	.	.	.	.	.	.	.	.	.	To
0	0	0	0	1	0	1	1	1	1	1	1	1	1	1	1	0BFF
0	0	0	0	1	1	0	0	0	0	0	0	0	0	0	0	0CFF
.	.	.	.	.	.	.	.	.	.	.	.	.	.	.	.	To
0	0	0	0	1	1	1	1	1	1	1	1	1	1	1	1	0FFF

If we observe A10 and A11, 2-to-4 decoder would be an obvious choice for CS signals. The implementation has been shown in Fig. 2.6. We can write the truth table as follows:

Table 2.4 Selection of RAM Chip

A11	A10	Device selected
0	0	RAM chip1
0	1	RAM chip2
1	0	RAM chip3
1	1	RAM chip4

Here, when we observe the memory map, there are no windowing between memory addresses and the foldback is also removed. Above that, address space is not wasted since the unused lines can be used in future by making use of higher decoder.

Input/Output (I/O) Interfacing

- One communicates with the microcomputer system via the I/O devices interfaced to it.
- The user can enter the program and data using the keyboard on the terminal and execute the program to obtain the results.
- Thus, I/O devices provide the efficient means of communication between the computer and the outside world.
- I/O devices are commonly called peripherals and include key board, CRT display, printers, disks etc.

- The characteristics of I/O devices are normally different from those of microcomputer. For example, the speed of operation of peripherals is usually slower compared to the microcomputer.

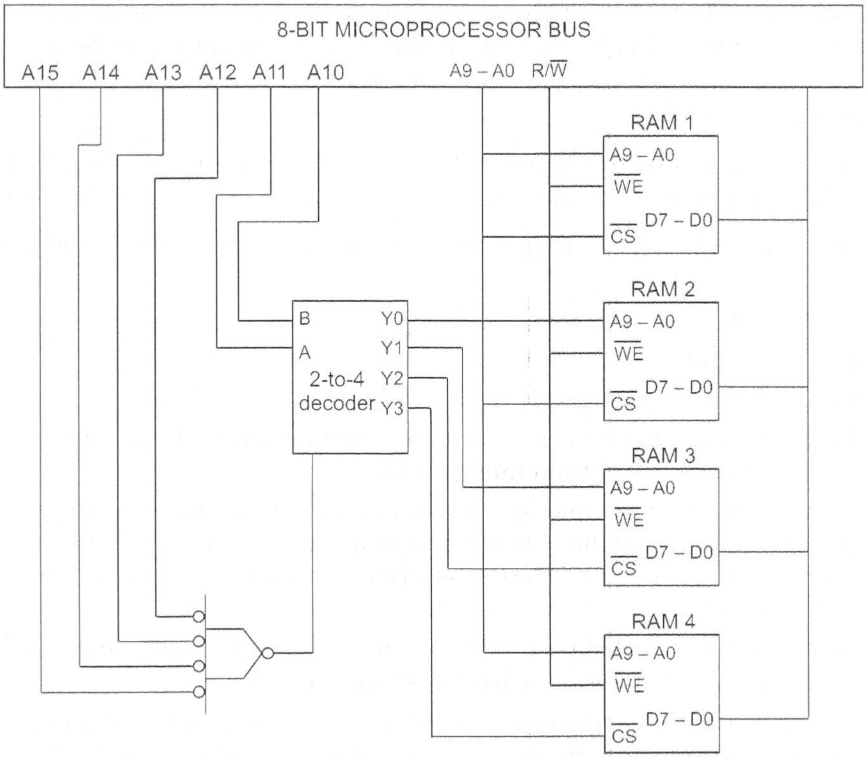

Fig. 2.6 Microprocessor and RAM Chip Connections

- Word length of the microcomputer may be different from the data format of the peripheral device.
- To make these characteristics compatible, interface hardware circuitry is used in between them. They provide all type of input/output transfers between microcomputer and peripherals by using an I/O bus which carries three types of signals: device address, data and command status.
- For 16- and 23-bit microprocessors, a separate intelligent I/O processor or data channel is provided to route all I/O transfers.
- To make 8-bit microprocessors inexpensive, a separate interface is provided with I/O device.
- The microprocessor uses I/O bus when it executes an I/O instruction.
- A typical I/O instruction has three fields
- When the microcomputer executes an I/O instruction, the control unit decodes the opcode field and identifies it as an I/O instruction.
- Then the microprocessor places the device address and command from the respective fields of the I/O instruction on to the I/O bus.

- The interfaces of various devices connected to this I/O bus decode this address, and appropriate interface is selected.
- The identified interface decodes the command line and determines the function to be performed. Typical functions include receiving data from an input device into the microprocessor or sending data to an output device from the microprocessor.

In a typical microcomputer system, the user gets involved with to types of I/O devices: **physical I/O and Virtual I/O**.

When the microcomputer has no operating system, the user must work directly with physical I/O devices and perform detailed I/O design.

There are three ways of transferring data between the microprocessor and a physical I/O device.

(*i*) Programmed I/O
(*ii*) Interrupt driven I/O
(*iii*) Direct Memory Access (DMA)

- The microprocessor executes a program to communicate with an external device via a register called I/O port for **programmed I/O.**
- An external device requests microprocessor to transfer data by activating a signal on the microprocessor's interrupt line during interrupt I/O. In response, the microprocessor executes a program called the **interrupt-service routine** to carry out the function desired by the external device.
- Data transfer between microcomputer's memory and an external device occurs without microprocessors involvement in **Direct Memory Access.**

For a microcomputer with an operating system, the user works with virtual I/O devices. The user does not have to be familiar with the characteristics of the physical I/O device. Instead the user performs data transfers between the microcomputer and the physical I/O device indirectly by calling the I/O routines provided by operating system using virtual I/O instructions.

Programmed I/O

- The microcomputer communicates with an external device via one or more registers called I/O ports.
- I/O ports are occasionally fabricated by the manufacturer in the same chip as the memory chip to achieve minimum chip count.
- I/O ports are usually two types
  (*i*) Each bit in the port can be individually configured as either input or output port
  (*ii*) For the other type, all bits in a port can be set up as either all parallel outputs or parallel inputs. Each port can be configured as an input or output port by another register called command or data direction register. The port contains the actual data. The command register says whether they are inputs or outputs.

In the first method, the command register is loaded with 0s and 1s to indicate how the corresponding port bits act. For example, consider the command register is loaded with 65H. Then the corresponding port acts as follows:

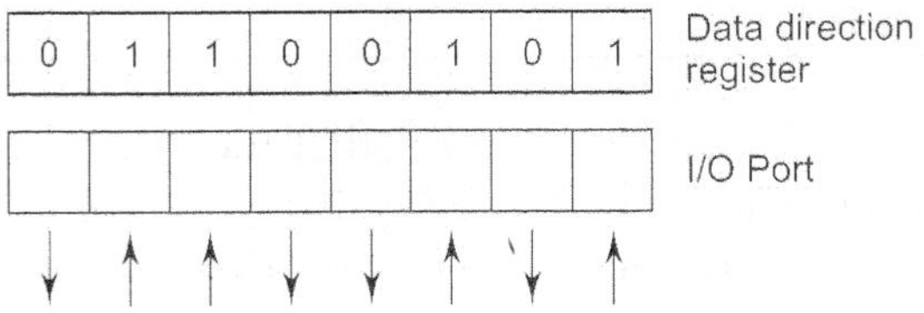

Fig. 2.7 Command Register

In the preceding example, since 65H is sent as output into data direction register, bits1, 3, 4, and 8 of the port are set up as outputs and bits 0, 2, 5, and 6 at set as inputs.

The following figure gives clear idea of this type of configuring I/O ports:

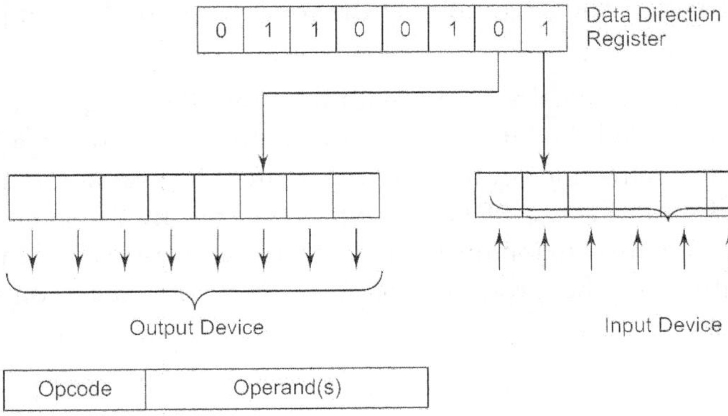

Fig. 2.8 Command Register

Standard I/O versus Memory-Mapped I/O

I/O ports are addressed using either standard I/O or memory mapped I/O. The standard I/O is also called isolated I/O.

It uses $IO/\overline{M}$ control pin on the microprocessor. Processor outputs high on this pin to indicate an I/O operation is taking place. A low on this pin indicates a memory operation. Address will be of 8-bits and hence can address up to 256 different devices. Here we can use only IN and OUT instructions.

In memory mapped I/O, the processor does not differentiate between I/O and memory. The processor uses a portion of memory address to represent I/O ports. The I/O ports are mapped into the processors main memory. Hence are called memory mapped I/O. Here we can use all instructions that are used to work with memory. The address bus width is same as that of microprocessor and hence decoding logic becomes bit complicated.

Unconditional and Conditional I/O

The processor can send data to the device at any time in unconditional I/O. The external device must always be ready for data transfer. A typical example is when the processor outputs seven segment code through an I/O port to drive a seven segment display connect to this I/O port.

In conditional I/O, the processor outputs data to an external device via handshaking. Data transfer occurs by the exchanging of the control signals between the processor and the external device. Data transfer takes place only when the device is ready.

2.4 INTERRUPT DRIVER I/O

A disadvantage of conditional programmed I/O is that the microcomputer needs to check the status bit (BUSY signal for the A/D converter) by waiting in a loop. This type of I/O transfer is dependent on the speed of the external device. For a slow device, this waiting may slow down the capability of the microprocessor to process other data. The interrupt I/O technique is efficient in this type of situation.

Interrupt I/O is a device-initiated I/O transfer. The external device is connected to a pin called the interrupt (INT) pin on the processor chip. When the device needs an I/O transfer with the microcomputer, it activates the interrupt pin of the processor chip. The microcomputer usually completes the current instruction and saves at least the contents of the current program counter on the stack.

The microcomputer then automatically loads an address into the program counter to branch to a subroutine like program called the interrupt service routine. This program is written by the user. The external device wants the microcomputer to execute this program to transfer data. The last instruction of the service routine is a RETURN, which is typically the same instruction used at the end of a subroutine. This instruction normally loads the address (saved in the stack before going to the service routine) in the program counter. Then, the microcomputer continues executing the main program.

Types of Interrupts

There are typically three types of interrupts: **external interrupts, traps** or **internal interrupts**, and **software interrupts.**

External interrupts are initiated through the microcomputer's interrupt pins by external devices such as A/D converters.

External interrupts can further be divided into two types: **maskable** and **nonmaskable.**

A **maskable interrupt** is enabled or disabled by executing instructions such as EI or DI. If the microcomputer's interrupt is disabled, the microcomputer ignores the maskable interrupt. Some processors, such as the Intel 8086, have an interrupt flag bit in the processor status register. When the interrupt is disabled, the interrupt flat bit is 1, so no maskable interrupts are recognized by the processor. The interrupt flag bit resets to zero when the interrupt is enabled.

The **nonmaskable interrupt** has higher priority than the maskable interrupt. If both maskable and nonmaskable interrupts are activated at the same time, the processor will service the nonmaskable interrupt first.

Internal interrupts, or traps, are activated internally by exceptional conditions such as overflow, division by zero, or execution of an illegal op-code. Traps are handled the same way as external interrupts. The user writes a service routine to take corrective measures and provide an indication to inform the user that an exceptional condition has occurred.

Many processors include **software interrupts**, or system calls. When one of these instructions is executed, the processor is interrupted and serviced similarly to external or internal interrupts. Software interrupt instructions are normally used to call the operating system. Software interrupt instructions allow the user to switch from user to supervisor mode.

Interrupt Address Vector

The technique used to find the starting address of the service routine (commonly known as the interrupt address vector) varies from one processor to another. With some processors, the manufacturers define the fixed starting address for each interrupt. Other manufacturers use an indirect approach by defining fixed locations where the interrupt address vector is stored.

Saving the Microprocessor Registers

When a processor is interrupted, it saves at least the program counter on the stack so that processor can return to the main program after executing the service routine. Some processors save only one or two registers, such as the program counter and status register.

Other processors save all microprocessor registers before going to the service routine. The user should know the specific registers the processor saves prior to executing the service routine. This will enable the user to use the appropriate return instruction at the end of the service routine to restore the original conditions upon return to the main program.

Interrupt Priorities

A processor is typically provided with one or more interrupt pins on the chip. Therefore, a special mechanism is necessary to handle interrupts from several devices that share on of these interrupt lines. There are two ways of servicing multiple interrupts: **polled** and **daisy chain techniques.**

Polled interrupts are handled by software and therefore are slower when compared with daisy chaining. The processor responds to an interrupt by executing one general service routine for all devices. The priorities of devices are determined by the order in which the routine polls each device. The processor checks the status of each device in the general service routine, starting with the highest priority device to service an interrupt. Once the processor determines the source of the interrupt, it branches to the service routine for the device.

In a **daisy chain priority system**, devices are connected in a daisy chain fashion to set up a priority system. Suppose one or more devices interrupt the processor. In response, the processor pushes at lease the PC and generates an interrupt acknowledge ($\overline{\text{INTA}}$) signal to the highest priority device. If this device has generated the interrupt, it will accept the $\overline{\text{INTA}}$. Otherwise, it will pass the $\overline{\text{INTA}}$ onto the next device until $\overline{\text{INTA}}$ is accepted. Once accepted, the device provides a means for the processor to find an interrupt address vector by using external hardware. The daisy chain priority scheme is based on mostly hardware and is therefore faster than the polled interrupt.

2.5 DIRECT MEMORY ACCESS (DMA)

Direct Memory Access (DMA) is a technique that transfers data between a microcomputer's memory and I/O device without involving the microprocessor. DMA is widely used in transferring large blocks of data between a peripheral device and the microcomputer's memory. The DMA technique uses a DMA controller chip for the data transfer operation. The main functions of a typical DMA controller are summarized as follows:

- The I/O devices request DMA operation via the DMA request line of the controller chip.
- The controller chip activates the microprocessor HOLD pin, requesting the CPU to release the bus.
- The processor sends $\overline{\text{HLDA}}$ (hold acknowledge) back to the DMA controller, indicating that the bus is disabled. The DMA controller places the current value of its internal registers, such as the address register and counter, on the system bus and sends a DMA acknowledge to the peripheral device. The DMA controller completes the DMA transfer.

There are three basic types of DMA: **block transfer, cycle stealing,** and **interleaved DMA.**

For **block transfer DMA**, the DMA controller chip takes the bus from the microcomputer to transfer data between the memory and I/O device. The microprocessor has no access to the bus until the transfer is completed. During this time, the microprocessor can perform internal operations that do not need the bus. This method is popular with microprocessors. Using this technique, blocks of data can be transferred.

Data transfer between the microcomputer memory and an I/O device occurs on a word-by-word basis with **cycle stealing**. Typically, the microprocessor clock is enabled by ANDing an INHIBIT signal with the system clock. The system clock has the same frequency as the microprocessor clock. The DMA controller controls the INHIBIT line. During normal operation, the INHIBIT line is HIGH, providing the microprocessor clock. When DMA operation is desired, the controller makes the INHIBIT line LOW for one clock cycle. The microprocessor is then stopped completely for the cycle. Data transfer between the memory and I/O takes place during this cycle. This method is called cycle stealing because the DMA controller takes away or steals a cycle without microprocessor recognition. Data transfer takes place over a period of time.

With **interleaved DMA**, the DMA controller chip takes over the system bus when the microprocessor is not using it. For example, the microprocessor does not use the bus while incrementing the program counter or performing an ALU operation. The DMA controller chip identifies these cycles and allows transfer of data between the memory and I/O device. Data transfer takes place over a period for time for this method.

2.6 INSTRUCTION AND DATA FLOW (WITH TIMING DIAGRAM)

When software is installed onto a modern day personal computer (most commonly from a CD-ROM, though other media or downloading from the internet is also common), code comprising the program and any associated files is stored on the hard drive.

This code comprises of a series of instructions for performing designated tasks, and data associated with these instructions. The CPU then executes the program from memory, processing each instruction in turn. Of course, in order to execute the instructions, it is necessary for the CPU to understand what the instruction is telling it to do. Therefore, recognition for instructions that could be encountered needs to be programmed into the processor.

Once the instruction has been recognized, and the actions that should be carried out are decided upon, the actions are then performed before the CPU proceeds on to the next instruction in memory. This process is called the 'instruction execution cycle'.

When a processor is executing a program, the program is in a machine language. However, programmers almost never write their programs directly into this form. While it may not have been originally written in this way, it is translated to a machine language at some point before execution so that it is understandable by the CPU. Machine language can be directly interpreted by the hardware itself, and is able to be easily encoded as a string of binary bits and sent easily via electrical signals.

The instruction set is a collection of pre-defined machine codes, which the CPU is designed to expect and be able to act upon when detected. Different processors have different instruction sets, to allow for greater features, easier coding, and to cope with changes in the actual architecture of the processor itself. Each machine code of an instruction set consists of two separate fields:

Opcode	Operand(s)

Fig. 2.9 Format of Instruction

The opcode is a short code which indicates what operation is expected to be performed. Each operation has a unique opcode. The operand, or operands, indicates where the data required for the operation can be found and how it can be accessed (the addressing mode). The length of a machine code can vary - common lengths vary from one to twelve bytes in size. The exact format of the machine codes is again CPU dependant.

Opcodes are also given mnemonics (short names) so that they can be easily referred to in code listings and similar documentation. For example, an instruction to store the contents of the accumulator in a given memory address could be given the binary opcode 000001, which may then be referred to using the mnemonic STA (short for STore Accumulator).

Once a program is in memory, it has to be executed. To do this, each instruction must be looked at, decoded and acted upon in turn until the program is completed. This is achieved by the use of what is termed the 'instruction execution cycle', which is the cycle by which each instruction in turn is processed. However, to ensure that the execution proceeds smoothly, it is also necessary to synchronise the activites of the processor.

To keep the events synchronised, the clock located within the CPU control unit is used. This produces regular pulses on the system bus at a specific frequency, so that each pulse is an equal time following the last. This clock pulse frequency is linked to the clock speed of the processor - the higher the clock speed, the shorter the time between pulses. Actions only occur when a pulse is detected, so that commands can be kept in time with each other across the whole computer unit.

The instruction execution cycle can be clearly divided into three different parts, which will now be looked at in more detail.

Fetch Cycle

The fetch cycle takes the address required from memory, stores it in the instruction register, and moves the program counter on one so that it points to the next instruction.

Decode Cycle

Here, the control unit checks the instruction that is now stored within the instruction register. It determines which opcode and addressing mode have been used, and as such what actions need to be carried out in order to execute the instruction in question.

Execute Cycle

Once the opcode is known, the execution cycle can occur. Different actions need to be carried out dependant on the opcode, with no two opcodes requiring the same actions to occur. However, there are generally four groups of different actions that can occur:

(*a*) Transfer of data between the CPU and memory.

(*b*) Transfer of data between the CPU and an input or output devices.

(*c*) Processing of data, possibly involving the use of the arithmetic and logic unit.

(*d*) A control operation, in order to change the sequence of subsequent operations. These can possibly be conditional, based on the values stored at that point within the flag register.

We have to examine the process of communication between the microprocessor and memory to understand the functions of various signals.

The first step in the communication process is reading from memory or fetching an instruction. This process is called fetch cycle. We need to understand timings of various signals in relation to the system clock. It can be better understood through timing diagram as shown below.

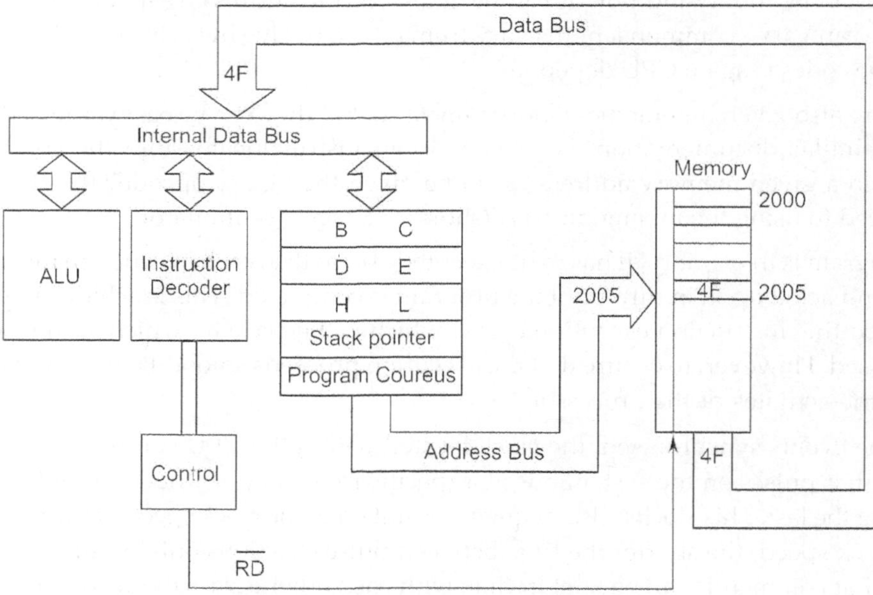

Fig. 2.10 Execution of Instruction

Consider the example of fetching the machine code of instruction MOV C, A (0100 1111 = 4F) stored at the address 2005.

Step 1: Program Counter places the 16-bit memory address on the address bus. In Timing diagram, during the T-state T_1 the higher order memory address 20H is placed on the address lines A_{15}–A_8, the lower order address 05H is placed on the bus AD_7–AD_0, and ALE signal goes high. The status signal $IO/\overline{M}$ goes low, indicating this is a memory-related operation.

Step 2: The control unit sends the control signal RD to enable the memory chip. The control signal $\overline{RD}$ is sent out during the clock period T_2, thus enabling the memory chip. This signal is active for two clock periods.

Step 3: The byte from the memory location is placed on the data bus. When the memory is enabled, the instruction byte (4F) is placed on the bus AD_7–AD_0 and transferred to the microprocessor. The RD signal causes 4F to be placed on the multiplexed bus and when RD goes high, it causes the bus to go into high impedance.

Step 4: The byte is placed in the instruction decoder of the microprocessor and the task is carried out according to the instruction.

The instruction decoder decodes the machine code and the contents of the accumulator are copied into register C. This task is performed during period T_4.

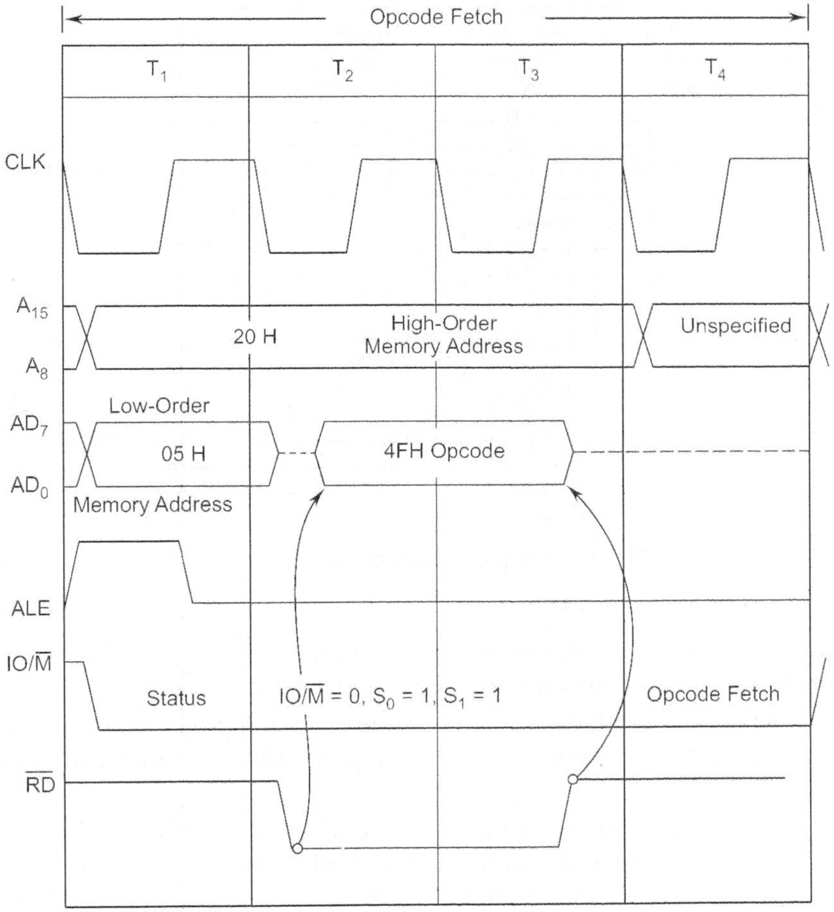

Fig. 2.11 Timing Diagram of Opcode Fetch

If we observe the timing diagram above, demultiplexing AD_7–AD_0 becomes apparent. It clearly shows that, lower order address (05H) is lost after the first clock period. This address needs to be latched and used for identifying the memory address. If the bus AD_7–AD_0 is used to identify the memory location 2005H, the address will change to 204FH after the first clock period.

The following Fig. 2.12 shows a schematic that uses a latch and ALE signal to demultiplex the bus. The bus AD_7–AD_0 is connected as the input to the latch 74LS373. The ALE signal is connected to the enable pin of the latch and the output control signal of latch is grounded.

ALE goes high during T_1. When ALE is high, the latch is transparent and output changes according to input. During T_1, the output of the latch is 05H. When ALE goes low, the data byte 05H is latched until next ALE. And the output of latch represents the lower order address bus A_7–A_0.

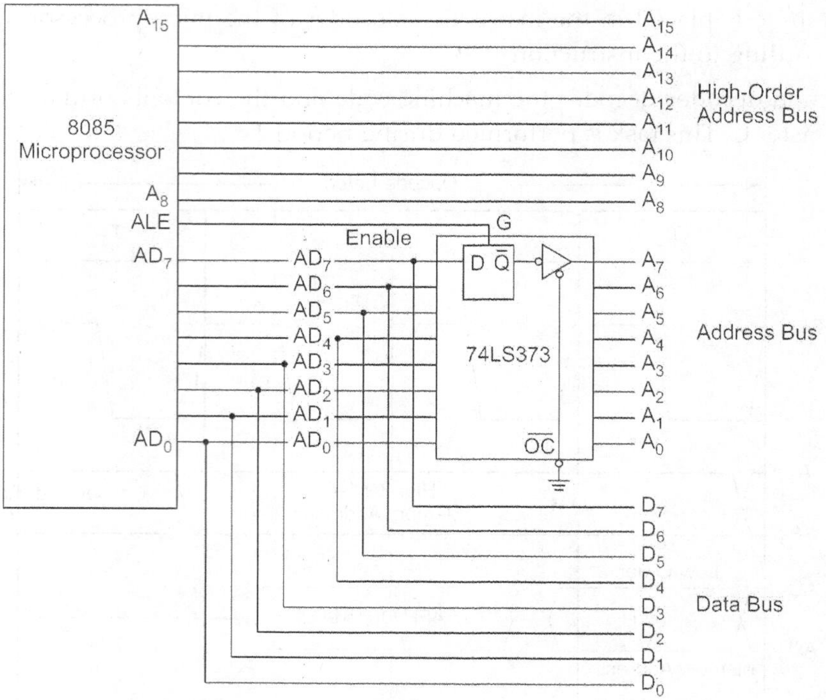

Fig. 2.12 Demultiplexing of Lower Order Address Bus

Now, we can define following terms:

(*a*) **Instruction Cycle:** It is the time required to complete the execution of an instruction. 8085 has instructions which consume one to six machine cycles.

(*b*) **Machine Cycle:** It is defined as the time required to complete one operation of accessing memory, I/O, or acknowledging an external request. This cycle may consists of three to six T-states

(*c*) **T-State:** It is defined as one subdivision of the operation performed in one clock period. These subdivisions are internal states synchronized with the system clock and each T-state is precisely equal to one clock period.

(*d*) **Tri-state logic** devices have three stages: logic 1, logic 0 and high impedance. The term TRI-STATE is a trade mark of National Semiconductor and is used to represent three logic states. A tri-state device has a third line called enable. When this line is activated, the device functions the same way as ordinary logic devices. When this third line is disabled, the logic device goes into high impedance state-as if it were disconnected from the system. Ordinarily, current is required to drive a device in logic 0 or logic 1 states. In high impedance state, practically no current is drawn from the system.

2.7 BASIC MICROPROCESSOR INSTRUCTIONS

Microprocessors perform various basic operations including data transfer instructions, arithmetic instructions, logic instructions, control transfer instructions and machine control instructions.

(*a*) Data Transfer Instructions

Data transfer instructions transfer data from one location designated as the source location to another location designated as the destination. The data transfer could take place from one register to another, from one memory location to another memory location, from a memory location to a register or from a register to a memory location, and so on. In fact, they are more correctly referred to as data movement operations as the contents of the source are not transferred but are copied into the destination register without modifying the contents of the source. It may be mentioned here that these operations do not affect the flags. Data transfer operations of the 8085 microprocessor are of three types, namely MOVE, LOAD and STORE:

MOV destination, source:	Copy data from the source to the destination location
LDA address:	Copy the data byte at the memory location specified by the 16-bit address into the accumulator
STA address :	Copy the data from the accumulator to the memory location specified by the 16-bit address.

(*b*) Arithmetic Instructions

Arithmetic instructions performed by microprocessors include addition, subtraction, multiplication, division, comparison, negation, increment and decrement. It may be mentioned here that most of the eight-bit microprocessors do not support multiplication and division operations. These operations are supported by the 16-bit and 32-bit microprocessors. The arithmetic operations supported by the 8085 microprocessor are addition, subtraction, increment and decrement operations. Examples are as follows:

ADD R:	Adds the contents of the register to the accumulator
ADI eight-bit:	Adds the eight-bit data to the accumulator
SUB R:	Subtracts the contents of the register from the accumulator
SUI eight-bit:	Subtracts eight-bit data from the contents of the accumulator
INR R:	Increments the contents of the register
DCR R:	Decrements the contents of the register

(*c*) Logic Instructions

Microprocessors can perform all the logic functions of hard-wired logic. The basic logic operations performed by all microprocessors are AND, OR, NOT and EXCLUSIVE-OR. Other logic operations include 'shift' and 'rotate' operations. All these operations are performed on a bit-for-bit basis on bytes or words. For instance, 11111111 AND 10111010 equals 10111010, and 11111111 OR 10111010 equals 11111111. Some microprocessors also perform bit-level instructions such as 'set bit', 'clear bit' and 'complement bit' operations. It may be mentioned that logic operations always clear the carry and overflow flags, while the other flags change to reflect the condition of the result.

The basic shift operations are the 'shift left' and 'shift right' operations. In the *shift left* operation, also known as the arithmetic shift left, all bits are shifted one position to the left, with the rightmost bit set to '0' and the leftmost bit transferred to the carry position in the status register. In the *shift right* operation, also known as logic shift right, all bits are shifted one bit position to the right, with the leftmost bit set to '0' and the rightmost bit transferred to the carry position in the status register.

If in the shift right operation the leftmost bit is left unchanged, it is called arithmetic shift right. In a 'rotate' operation, the bits are circulated back into the register. Carry may or may not be included.

As an illustration, in a 'rotate left' operation without carry, the leftmost bit goes to the rightmost bit position, and, in a 'rotate right' with carry included, the rightmost bit goes to the carry position and the carry bit takes the position of the leftmost bit.

Examples of logic instructions performed by the 8085 microprocessor include the following:

ANA R/M: Logically AND the contents of the register/memory with the contents of the accumulator

ANI eight-bit: Logically AND the eight-bit data with the contents of the accumulator

ORA R/M: Logically OR the contents of the register/memory with the contents of the accumulator

ORI eight-bit: Logically OR the eight-bit data with the contents of the accumulator

XRA R/M: Logically EXCLUSIVE-OR the contents of the register memory with the contents of the accumulator

XRI eight-bit: Logically EXCLUSIVE-OR the eight-bit data with the contents of the accumulator

CMA: Complement the contents of the accumulator

RLC: Rotate each bit in the accumulator to the left position

RRC: Rotate each bit in the accumulator to the right position

(*d*) Control Transfer or Branch or Program Control Instructions

Microprocessors execute machine codes from one memory location to the next, that is, they execute instructions in a sequential manner. Branch instructions change the flow of the program either unconditionally or under certain test conditions. Branch instructions include 'jump', 'call', 'return' and 'interrupt'.

'Jump' instructions are of two types, namely 'unconditional jump' instructions and 'conditional jump' instructions. If the microprocessor is so instructed as to load a new address in the program counter and start executing instructions at that address, it is termed an unconditional jump. In the case of a conditional jump, the program counter is loaded with a new instruction address only if and when certain conditions are established by the microprocessor after reading the appropriate status register bits. 'Call' instructions transfer the flow of the program to a subroutine. The 'call' instruction differs from the 'jump' instruction as 'call' saves a return address (the address of the program counter plus one) on the stack. The 'return' instruction returns control to the instruction whose address was stored in the stack when the 'call' instruction was encountered. 'Interrupt' is a hardware-generated call (externally driven from a hardware signal) or a software-generated call (internally derived from the execution of an instruction or by some internal event). Examples of transfer control instructions of the 8085 microprocessor are as follows:

JMP 16-bit address: Change the program sequence to the location specified by the 16-bit address

JZ 16-bit address: Change the program sequence to the location specified by the 16-bit address if a zero flag is set

JC 16-bit address: Change the program sequence to the location specified by the 16-bit address if a carry flag is set

CALL 16-bit address: Change the program sequence to the location of the subroutine specified by the 16-bit address

RET: Return to the calling program

(*e*) Machine Control Instructions

Machine control instructions include HALT and NOP instructions. Machine control instructions performed by the 8085 microprocessor include the following:

HLT: Stop processing and wait

NOP: No operation

2.8 ADDRESSING MODES

Microprocessors perform operations on data stored in the register or memory. These data are specified in the operand field of the instruction. The data can be specified in various ways as a direct data value or stored in some register or memory location, and so on. These are referred to as the *addressing modes* of the microprocessor. In other words, the addressing mode as expressed in the instruction tells us how and from where the microprocessor can get the data to act upon. Addressing modes are of direct relevance to compiler writers and to programmers writing the code in assembly language.

Different microprocessor architectures provide a variety of addressing modes. The most commonly used addressing modes are absolute, immediate, register direct, register indirect, indexed, program counter relative, implicit and relative addressing modes. They account for more than 90 % of the total addressing modes.

Absolute or Memory Direct Addressing Mode

In *absolute addressing mode,* the data are accessed by specifying their address in the memory [Fig. 2.13 (a)]. This mode is useful for accessing fixed memory locations, such as memory mapped I/O devices. For example, the instruction MOV A, 30H in the 8085 microprocessor moves the contents of memory location 30H into the accumulator [Fig. 2.13 (b)]. In this case the accumulator has the value 07H.

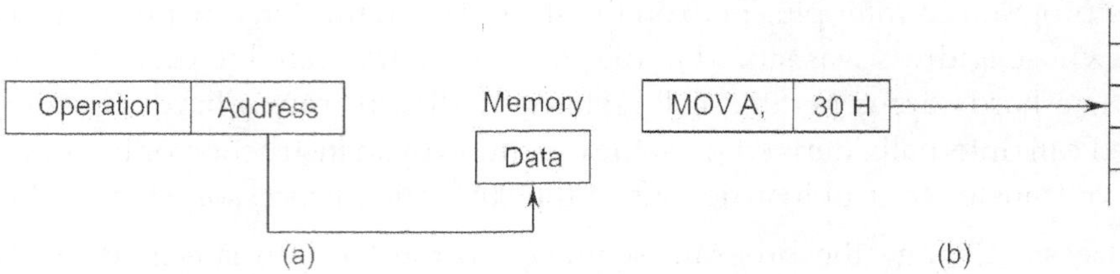

Fig. 2.13 Absolute Addressing Mode

Immediate Addressing Mode

In *immediate addressing mode* the value of the operand is held within the instruction itself (Fig. 2.14). This mode is useful for accessing constant values in a program. It is faster than the absolute addressing mode and requires less memory space. For example, the instruction MVI A, #30H moves the data value 30H into the accumulator. The sign # in the instruction tells the assembler that the addressing mode used is immediate.

Fig. 2.14 Immediate Addressing Mode

Register Direct Addressing Mode

In *register direct addressing mode,* data are accessed by specifying the register name in which they are stored [Fig. 2.15 (a)]. Operations on registers are very fast, and hence instructions in this mode require less time than absolute addressing mode instructions. As an example, the instruction MOV A, R1 in the 8051 microprocessor moves the contents of register R1 into the accumulator [Fig. 2.15 (b)]. The contents of the accumulator after the instruction are 07H.

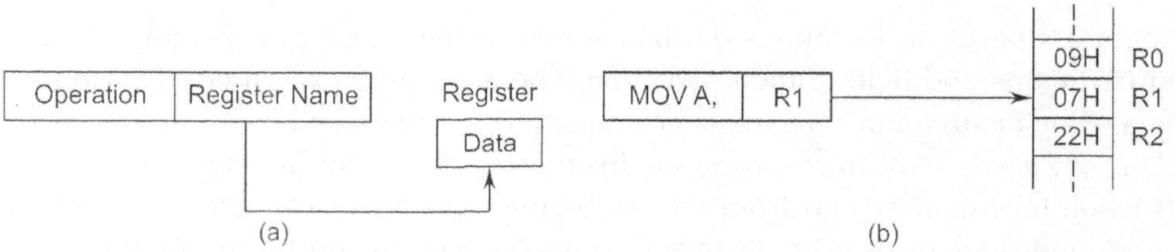

Fig. 2.15 Register Direct Addressing Mode

Register Indirect Addressing Mode

In all the modes discussed so far, either the value of the data or their location is directly specified. The *indirect addressing mode* uses a register to hold the actual address where the data are stored. That is, in this case the memory location of the data is stored in a register [Fig. 2.16 (a)]. In other words, in indirect addressing mode, the address is specified indirectly and has to be looked up.

This addressing mode is useful when implementing the pointer data type of high-level language. In the 8085 microprocessor, the R0 and R1 registers are used as an eight-bit index and the DPTR as a 16-bit index. The mnemonic symbol used for indirect addressing is @. As an example, the instruction MOV A, @R0 moves the contents of the memory location whose address is stored in R0 into the accumulator. The value of the accumulator in this example is 07H [Fig. 2.16 (b)]. This addressing mode can also be enhanced with an offset for accessing data structures in data space memory. This is referred to as *register indirect* with displacement. As an example, the instruction MOVC A, @A+DPTR copies the code byte at the memory address formed by adding the contents of A and DPTR to A.

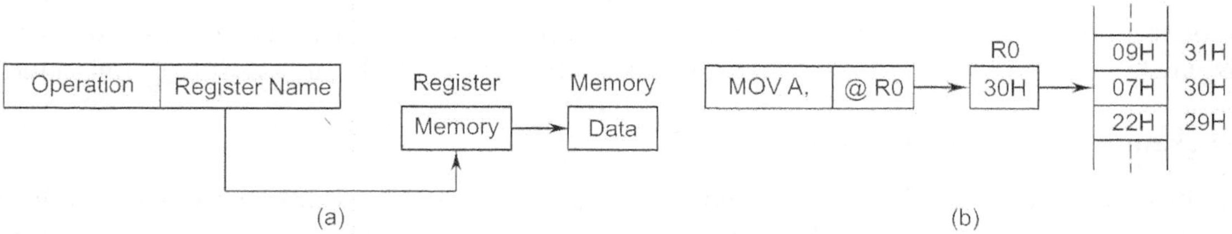

Fig. 2.16 Register Indirect Addressing Mode.

Indexed Addressing Mode

In the *indexed addressing mode,* the address is obtained by adding the contents of a register to a constant (Fig. 2.17). The instruction 'move the contents of accumulator A to the memory location whose address is given by the contents of register plus constant is an example of indexed addressing. The indexed addressing mode is useful whenever the absolute location of the data is not known until the program is running. This addressing mode is used to access a continuous table or array of data items stored in memory. The content of the constant gives the starting address, while the contents of the register determine the element of the array or table to be accessed. If the program counter is used in the indexed addressing mode, it is known as the program counter relative addressing mode.

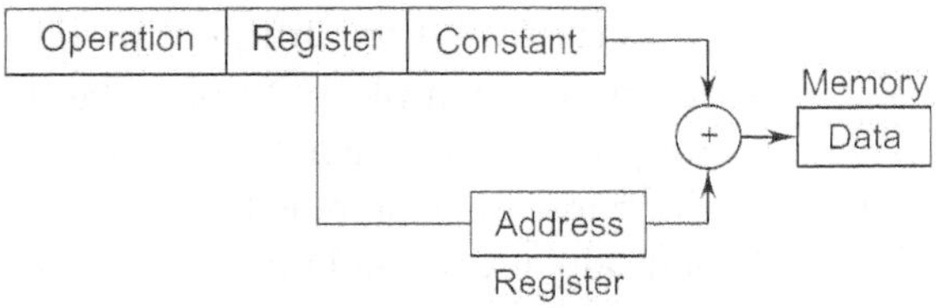

Fig. 2.17 Indexed Addressing Mode

Implicit Addressing Mode and Relative Addressing Mode

In *implicit addressing mode,* no operand is used in the instruction and the location of the operand is obvious from the instruction itself. Examples include 'clear carry flag', 'return from subroutine' and so on.

The relative addressing mode is used for 'jump' and 'branch' instructions only. In this, a displacement is added to the address in the program counter and the next instruction is fetched from the new address in the program counter. This mode is particularly useful in connection with conditional jumps.

2.9 PERIPHERAL DEVICES

Microprocessors and peripheral devices provide a complete solution in increasingly complex application environments. A peripheral device typically belongs to the category of MSI logic devices. This section gives an introduction to the popular peripheral devices that are used along with the microprocessor in a microcomputer system. The different peripheral devices used in a microcomputer system include a programmable counter/timer, a programmable peripheral interface (PPI), EPROM, RAM, a programmable interrupt controller (PIC), a direct memory access (DMA) controller, a programmable communication interface – a universal synchronous/asynchronous receiver/transmitter (USART), a math coprocessor, a programmable keyboard/display interface, a CRT controller, a floppy disk controller and clock generators and transceivers.

Programmable Timer/Counter

The programmable timer/counter is used for the generation of an accurate time delay for event counting, rate generation, complex waveform generation applications and so on. Examples of programmable timer/counter devices include Intel's 8254 and 8253 family of devices. Intel 8254 contains three 16-bit counters that can be programmed to operate in several different modes. Some of the functions common to microcomputers and implementable with 8254 are a real-time clock, an event counter, a digital one-shot, programmable rate generator, a square-wave generator, a binary rate multiplier, a complex waveform generator and a complex motor controller.

Programmable Peripheral Interface

Programmable peripheral interface (PPI) devices are used to interface the peripheral devices with the microprocessors. 8255 PPI is a widely used programmable parallel I/O device. 8255 can be programmed to transfer data under various conditions, from simple I/O to interrupt I/O. It can function in bit reset (BSR) mode or I/O mode. In I/O mode it has three ports, namely port A, port B and port C. The I/O mode is further divided into three different modes, namely mode 0, mode 1 and mode 2. In mode 0, all ports function as simple I/O ports. Mode 1 is a handshake mode whereby port A and/or B use bits from port C as handshake signals. In mode 2, port A can be set up for bidirectional data transfer using handshake signals from port C, and port B can be set up either in mode 0 or in mode 1. In BSR mode, individual bits in port C can be set or reset.

Programmable Interrupt Controller

A programmable interrupt controller (PIC) is a device that allows priority levels to be assigned to its interrupt outputs. It functions as an overall manager in an interrupt-driven system environment. When the device has multiple interrupt outputs, it will assert them in the order of their relative priority.

Common modes of a PIC include hard priorities, rotating priorities and cascading priorities. Intel 8259 is a family of programmable interrupt controllers (PICs) designed and developed for use with the Intel 8085 and Intel 8086 microprocessors. The family originally consisted of the 8259, 8259A, and 8259B PICs, although a number of manufacturers make a wide range of compatible chips today.

It handles up to eight vectored priority interrupts for the CPU. It is designed to minimize the software and real-time overhead in handling multi-level priority interrupts. It accepts requests from peripheral equipment, determines which of the incoming requests is of the highest priority, ascertains whether an incoming request has a higher priority value than the level currently being serviced and issues an interrupt to the CPU on the basis of this determination.

DMA Controller

In a direct memory access (DMA) data transfer scheme, data are transferred directly from an I/O device to memory, or vice versa, without going through the CPU. The DMA controller is used to control the process of data transfer. Its primary function is to generate, upon a peripheral request, a sequential memory address that will allow the peripheral to read or write data directly to or from memory. One of the popular known programmable DMA controllers is Intel's 8257. It is a four-channel direct memory access (DMA) controller. It is specifically designed to simplify the transfer of data at high speeds for microcomputer systems. It has a priority logic that resolves the peripheral requests and issues a composite hold request to the CPU. It maintains the DMA cycle count for each channel and outputs a control signal to notify the peripheral that the programmed number of DMA cycles is completed.

Programmable Communication Interface

Programmable communication interfaces (PCIs) are interface devices that are used for data communication applications with microprocessors. They basically convert the data from the microprocessor into a format acceptable for communication and also convert the incoming data into a format understood by the microprocessor.

8251 is a PCI device designed for Intel's 8085, 8086 and 8088 microprocessors and is used in serial communication applications. It is basically a universal synchronous/asynchronous receiver/transmitter (USART) that accepts data characters from the CPU in parallel format and then converts them into a continuous serial data stream for transmission.

Simultaneously, it can receive a serial data stream and convert it into parallel data characters for the CPU. The USART will signal the CPU whenever it can accept a new character for transmission or whenever it has received a character for the CPU.

Math Coprocessor

Math coprocessors are special-purpose processing units that assist the microprocessor in performing certain mathematical operations. The arithmetic operations performed by the coprocessor are floatingpoint operations, trigonometric, logarithmic and exponential functions and so on. Examples include Intel's 8087, 80287, etc. The 8087 numeric coprocessor provides the instructions and data types needed for high-performance numeric application, providing up to 100 times the performance of a CPU alone.

Another widely used math coprocessor is 80287. The 80287 numeric processor extension (NPX) provides arithmetic instructions for a variety of numeric data types in 80286 systems. It also executes numerous built-in transcendental functions (*e.g.*, tangent and log functions).

Programmable Keyboard/Display Interface

Programmable keyboard/display interfaces are devices used for interfacing the keyboard and the display to the microprocessor. The keyboard section of the device debounces the keyboard entries and provides data to the microprocessor in the desired format. The display section converts the data output of the microprocessor into the form desired by the display device in use.

8279 is a general-purpose programmable keyboard and display I/O interface device designed for use with Intel microprocessors. The keyboard portion can provide a scanned interface to a 64-contact key matrix. Keyboard entries are debounced and strobbed in eight-character FIFO. If more than eight characters are entered, overrun status is set. Key entries set the interrupt output line to the CPU. The display portion provides a scanned display interface for LED, incandescent and other popular display technologies. Both numeric and alphanumeric segment displays may be used. The 8279 has a 16×8 display RAM.

Programmable CRT Controller

The programmable CRT controller is a device to interface CRT raster scan displays with the microprocessor system. Its primary function is to refresh the display by buffering the information from the main memory and keeping track of the display position of the screen. One of the commonly used programmable CRT controllers is Intel's 8275H. It allows a simple interface to almost any raster scan CRT display with minimum external hardware and software overheads. The number of display characters per row and the number of character rows per frame are software programmable.

Floppy Disk Controller

The floppy disk controller is used for disk drive selection, head loading, the issue of read/write commands, data separation and serial-to-parallel and parallel-to-serial conversion of data. Examples of floppy disk controllers include Intel's 82078, 82077 and 8272.

Clock Generator

The clock generator is a circuit that produces a timing signal for synchronization of the circuit's operation. Examples of clock generators used in microprocessor systems include 8284 and 82284.

8284 generates the system clock for the 8086 and 8088 processors. It requires a crystal or a TTL signal source for producing clock waveforms. It provides local READY and MULTIBUS READY synchronization.

82284 is a clock generator/driver that provides clock signals for the 80286 processor and support components. It also contains logic to supply READY to the CPU from either asynchronous or synchronous sources and synchronous RESET from an asynchronous input with hysteresis. The 82284 is packaged in 18-pin DIP and contains a crystal-controlled oscillator, an MOS clock generator, a peripheral clock generator, multibus ready synchronization logic and system reset generation logic.

Octal Bus Transceiver

Bus transceivers are devices with a high-output drive capability for interconnection with data buses.

In a microprocessor-based system, they provide an interface between the microprocessor bus and the system data bus. 8286 is an eight-bit bipolar transceiver with a three-state output that is used in a wide variety of buffering applications in microcomputer systems. It comes in a 20-pin DIP package.

2.10 ARCHITECTURAL ADVANCEMENT OF MICROPROCESSOR

In the previous sections, the basics of the workings and architecture of the central processing unit has been explained. There has been a general look at simple processor architecture, an explanation of the method by which instructions are executed, and how the various different addressing modes affect how the CPU processes instructions. However, modern CPUs are very rarely as simple as the ones that have been discussed thus far.

While the information covered up to this point is still applicable and relevant to the majority of microprocessors, many refinements to the workings and architecture have also been implemented.

(*a*) **Pipelining :** Up until this point in chapter, we have assumed that the processor is only able to process one instruction at a time. All examples have shown an instruction having to be executed in full before the next one can be started on. However, this is not how modern CPUs work. Pipelining is the name given to the process by which the processor can be working on more than one instruction at once.

The simplest way to approach pipelining is to consider the three stage fetch, decode and execute instruction execution cycle outlined earlier. There are times during each of these subcycles of the main cycle where the main memory is not being accessed, and the CPU could be considered 'idle'. The idea, therefore, is to begin the fetch stage for a second instruction while the first stage is being decoded. Then, when instruction one is being executed and instruction two is being decoded, a third instruction can be fetched.

While pipelining can severely cut the time taken to execute a program, there are problems that cause it to not work as well as it perhaps should. The three stages of the instruction execution process do not necessarily take an equal amount of time, with the time taken for 'execute' being generally longer than 'fetch'. This makes it much harder to synchronise the various stages of the different instructions. Also, some instructions may be dependent on the results of other earlier instructions. This can arise when data produced earlier needs to be used, or when a conditional branch based on a previous outcome is used.

One of the simplest ways in which the effects of these problems can be reduced is by breaking the instruction execution cycle into stages that are more likely to be of an equal duration. For example, the diagram below shows how the cycle can be broken down into six stages rather than three:

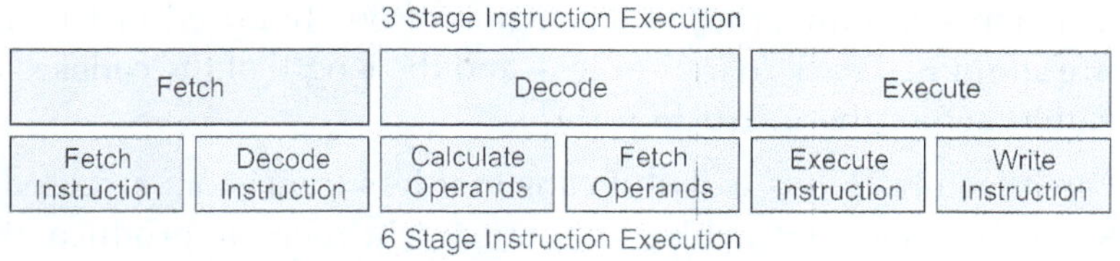

Fig. 2.18 Pipelining Process

However, while this may solve some of the problems outlined above, it is not without creating further problems of its own. Firstly, it is not always the case than an instruction will use all six of these stages. Simple load instructions, for example, will not require the use of the final 'write operand' stage, which would possibly upset the synchronisation.

There is also the matter of potential conflicts within the memory system, as three of the above stages (fetch instruction, fetch operands, write operand) require access to the memory. Many memory management systems would not allow three seperate instructions to be accessing the memory at once, and hence the pipelining would not be as beneficial as it would first seem.

On top of this, the problem of conditional branching and result dependant instructions also occurs. This means that the processor needs to be designed well in order to cope with these potential interruptions to the flow of data. As you can tell, there are many issues which need to be taken into consideration relating to the technique of pipelining. While it is a powerful technique for the purpose of increasing CPU performance, it does require careful design and consideration in order to achieve the best possible results.

(*b*) **CISC and RISC :** Years of development have been undertaken into improving the architecture of the central processing unit, with the main aim of improving performance. Two competing architectures were developed for this purpose, and different processors conformed to each one. Both had their strengths and weaknesses, and as such also had supporters and detractors.

CISC: Complex Instruction Set Computers

Earlier developments were based around the idea that making the CPU more complex and supporting a larger number of potential instructions would lead to increased performance. This idea is at the root of CISC processors, such as the Intel x86 range, which have very large instruction sets reaching up to and above three hundred separate instructions. They also have increased complexity in other areas, with many more specialized addressing modes and registers also being implemented, and variable length of the instruction codes themselves.

Performance was improved here by allowing the simplification of program compilers, as the range of more advanced instructions available led to less refinements having to be made at the compilation process. However, the complexity of the processor hardware and architecture that resulted can cause such chips to be difficult to understand and program for, and also means they can be expensive to produce.

RISC: Reduced Instruction Set Computers

In opposition to CISC, the mid-1980s saw the beginnings of the RISC philosophy. The idea here was that the best way to improve performance would be to simplify the processor workings as much as possible. RISC processors, such as the IBM PowerPC processor, have a greatly simplified and reduced instruction set, numbering in the region of one hundred instructions or less. Addressing modes are simplified back to four or less, and the length of the codes is fixed in order to allow standardization across the instruction set.

Changing the architecture to this extent means that less transistors are used to produce the processors. This means that RISC chips are much cheaper to produce than their CISC counterparts. Also the reduced instruction set means that the processor can execute the instructions more quickly, potentially allowing for greater speeds. However, only allowing such simple instructions means a greater burden is placed upon the software itself. Less instructions in the instruction set means a greater emphasis on the efficient writing of software with the instructions that are available. Supporters of the CISC architecture will point out that their processors are of good enough performance and cost to make such efforts not worth the trouble.

Table 2.5 Comparison of CISC and RISC

CISC		RISC
Large (100 to 300)	**Instruction Set**	Small (100 or less)
Complex (8 to 20)	**Addressing Modes**	Simple (4 or less)
Specialised	**Instruction Format**	Simple
Variable	**Code Lengths**	Fixed
Variable	**Execution Cycles**	Standard for most
Higher	**Cost/CPU Complexity**	Lower
Compilation	**Simplifies**	Processor design
Processor design	**Complicates**	Software

Looking at the most modern processors, it becomes evident that the whole rivalry between CISC and RISC is now not of great importance. This is because the two architectures are converging closer to each other, with CPUs from each side incorporating ideas from the other. CISC processors now use many of the same techniques as RISC ones, while the reduced instruction sets of RISC processors contain similar numbers of instructions to those found in certain CISC chips. However, it is still important that you understand the ideas behind these two differing architectures, and why each design path was chosen.

2.11 MODERN ARCHITECTURES

Cache Memory

This is a small amount of high-speed memory used specifically as a fast and effective method of storage for commonly used instructions. Most programs end up accessing the same data and instructions over and over again at some point in their execution. Placing these in higher speed storage, such as a cache, provides a great improvement in the time taken for processing over continual accessing from the main memory at a slower speed.

Home computer processors traditionally have implemented the cache directly into their architecture, in what is known as a 'Level 1' cache. The most modern CPUs also make use of external caches, which are referred to as 'Level 2' cache and much larger in size than 'Level 1' caches. More recent processors have larger caches - for instance, the Intel 486 had a cache of only eight kilobytes, while the Pentium II used multiple stores totalling up to two megabytes of storage space.

Specialised Instruction Set Extensions

The most commonly known extensions to the traditional CPU instruction set are Intel's MMX and AMD's 3D technology. These both come into use when the processor is asked to perform operations involving graphics, audio and video, and consist of a number of specific instructions which are specialized to perform the short repetitive tasks that make up the large majority of multimedia processing. These extensions use SIMD (Single Instruction, Multiple Data) instructions in order to greatly reduce the time taken, as such instructions perform their operations to multiple pieces of data at the same time.

MMX makes use of fifty-seven SIMD instructions, while the Pentium 4 raises this number to one hundred and forty-four. This includes further extensions to improve operations relating to internet-related activity, such as the streaming of music and video files. The improved 3D technology found in the AMD Athlon processor also contains SIMD instructions for this purpose. Such extensions ultimately enhance the performance of the processor in activites relating to gaming, multimedia applications, and use of the internet and other forms of communication.

Hyperthreading

Hyperthreading is a new technology, introduced by Intel with their most recent Pentium 4 processors. It works by using what is known as 'simultaneous multithreading' to make the single processor appear to the computer operating system as multiple logical processors. This enables the CPU via use of shared hardware resources to execute multiple seperate parts of a program (or 'threads') at the same time.

This technology does not provide the same performance increase as actual seperate processors would do, but provides a considerable boost for less cost and power consumption than said multiple processors would require. Current processors such as the aforementioned Pentium 4 currently split the CPU into two logical processors. Intel are currently working on further advancements which will enable splitting higher numbers of threads to be simultaneously executed.

2.12 PROCESSOR SELECTION

For all but a very few applications, more than one right solution is possible because several microprocessors can meet the requirements. As with most real-world engineering decisions, the selection consists of a series of tradeoffs between cost and functionality. The specific selection process will depend on the complexity of the finished product, but the following items must be taken into consideration:

1. Number of I/O pins required
2. Interfaces required
3. Memory requirements
4. Number of interrupts required
5. Real-time considerations
6. Development environment
7. Processing speed required
8. ROMability
9. Memory architecture
10. Power requirements
11. Environmental requirements
12. Life cycle costs
13. Operator training/competence
14. The "real" requirements

2.13 BASIC PROCESSOR DESIGN

The operations of most modern computers are modeled based on the Von Neumann machine. The microprocessor executes each instruction in a series of small steps.

1. Fetch the instruction from memory into the instruction register
2. Change the program counter to point to the following instruction.
3. Determine the type of instruction fetched.
4. If the instruction uses a word in memory, determine where it is.
5. Fetch the word, into a register if needed.
6. Execute the instruction
7. Go to step 1.

This sequence is referred to as the fetch-decode-execute cycle, central to the operation of most modern processors.

Producing a faster processor is usually pursued through the following strategies:

Direct Execution by Hardware

All common instructions are directly executed in hardware, not interpreted by microinstructions thus making the common instructions execute at high speed.

Easy Instruction Decoding

A critical limit on the rate of issue of instructions is decoding individual instructions to determine the required resources. In order to aid this process, instructions should be made regular and have fixed length with a small number of fields.

Higher Clock Rate

All else being equal, the higher the clock rate, a processor can process the more instructions in a given time. So it is in the last decade we have seen new processors coming out with higher and higher clock rate. While 25Mhz was said to have 'rocket like speed' in 1988, nowadays 1GHz or more can be expected from top-of-the-line models, up to 1.7 GHz for Pentium 4 from Intel.

Actually, achieving a high clock rate has little to do with advancement in processor architecture. It is primarily determined by the currently available fabrication technology, which is usually specified in terms of how far apart are the traces in an integrated circuit. Currently, most companies are moving their production facilities from 0.18 micron to the 0.13-micron production line. To understand the relation between achievable clock rate and trace width, remember that the speed of electricity in a copper wire is limited. Assuming it is about 20cm per nanosecond, then a CPU with 1GHz clock will have 1ns time between cycles, which is just enough time to travel 20cm (we are ignoring lost time due to the time needed to switch gates polarity etc.). It should now be clear that CPUs on which the transistor elements are more tightly packed can accomplish the same amount of work in less time than otherwise. In short, the smaller the traces are, the higher the clock rate can be.

Another way to decrease the distance between components and therefore increase the speed of communication between them is to provide multiple levels of metal for interconnection. Current processors use between 4 and 6 *layers* of metal interconnections. Its down side is it is more expensive to fabricate.

Additional benefit that comes from a smaller trace and more layers is a higher *chip density*: more transistors can be fitted inside the same chip area. Chips designers gladly use this available area to add more functionality into the chip, such as bigger caches, more registers, or more complex logic supports.

On the other hand, the combination of smaller traces, more transistors, and high clock rate give rise to a more severe heat problem. To combat this (and prevent a CPU meltdown), most CPU nowadays requires huge heat sinks and a dedicated 'CPU fan'.

More effectively, to produce less heat, current processors are designed to work at a lower voltage than their predecessor, 2.0 or even 1.6 volt compared to the 5.0 or 3.3 volt of older processors. But going down in voltage also produces a new problem: it is more difficult to detect the difference between logical 0 and logical 1. Which means the electrical tolerances for system components are more sensitive (equates to costlier system). As a way out of this, some processors work with 'dual-voltage', a lower internal one for the CPU core and a higher external one for I/O.

Higher Instructions Per Cycle Rate

This is where most of the new ideas in CPU architecture come into being. The basic idea is to get more program instructions done in a fixed number of clock cycles. Alternately, the same goal can be expressed as "lowering the average CPI" (Cycles Per Instruction).

Maximize Instruction Issue Rate

Parallelism plays a critical role in improving performance. Processors should be able to process multiple instructions simultaneously, thus large number of slow instructions can be issued in a short time, improving the processors MIPS rating.

2.14 TYPICAL MICROPROCESSOR DEVELOPMENT SCHEMES

To develop applications on a microprocessor, some basic tools are essential:

(*a*) A development system or cross compiler
(*b*) A PROM programmer
(*c*) Debug hardware.

In the prehistoric days of systems (before the IBM PC), the standard development system consisted of a computer from the company that sold the microprocessor ICs and a PROM programmer. The development systems were expensive, slow, and limited to developing software only for that manufacturer's parts. Some third-party companies had development systems as well. These also were expensive and slow but could often be upgraded (at a huge cost) with hardware that would permit software development for more than one manufacturer's parts.

It is unarguable that the standardization of the business world around the IBM PC and its derivatives has been a real advantage to the embedded systems developer. Most manufacturers of microprocessor ICs now provide development software instead of systems for their parts. These cross-compilers run on a PC to compile or assemble code for the manufacturer's microprocessor. Some manufacturers even give away some sort of development tools (usually an assembler) to potential customers on the premise that they are in the IC business, not the software business. It is unknown how many microprocessor selections have been made based on the availability of these free tools, but the number must be large.

Many IC manufacturers still provide complete development systems for their parts. These are usually PCs with the manufacturer's software included, and they constitute a complete development environment.

Some will argue that the PROM programmer no longer is an essential development tool, and they are right. If the project is to be developed in RAM or on an embedded PC or on a flash-based, downloadable processor, a PROM programmer is not needed.

However, some projects still are developed in an environment in which parts have to be erased and programmed every time the code changes; for those developers, a PROM programmer is needed.

When the development system consists of a cross compiler and a PROM programmer and little else, debugging is simple, although often tedious. The code is run, the operation of the system is observed, and the code is examined to see why things do not work. This process is repeated until all bugs are found. For some systems, especially at small companies, this stare-at-the code method is still used and works well. This method becomes less and less attractive as system complexity grows and development schedules shrink.

The next level of debug is a monitor program, sometimes called a debuggm. This simple program resides in the system and provides commands to examine and alter memory, download code, and insert breakpoints into the code.

A breakpoint is an unconditional branch instruction that takes the code back to the monitor program, where the registers and memory may be examined. Monitor programs require some kind of terminal, and the monitor program itself takes up considerable memory, so they typically are not used with very simple microprocessor.

As microprocessors become more complex, debugging the completed system becomes more difficult. Many designers, especially at large companies, use an emulator for system debugging. The emulator has a probe that replaces the microprocessor IC in the system (the target) and is supposed to run exactly the same as the target part. However, the emulator allows the engineer to insert breakpoints into the code so that the microprocessor's operation can be stopped. While stopped, the memory, internal registers, and other information about the microprocessor can be examined, the same as with a monitor program. In a simple emulator, the breakpoint is typically a specific address-for example at the instruction in the pool timer that turns on the motor relay. More sophisticated emulators have additional hardware that allows breakpoints when specific values are written or read to or from memory, when a specific sequence of instructions is executed, or for many other causes.

One drawback to emulators is their cost. Ranging from a few hundred dollars for a simple microprocessor (such as the Intel 8031 family) to several thousand dollars for a more complicated IC, the cost is often prohibitive.

As microprocessors have grown even faster, their speed has outpaced the emulator industry's ability to keep up. In addition, the use of more powerful processors for applications often means some CPU horsepower is left over after the application is developed. Many developers have moved away from emulators and back to the monitor or debugger programs. These now more-sophisticated programs take advantage of leftover CPU capacity to provide event tracing, throughput measurement, code histograms that show how much time the CPU spends in each section of code, and other powerful debugging information.

REVIEW QUESTIONS

1. What are the different generations of microprocessor evolution?
2. What are different types of microprocessor?
3. Compare the different types of microprocessor.
4. Draw the block diagram of microprocessor architecture and explain operation of its components.
5. What do you understand by addressing modes? Explain different types of addressing modes.
6. What do you understand by interrupts? What are different types of interrupts?
7. What are different data transfer schemes?
8. What do you understand by instruction? Classify them according to operation.

3

8085 8-Bit Microprocessors

- Single +5V Power Supply with 10% Voltage Margins
- 3 MHz, 5 MHz and 6 MHz Selections Available
- 20% Lower Power Consumption than 8085A for 3MHz and 5 MHz
- 1.3 μs Instruction Cycle (8085AH); 0.8 μs (8085AH-2); 0.67 μs (8085AH-1)
- 100% software Compatible with 8080A
- On-Chip Clock Generator (with External Crystal, LC or RC Network)
- On-chip system Controller; Advanced Cycle Status Information Available for Large System Control
- Four Vectored Interrupt Inputs (One is Non-Maskable) Plus an 8080A-compatible interrupt.
- Serial In/Serial Out Port
- Decimal, Binary and Double Precision Arithmetic
- Direct Addressing Capability to 64K Bytes of Memory.
- Available in 40-Lead Cerdip and Plastic Packages.

The Intel 8085AH is a complete 8-bit parallel Central Processing Unit (CPU) implemented in N-channel, depletion load, silicon gate technology (HMOS). Its instruction set is 100% software compatible with the 8080A microprocessor, and it is designed to improve the present 8080A's performance by higher system speed. Its high level of system integration allows a minimum system of three IC's [8085AH(CPU), 8156H (RAM/IO) and 8755A (EPROM/IO)] while maintaining total system expandability. The 8085AH-2 and 8085AH-1 are faster versions of the 8085AH.

The 8085AH incorporates all of the features that the 8224 (clock generator) and 8228 (system controller) provided for the 8080A, thereby offering a higher level of system integration.

The 8085AH uses a multiplexed data bus. The address is split between the 8-bit address bus and the 8-bit data bus. The on-chip address latches of 8155H/8156H/8755A memory products allow a direct interface with the 8085AH.

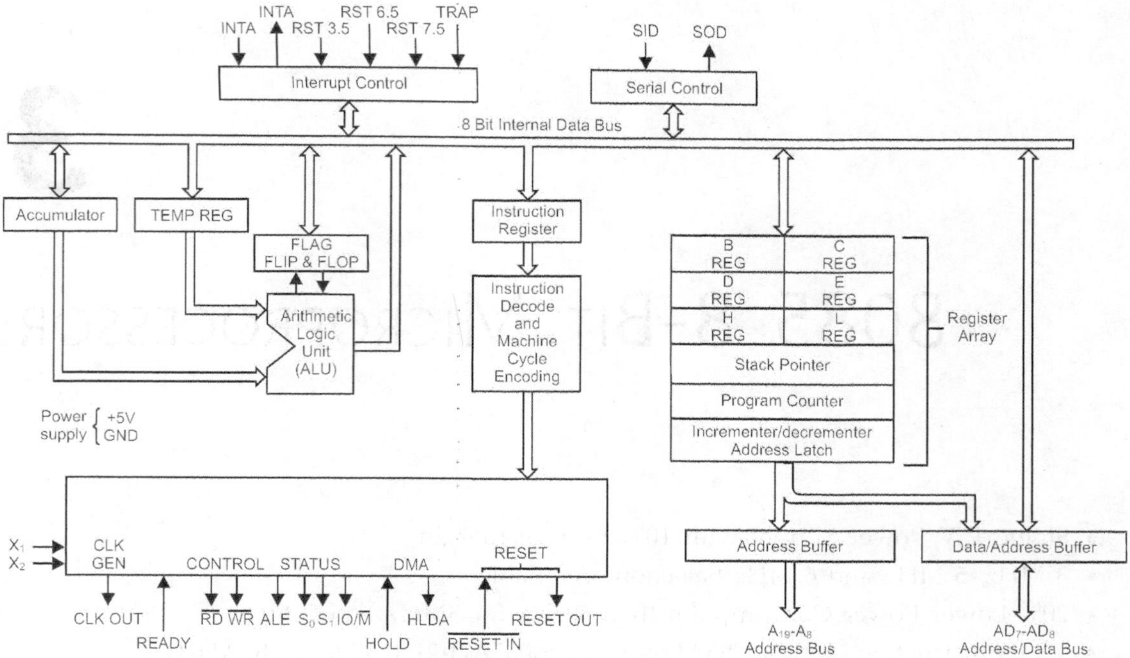

Fig. 3.1 8085AH CPU Functional Block Diagram

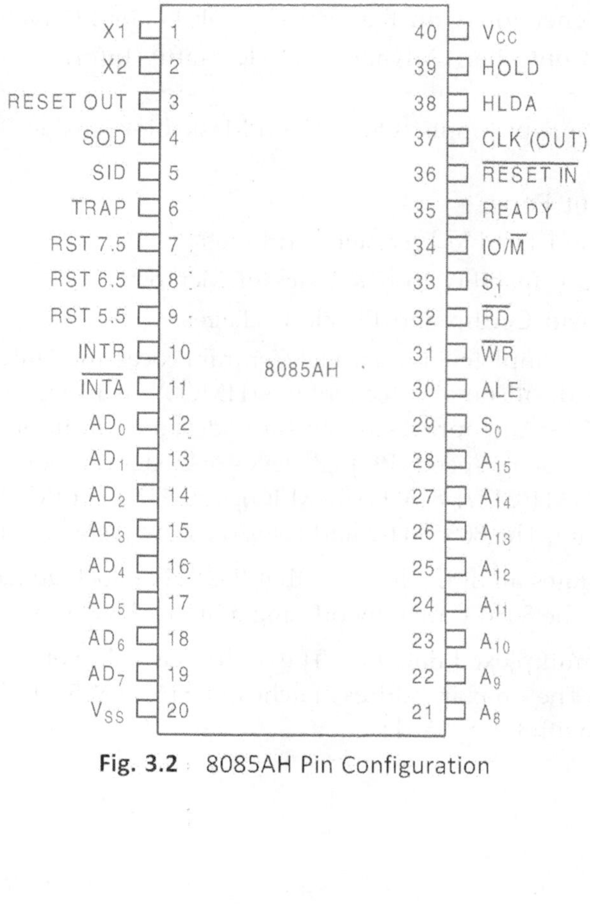

Fig. 3.2 8085AH Pin Configuration

3.1 PIN DESCRIPTION

Table 3.1

<table>
<tr><th>Symbol</th><th>Type</th><th>Name and Function</th></tr>
<tr><td>A8-A15</td><td>O</td><td>ADDRESS BUS: The most significant 8 bits of memory address or the 8 bits of the I/O address, 3-stated during Hold and Halt modes and during RESET.</td></tr>
<tr><td>AD0-7</td><td>I/O</td><td>MULTIPLEXED ADDRESS/DATA BUS: Lower 8 bits of the memory address (or I/O address) appear on the bus during the first clock cycle (T state) of a machine cycle. It then becomes the data bus during the second and third clock cycles.</td></tr>
<tr><td>ALE</td><td>O</td><td>ADDRESS LATCH ENABLE: It occurs during the first clock state of a machine cycle and enables the address to get latched into the on-chip latch of peripherals. The falling edge of ALE is set to guarantee setup and hold times for the address information. The falling edge of ALE can also be used to strobe the status information. ALE is never 3-stated.</td></tr>
<tr><td>S0, S1
and IO/$\overline{M}$</td><td>O</td><td>MACHINE CYCLE STATUS:
<table>
<tr><th>IO/$\overline{M}$</th><th>S1</th><th>S0</th><th>Status</th></tr>
<tr><td>0</td><td>0</td><td>1</td><td>Memory write</td></tr>
<tr><td>0</td><td>1</td><td>0</td><td>Memory read</td></tr>
<tr><td>1</td><td>0</td><td>1</td><td>I/O write</td></tr>
<tr><td>1</td><td>1</td><td>0</td><td>I/O read</td></tr>
<tr><td>0</td><td>1</td><td>1</td><td>Opcode fetch</td></tr>
<tr><td>1</td><td>1</td><td>1</td><td>Interrupt Acknowledge</td></tr>
<tr><td>*</td><td>0</td><td>0</td><td>Halt</td></tr>
<tr><td>*</td><td>X</td><td>X</td><td>Hold</td></tr>
<tr><td>*</td><td>X</td><td>X</td><td>Reset</td></tr>
</table>
* = 3-state (high impedance)
X = unspecified
S1 can be used as an advanced R/$\overline{W}$ status. IO/$\overline{M}$, S0 and S1 become valid at the beginning of a machine cycle and remain stable throughout the cycle. The falling edge of ALE may be used to latch the state of these lines.</td></tr>
<tr><td>$\overline{RD}$</td><td>O</td><td>READ CONTROL: A low level on $\overline{RD}$ indicates the selected memory or I/O device is to be read and that the Data Bus is available for the data transfer, 3-stated during Hold and Halt modes and during RESET.</td></tr>
<tr><td>$\overline{WR}$</td><td>O</td><td>WROTE CONTROL: A low level on $\overline{WR}$ indicates the data on the Data Bus is to be written into the selected memory or I/O location. Data is set up at the trailing edge of $\overline{WR}$. 3-stated during Hold and Halt modes and during RESET.</td></tr>
<tr><td>READY</td><td>I</td><td>READY: If READY is high during a read or write cycle, it indicates that the memory or peripheral is ready to send or receive data. If READY is low, the CPU will wait an integral number of clock cycles for READY to go high before completing the read or write cycle. READY must conform to specified setup and hold times.</td></tr>
</table>

HOLD	I	**HOLD:** Indicates that another master is requesting the use of the address and data buses. The CPU, upon receiving the hold request, will relinquish the use of the bus as soon as the completion of the current bus transfer. Internal processing can continue. The processor can regain the bus only after the HOLD is removed. When the HOLD is acknowledged, the address, Data $\overline{\text{RD}}$, $\overline{\text{WR}}$, and IO/M lines are 3-stated.
HLDA	O	**HOLD ACKNOWLEDGE:** Indicates that the CPU has received the HOLD request and that it will relinquish the bus in the next clock cycle. HLDA goes low after the Hold request is removed. The CPU takes the bus one half clock cycle after HLDA goes low.
INTR	I	**INTERRUPT REQUEST:** Is used as a general purpose interrupt. It is sampled only during the next to the last clock cycle of an instruction and during Hold and Halt states. If it is active, the Program Counter (PC) will be inhibited from incrimenting and an $\overline{\text{INTA}}$ will be issued. During this cycle a RESTART or CALL instruction can be inserted to jump to the interrupt service routine. The INTR is enabled and disabled by software. It is disabled by Reset and immediately after an interrupt is accepted.
$\overline{\text{INTA}}$	O	**INTERRUPT ACKNOWLEDGE:** Is used instead of (and has the same timing as) $\overline{\text{RD}}$ during the Instruction cycle after an INTR is accepted. It can be used to activate an 8259A Interrupt chip or some other interrupt port.
RST 5.5 RST 6.5 RST 7.5	I	**RESTART INTERRUPTS:** These three inputs have the same timing as INTR except they cause and internal RESTART to be automatically inserted. The priority of these interrupt is ordered as shown in Table 3.2. These interrupts have a higher priority than INTR. In addition, they may be individually masked out using the SIM instruction.
TRAP	I	**TRAP:** Trap interrupt is a non-maskable RESTART interrupt. It is recognized at the same time as INTR or RST 5.5-7.5. It is unaffected by any mask or Interrupt Enable. It has the highest priority of any interrupt. (See Table 3.2)
$\overline{\text{RESET IN}}$	I	**RESET IN:** Sets the Program Counter to zero and resets the Interrupt Enable and HLDA flip-flops. The data and address buses and the control lines are 3-stated during RESET and because of the asynchronous nature of RESET, the processor's internal registers and flags may be altered by RESET with unpredictable results. $\overline{\text{RESET IN}}$ is a Schmitt-triggered input, allowing connection to an R-C network for power-on RESET delay. Upon power-up, $\overline{\text{RESET IN}}$ must remain low for at least 10 ms after minimum Vcc has been reached. For proper reset operation after the power-up duration, $\overline{\text{RESET IN}}$ should be kept low a minium of three clock periods. The CPU is held in the reset condition as long as $\overline{\text{RESET IN}}$ is applied.

RESET OUT	O	**RESET OUT:** Reset Out indicates CPU is being reset. Can be used as a system reset. The signal is synchronized to the processor clock and lasts an integral number of clock periods.
X1, X2	I	**X1 and X2:** Are connected to a crystal, LC or RC network to drive the internal clock generator. X1 can also be an external clock input from a logic gate. The input frequency is divided by 2 give the processor's internal operating frequency.
CLK	O	**CLOCK:** Clock output for use as system clock. The period of CLK is twice the X1, X2 input period.
SID	I	**SERIAL INPUT DATA LINE:** The data on this line is loaded into accumulator bit 7 whenever a RIM instruction is executed.
SOD	O	**SERIAL OUTPUT DATA LINE:** The output SOD is set or reset as specified by the SIM instruction.
VCC		**POWER:** +5 volt supply.
VSS		**GROUND:** Reference.

Table 3.2 Interrupt Priority, Restart Address and Sensitivity

Name	Priority	Address Branched When Interrupt Occurs	Type Trigger
TRAP	1	24H	Rising Edge AND High Level until Sampled
RST 7.5	2	3CH	Rising Edge (Latched)
RST 6.5	3	34H	High Level until sampled
RST 5.5	4	2CH	High Level until Sampled
INTR	5	(Note)	High Level until Sampled

Notes:

1. The processor pushes the PC on the stack before branching to the indicated address.
2. The address branched to depends on the instruction provided to the CPU when the interrupt is acknowledged.

3.2 FUNCTIONAL DESCRIPTION

The 8085AH is a complete 8-bit parallel central processor. It is designed with N-channel, depletion load, silicon gate technology (HMOS), and requires a single +5V supply. Its basic clock speed is 3MHz (8085AH), 5MHz (8085AH-2) or 6 MHz (8085-AH-1), thus improving on the present 8080A's performance with higher system speed. Also it is designed to fit into a minimum system of three IC's: The CPU (8085AH), a RAM/IO (8156H), and an EPROM/IO chip (8755A).

The 8085AH has twelve addressable 8-bit registers. Four of them can function only as two 16-bit register pairs. Six others can be used interchangeably as 8-bit registers or as 16-bit register pairs. The 8085AH register set is as follows:

Mnemonic	Register	Contents
ACC or A	Accumulator	8-Bits
PC	Program Counter	16-Bit Address
BC, DE, HL	General-Purpose Registers; data pointer (HL)	8-Bits × 6 or 16-Bits × 3
SP	Stack Pointer	16-Bits Address
Flags or F	Flag Register	5 Flags (8-Bit Space)

The 8085AH uses a multiplexed Data Bus. The address is split between the higher 8-bit Address Bus and the lower 8-bit Address/Data Bus. During the first T state (clock cycle) of a machine cycle the low order address is sent out on the Address/Data bus. These lower 8 bits may be latched externally by the Address Latch Enable signal (ALE). During the rest of the machine cycle the data bus is used for memory or I/O data.

The 8085AH provides $\overline{RD}$, $\overline{WR}$, S0, S1 and IO/$\overline{M}$ signals for bus control. An Interrupt Acknowledge signal ($\overline{INTA}$) is also provided. HOLD and all Interrupts are synchronized with the processor's internal clock. The 8085AH also provides Serial Input Data (SID) and Serial Output Data (SOD) lines for simple serial interface.

In addition to these features, the 8085AH has three maskable, vector interrupt pins, one nonmaskable TRAP interrupt, and a bus vectored interrupt, INTR.

3.2.1 Interrupt and Serial I/O

The 8085AH has 5 interrupt inputs: INTR, RST 5.5,RST 6.5, RST 7.5, and TRAP. Each of the three RESTART inputs, 5.5, 6.5, and 7.5, has programmable mask. TRAP is also a RESTART interrupt but it is non-maskable.

The three maskable interrupt cause the internal execution of RESTART (saving the program counter in the stack and branching to the RESTART address) if the interrupts are enabled and if the interrupt mask is not set. The nonmaskable TRAP causes the internal execution of a RESTART vector independent of the state of the interrupt enable or masks. (See Table 3.2)

There are two different types of inputs in the restart interrupts. RST 5.5 and RST 6.5 are high level-sensitive like INTR and are recognized with the same timing as INTR. RST 7.5 is rising edge-sensitive.

For RST 7.5, only a pulse is required to set an internal flip-flop which generates the internal interrupt request (a normally high level signal with a low going pulse is recommended for highest system noise immunity). The RST 7.5 request flip-flop remains set until the request is serviced. Then it is reset automatically. This flip-flop may also be reset by using the SIM instruction or by issuing a $\overline{\text{RESET IN}}$ to the 8085AH. The RST 7.5 internal flip-flop will be set by a pulse on the RST 7.5 pin even when the RST 7.5 interrupt is masked out.

The status of the three RST interrupt masks can only be affected by the SIM instruction and $\overline{\text{RESET IN}}$.

The interrupts are arranged in a fixed priority that determines which interrupt is to be recognized if more than one is pending as follows: TRAP–highest priority, RST 7.5, RST 6.5, RST 5.5, INTR–lowest priority. This priority scheme does not take into account the priority of a routine that was started by a higher priority interrupt. RST 5.5 can interrupt an RST 7.5 routine if the interrupts are re-enabled before the end of the RST 7.5 routine.

The TRAP interrupt is useful for catastrophic events such as power failure or bus error. The TRAP input is recognized just as any other interrupt but has the highest priority. It is not affected by any flag or mask. The TRAP input is both edge and level sensitive. The TRAP input must go high and remain high until it is acknowledged. It will not be recognised again until, it goes low, then high again. This avoids any false triggering due to noise or logic glitches. Note that the servicing of any interrupt (TRAP, RST 7.5, RST 6.5, RST 5.5, INTR) disables all future interrupts (except TRAPs) until an EI instruction is executed.

The TRAP interrupt is special in that it disables interrupts, but preserves the previous interrupt enable status. Performing the first RIM instruction following a TRAP interrupt allows you to determine whether interrupts were enabled or disabled prior to the TRAP. All subsequent RIM instructions provide current interrupt enable status. Performing a RIM instruction following INTR, or RST 5.5-7.5 will provide current Interrupt Enable status, revealing that interrupts are disabled.

The serial I/O system is also controlled by the RIM and SIM instruction. SID is read by RIM, and SIM sets the SOD data.

3.2.2 Driving the X1 and X2 Inputs

You may drive the clock inputs of the 8085AH, 8085AH-2, or 8085AH-1 with a crystal, an LC tuned circuit, an RC network, or an external clock source. The crystal frequency must be at least 1 MHz, and must be twice the desired internal clock frequency: hence, the 8085AH is operated with a 6 MHz crystal (for 3 MHz clock), the 8085AH-2 operated with a 10 MHz crystal (for 5 MHz clock), and the 8085AH-1 can be operated with a 12 MHz crystal (for 6MHz clock). If a crystal is used, it must have the following characteristics:

Parallel resonance at twice the clock frequency desired

C_L (load capacitance) $\leq$ 30pF

C_S (Shunt capacitance) $\leq$ 7pF

R_S (equivalent shunt resistance) $\leq$ 75Ω

Drive level: 10 mW

Frequency tolerance: ± 0.005% (suggested)

Note the use of the 20pF capacitor between X2 and ground. This capacitor is required with crystal frequencies below 4 MHz to assure oscillator startup at the correct frequency. A parallel-resonant LC circuit may be used as the frequency-determining network for the 8085AH, providing that its frequency tolerance of approximately ± 10% is acceptable. The components are chosen from the formula:

$$f = \frac{1}{2\pi\sqrt{L\left(C_{ext} + C_{int}\right)}}$$

To minimize variations in frequency, it is recommended that you choose a value for Cext that is at least twice that of Cint. or 30pF. The use of an LC circuit is not recommended for frequencies higher than approximately 5 MHz.

An RC circuit may be used as the frequency-determining network for the 8085AH if maintaining a precise clock frequency is of no importance. Variations in the on-chip timing generation can cause a wide variation in frequency when using the RC mode. Its advantage is its low component cost. The driving frequency generated by the circuit shown is approximately 3 MHz. It is not recommended that frequencies greatly higher or lower than this be attempted.

Fig. 3.3 shows the recommended clock driver circuits.

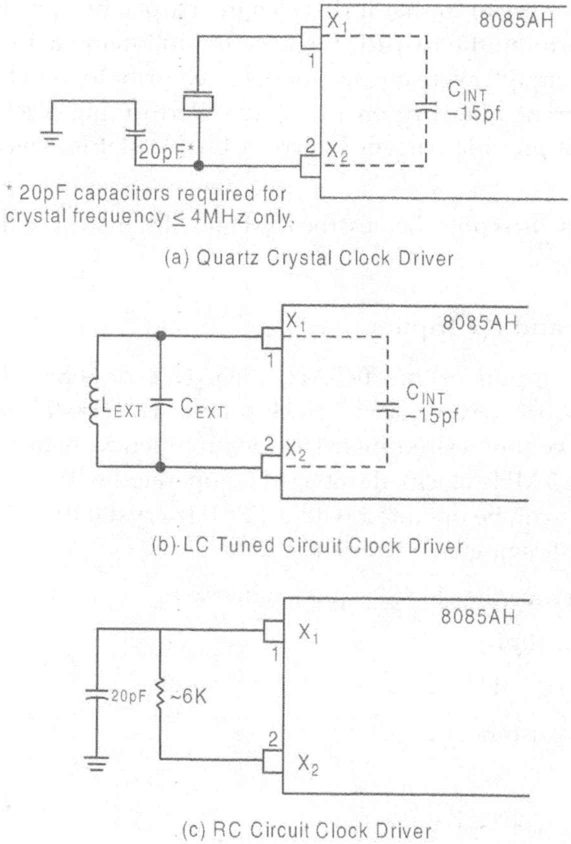

Fig. 3.3 Clock Drive Circuits

3.2.3 The ALU

The Arithmetic and logic unit performs the all arithmetic and logical operation. It includes accumulators and fine flags with temporary register.

- All the operations are performed and final result is always in accumulator.
- In temporary registers; the temporary storage of data during the operations takes place; that helps the accumulator to perform operation.
- Flags are used some time to perform some conditional operation on the basis of conditions that are caused due to operation done by execution of instruction, 8085 have 8-bit flag register in which 5 active flags.

7	6	5	4	3	2	1	0
S	Z	V	AC	V	P	V	CY

V = Undefined

Fig. 3.4 Flags of 8085

The explanation of flags are given below:

(*a*) **S-Sign flag.** Sign bit is used to show the status of number; If it is set '1' means number is negative; If it is set '0' means number is positive;

First bit is used for sign presentation and remaining 7 bits are used for magnitude representation.

(*b*) **Z-Zero flag.** If the result of a arithmatic operation is 'zero', then it is set '1'.

(*c*) **AC-Auxiliary carry flag.** If in a arithmatic operation, a carry is generated at 4th bits: and passed to 5th bit; then it is set. This is generally used for BCD operation.

(*d*) **P-Parity flag.** If after the execution of an arithmetic operation, result has even number of one (called even parity) the flag is set; otherwise for odd number of ones (called odd parity) It is Reset.

(*e*) **CY-Carry flag.** If an arithmatic operation generate a carry; then it is set.

3.2.4 General Purpose Register

It has 8 general purpose register each of 8-bits; If we want to perform 16-bit operation: they can be used in pairs such as BC, DE and HL.

3.2.5 Programme Counter

Programme counter contains the address of next instruction to be executed so that micro-processor knows that after completion of present instruction that has to execute the instruction that address is contained by program counter. The same work is performed by stack pointer but now for stack operation (discussed in detailed later).

This minimum system, using the standard I/O technique is as shown in Fig. 3.5.

The 8085AH CPU can also interface with the standard memory that does *not* have the multiplexed address/data bus. It will require a simple 8-bit latch as shown in Fig. 3.6.

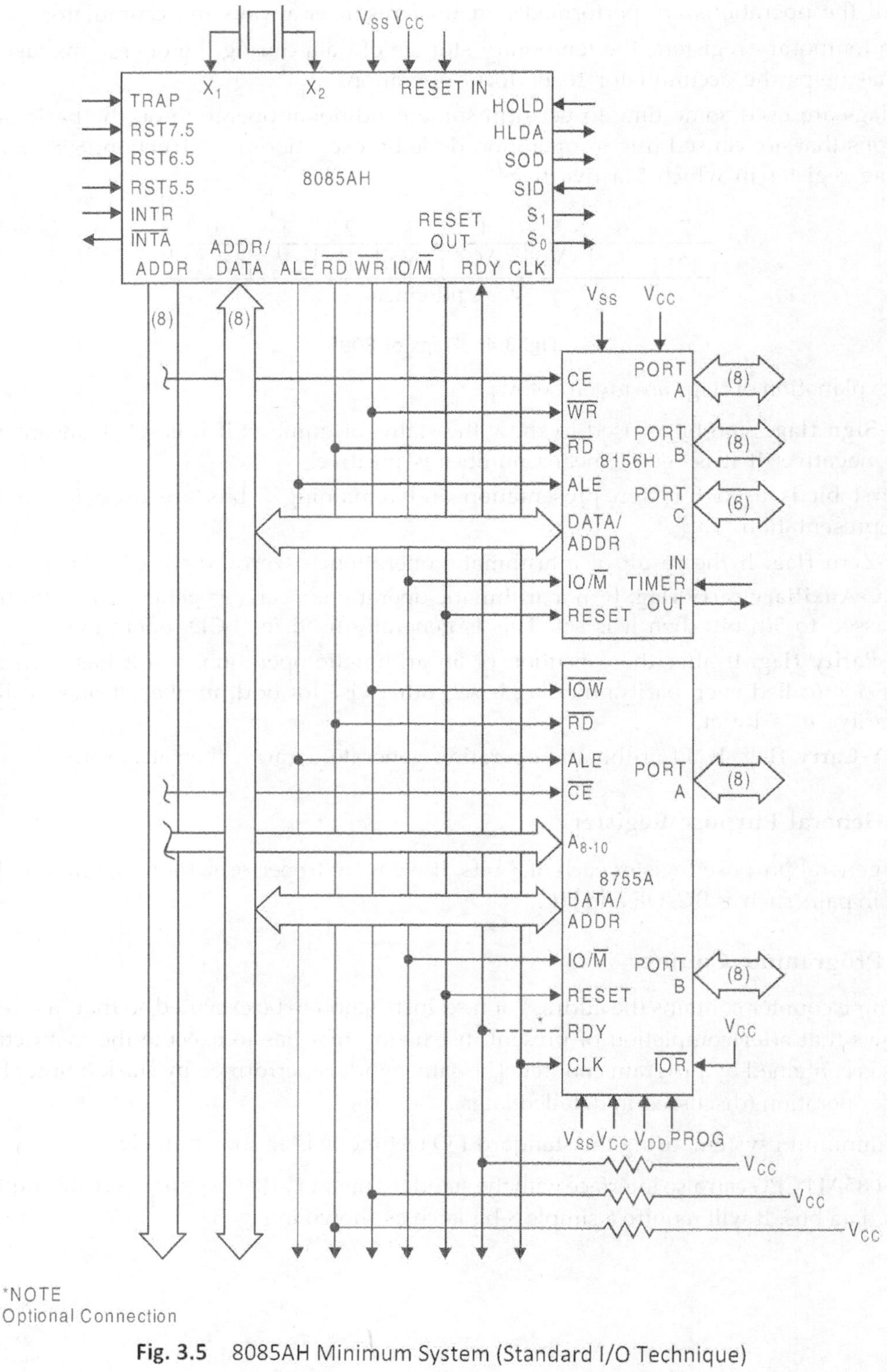

Fig. 3.5 8085AH Minimum System (Standard I/O Technique)

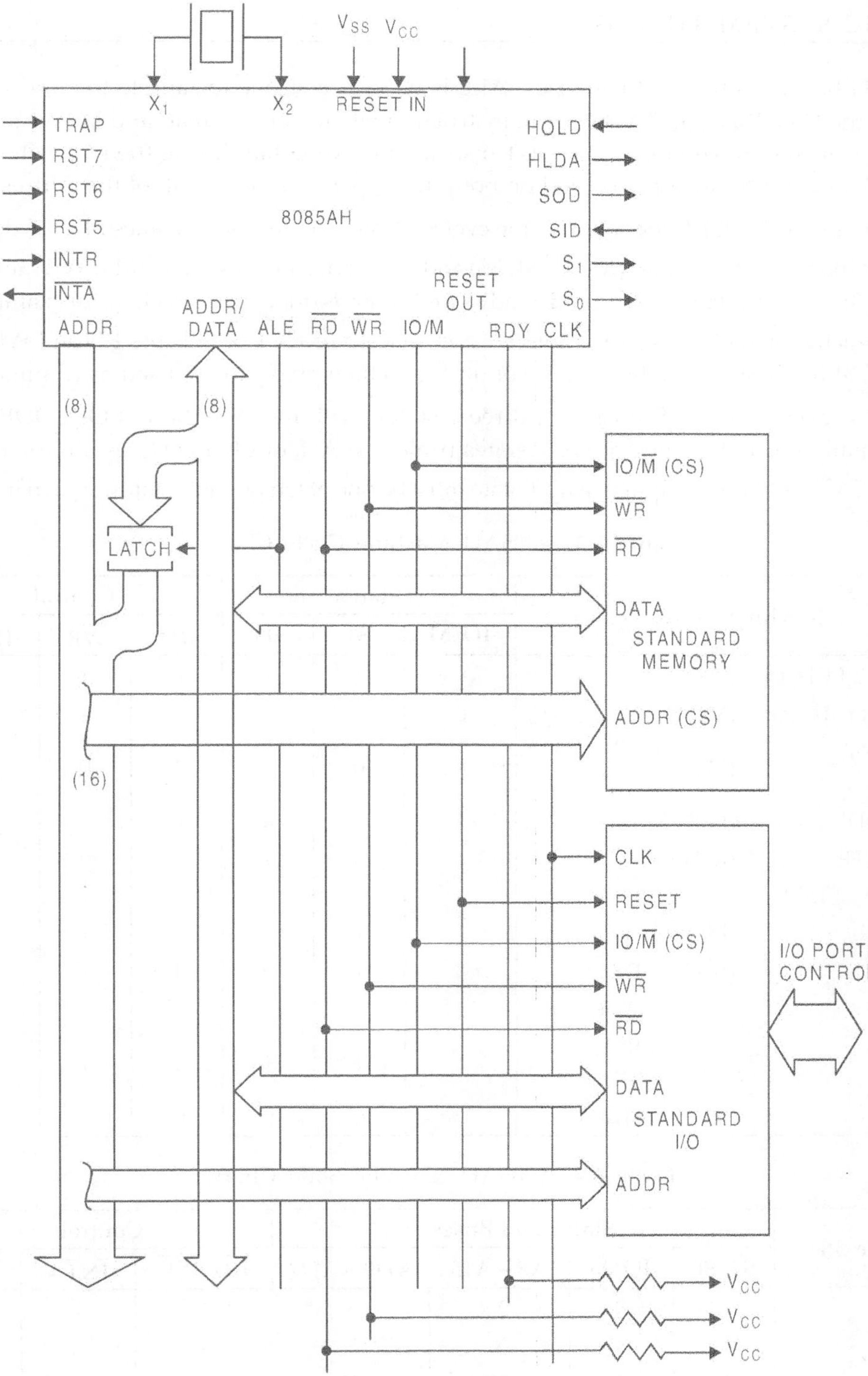

Fig. 3.6 8085 System (Using Standard Memories)

3.3 BASIC SYSTEM TIMING

The 8085AH has a multiplexed Data Bus. ALE is used as a strobe to sample the lower 8-bits of address on the Data Bus. Fig. 3.7 shows an instruction fetch, memory read and I/O write cycle (as would occur during processing of the OUT instruction). Note that during the I/O write and read cycle that the I/O port address is copied on both the upper and lower half of the address.

There are seven possible types of machine cycles. Which of these seven takes place is defined by the status of the three status lines (IO/$\overline{\text{M}}$, S1, S0) and the three control signals ($\overline{\text{RD}}$, $\overline{\text{WR}}$, and $\overline{\text{INTA}}$). (See Table 3.3) The status lines can be used as advanced controls (for device selection, for example), since they become active at the T1 state, at the outset of each machine cycle. Control lines $\overline{\text{RD}}$ and $\overline{\text{WR}}$ become active later, at the time when the transfer of data is to take place, so are used as command lines.

A machine cycle normally consists of three T states, with the exception of OPCODE FETCH, which normally has either four or six T states (unless WAIT or HOLD states are forced by the receipt of $\overline{\text{READY}}$ or HOLD inputs). Any T state must be one of ten possible states, shown in Table 3.4

Table 3.3 8085AH Machine Cycle Chart

Machine Cycle			Status			Control		
			IO/$\overline{\text{M}}$	S1	S0	$\overline{\text{RD}}$	$\overline{\text{WR}}$	$\overline{\text{INTA}}$
OPCODE FETCH	(OF)		0	1	1	0	1	1
MEMORY READ	(MR)		0	1	0	0	1	1
MEMORY WRITE	(MW)		0	0	1	1	0	1
I/O READ	(IOR)		1	1	0	0	1	1
I/O WRITE	(IOW)		1	0	1	1	0	1
AKNOWLEDGE OF INTR	(INA)		1	1	1	1	1	0
BUS IDLE	(BI):	DAD	0	1	0	1	1	1
		ACK. OF RST, TRAP	1	1	1	1	1	1
		HALT	TS	0	0	TS	TS	1

Table 3.4 8085AH Machine State Chart

Machine State	Status and Buses				Control		
	S1, S0	IO/$\overline{\text{M}}$	A8 – A15	AD0 – AD7	$\overline{\text{RD}}$, $\overline{\text{WR}}$	$\overline{\text{INTA}}$	ALE
T1	X	X	X	X	1	1	1*
T2	X	X	X	X	X	X	0
TWAIT	X	X	X	X	X	X	0
T3	X	X	X	X	X	X	0
T4	1	0 †	X	TS	1	1	0

T5	1	0 †	X	TS	1	1	0
T6	1	0 †	X	TS	1	1	0
TRESET	X	TS	TS	TS	TS	1	0
THALT	0	TS	TS	TS	TS	1	0
THOLD	X	TS	TS	TS	TS	1	0

0 = Logic "0" TS = High Impedance

0 = Logic "1" X = Unspecified

*ALE not generated during 2nd and 3rd machine cycles of DAD instruction.

† $IO/\overline{M}$ = 1 during T4 – T6 of INA machine cycle.

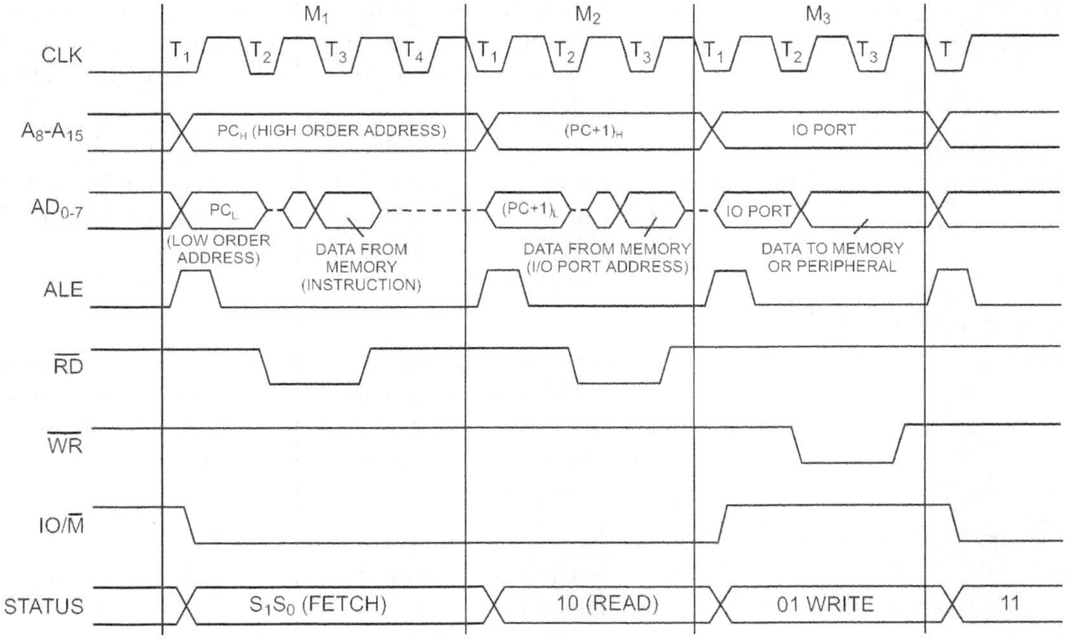

Fig. 3.7 8085AH Basic System Timing

3.4 8085 INSTRUCTION FORMAT

An Instruction is a command to the microprocessor to perform a given task on specified data.

Each instruction has two parts—one is the task to be performed called operation code 'or' opcode. And the second is the data to be operated on, called operand. The operand ('or' data) can be specified in various way (called Addressing mode). This can be data 'or' address of that particular data.

3.4.1 Instruction Details

Data Transfer Instructions

Opcode	Operand	Description
Copy from source to destination		
MOV	Rd, Rs M, Rs Rd, M	This instruction copies the contents of the source register into the destination register; the contents of the source register are not altered. If one of the operands is a memory location, its location is specified by the contents of the HL registers. *Example:* MOV B, C or MOV B, M
Move immediate 8-bit		
MVI	Rd, data M, data	The 8-bit data is stored in the destination register or memory. If the operand is a memory location, its location is specified by the contents of the HL registers. *Example:* MVI B, 57H OR MVI M, 57H
Load accumulator		
LDA	16-bit address	The contents of a memory location, specified by a 16-bit address in the operand, are copied to the accumulator. The contents of the source are not altered. *Example:* LDA 2034H
Load accumulator indirect		
LDAX	B/D Reg. pair	The contents of the designated register pair point to a memory location. This instruction copies the contents of that memory location into the accumulator. The contents of either the register pair or the memory location are not altered. *Example:* LDAX B
Load register pair immediate		
LXI	Reg. pair, 16-bit data	The instruction loads 16-bit data in the register pair designated in the operand. *Example:* LXI H, 2034H OR LXI H, XYZ
Load H and L registers direct		
LHLD	16-bit address	The instruction copies the contents of the memory location pointed out by the 16-bit address into register L and copies the contents of the next memory location into register H. The contents of source memory locations are not altered. *Example:* LHLD 2040H
Store accumulator direct		
STA	16-bit address	The contents of the accumulator are copied into the memory location specified by the operand. This is a 3-byte instruction, the second byte specifies the low-order address and the third byte specifies the high-order address. *Example:* STA 4350H

Store accumulator indirect		
STAX	Reg. pair	The contents of the accumulator are copied into the memory location specified by the contents of the operand (register pair). The contents of the accumulator are not altered. *Example:* STAX B
Store H and L registers direct		
SHLD	16-bit address	The contents of register L are stored into the memory location specified by the 16-bit address in the operand and the contents of H register are stored into the next memory location by incrementing the operand. The contents of registers HL are not altered. This is a 3-byte instruction, the second byte specifies the low-order address and the third byte specifies the high-order address. *Example:* SHLD 2470H
Exchange H and L with D and E		
XCHG	none	The contents of register H are exchanged with the contents of register D, and the contents of register L are exchanged with the contents of register E. *Example:* XCHG
Copy H and L registers to the stack pointer		
SPHL	none	The instruction loads the contents of the H and L registers into the stack pointer register, the contents of the H register provide the high-order address and the contents of the L register provide the low-order address. The contents of the H and L registers are not altered. *Example:* SPHL
Exchange H and L with top of stack		
XTHL	none	The contents of the L register are exchanged with the stack location pointed out by the contents of the stack pointer register. The contents of the H register are exchanged with the next stack location (SP + 1); however, the contents of the stack pointer register are not altered. *Example:* XTHL
Push register pair onto stack		
PUSH	Reg. pair	The contents of the register pair designated in the operand are copied onto the stack in the following sequence. The stack pointer register is decremented and the contents of the high-order register (B, D, H, A) are copied into that location. The stack pointer register is decremented again and the contents of the low-order register (C, E, L, flags) are copied to that location. *Example:* PUSH B or PUSH A

Pop off stack to register pair		
POP	Reg. pair	The contents of the memory location pointed out by the stack pointer register are copied to the low-order register (C, E, L, status flags) of the operand. The stack pointer is incremented by 1 and the contents of that memory location are copied to the high-order register (B, D, H, A) of the operand. The stack pointer register is again incremented by 1. *Example:* POP H OR POP A
Output data from accumulator to a port with 8-bit address		
OUT	8-bit port address	The contents of the accumulator are copied into the I/O port specified by the operand. *Example:* OUT F8H
Input data to accumulator from a part with 8-bit address		
IN	8-bit port address	The contents of the input port designated in the operand are read and loaded into the accumulator. *Example:* IN 8CH

Arithmetic Instructions

Opcode	**Operand**	**Description**
Add register or memory to accumulator		
ADD	R M	The contents of the operand (register or memory) are added to the contents of the accumulator and the result is stored in the accumulator. If the operand is a memory location, its location is specified by the contents of the HL registers. All flags are modified to reflect the result of the addition. *Example:* ADD B or ADD M
Add register to accumulator with carry		
ADC	R M	The contents of the operand (register or memory) and the Carry flag are added to the contents of the accumulator and the result is stored in the accumulator. If the operand is a memory location, its location is specified by the contents of the HL registers. All flags are modified to reflect the result of the addition. *Example:* ADC B or ADC M
Add immediate to accumulator		
ADI	8-bit data	The 8-bit data (operand) is added to the contents of the accumulator and the result is stored in the accumulator. All flags are modified to reflect the result of the addition. *Example:* ADI 45H
Add immediate to accumulator with carry		
ACI	8-bit data	The 8-bit data (operand) and the Carry flag are added to the contents of the accumulator and the result is stored in the accumulator. All flags are modified to reflect the result of the addition. *Example:* ACI 45H

Add register pair to H and L registers

DAD	Reg. pair	The 16-bit contents of the specified register pair are added to the contents of the HL register and the sum is stored in the HL register. The contents of the source register pair are not altered. If the result is larger than 16-bits, the CY flag is set. No other flags are affected. *Example:* DAD H

Subtract register or memory from accumulator

SUB	R M	The contents of the operand (register or memory) are subtracted from the contents of the accumulator, and the result is stored in the accumulator. If the operand is a memory location, its location is specified by the contents of the HL registers. All flags are modified to reflect the result of the subtraction. *Example:* SUB B or SUB M

Subtract source and borrow from accumulator

SBB	R M	The contents of the operand (register or memory) and the Borrow flag are subtracted from the contents of the accumulator and the result is placed in the accumulator. If the operand is a memory location, its location is specified by the contents of the HL registers. All flags are modified to reflect the result of the subtraction. *Example:* SBB B or SBB M

Subtract immediate from accumulator

SUI	8-bit data	The 8-bit data (operand) is subtracted from the contents of the accumulator and the result is stored in the accumulator. All flags are modified to reflect the result of the subtraction. *Example:* SUI 45H

Subtract immediate from accumulator with borrow

SBI	8-bit data	The 8-bit data (operand and the Borrow flag are subtracted from the contents of the accumulator and the result is stored in the accumulator. All flags are modified to reflect the result of the subtraction. *Example:* SBI 45H

Increment register or memory by 1

INR	R M	The contents of the designated register or memory) are incremented by 1 and the result is stored in the same place. If the operand is a memory location, its location is specified by the contents of the HL registers. *Example:* INR B or INR M

Increment register pair by 1

INX	R	The contents of the designated register pair are incremented by 1 and the result is stored in the same place. *Example:* INX H

Decrement register or memory by 1.		
DCR	R M	The contents of the designated register or memory are decremented by 1 and the result is stored in the same place. If the operand is a memory location, its location is specified by the contents of the HL registers. *Example:* DCR B or DCR M
Decrement register pair by 1		
DCX	R	The contents of the designated register pair are decremented by 1 and the result is stored in the same place. *Example:* DCX H
Decimal adjust accumulator		
DAA	none	The contents of the accumulator are changed from a binary value to two 4-bit binary coded decimal (BCD) digits. This is the only instruction that uses the auxiliary flag to perform the binary to BCD conversion, and the conversion procedure is described below. S, Z, AC, P, CY flags are altered to reflect the results of the operation. If the value of the low-order 4-bits in the accumulator is greater than 9 or if AC flag is set, the instruction adds 6 to the low-order four bits. If the value of the high-order 4-bits in the accumulator is greater than 9 or if the Carry flag is set, the instruction adds 6 to the high-order four bits. *Example:* DAA

Branching Instructions

Opcode	Operand	Description
Jump unconditionally		
JMP	16-bit address	The program sequence is transferred to the memory location specified by the 16-bit address given in the operand. *Example:* JMP 2034H or JMP XYZ
Jump conditionally		
Operand:	16-Bit address	The program sequence is transferred to the memory location specified by the 16-bit address given in the operand based on the specified flag of the PSW as described below. *Example:* JZ 2034H or JZ XYZ

Opcode	*Description*	*Flag Status*
JC	Jump on Carry	CY = 1
JNC	Jump on no Carry	CY = 0
JP	Jump on positive	S = 0
JM	Jump on minus	S = 1
JZ	Jump on zero	Z = 1
JNZ	Jump on no zero	Z = 0
JPE	Jump on parity even	P = 1
JPO	Jump on parity odd	P = 0

Unconditional subroutine call

CALL 16-bit address The program sequence is transferred to the memory location specified by the 16-bit address given in the operand. Before the transfer, the address of the next instruction after CALL (the contents of the program counter) is pushed onto the stack.

Example: CALL 2034H or CALL XYZ

Call conditionally

Operand: 16- bit address The program sequence is transferred to the memory location specified by the 16-bit address given in the operand based on the specified flag as described below. Before the transfer, the address of the next instruction after the call (the contents of the program counter) is pushed onto the stack.

Example: CZ 2034H or CZ XYZ

Opcode	*Description*	*Flag Status*
CC	Call on Carry	CY = 1
CNC	Call on no Carry	CY = 0
CP	Call on positive	S = 0
CM	Call on minus	S = 1
CZ	Call on zero	Z = 1
CNZ	Call on no zero	Z = 0
CPE	Call on parity even	P = 1
CPO	Call on parity odd	P = 0

Return from subroutine unconditionally

RET none The program sequence is transferred from the subroutine to the calling program. The two bytes from the top of the stack are copied into the program counter, and program execution begins at the new address.

Example: RET

Return from subroutine conditionally

Operand: none The program sequence is transferred from the subroutine to the calling program based on the specified flag as described below. The two bytes from the top of the stack are copied into the program counter, and program execution begins at the new address.

Example: RZ

Opcode	*Description*	*Flag Status*
RC	Return on Carry	CY = 1
RNC	Return on no carry	CY = 0
RP	Return on positive	S = 0
RM	Return on minus	S = 1
RZ	Return on zero	Z = 1
RNZ	Return on no zero	Z = 0
RPE	Return on parity even	P = 1
RPO	Return on parity odd	P = 0

Load program counter with HL contents		
PCHL	none	The contents of registers H and L are copied into the program counter. The contents of H are placed as the high-order byte and the contents of L as the low-order byte. *Example:* PCHL
Restart		
RST	0-7	The RST instruction is equivalent to a 1-byte call instruction to one of eight memory locations depending upon the number. The instructions are generally used in conjunction with interrupts and inserted using external hardware. However these can be used as software instructions in a program to transfer program execution to one of the eight locations. The addresses are:

Instruction	*Restart Address*
RST 0	0000H
RST 1	0008H
RST 2	0010H
RST 3	0018H
RST 4	0020H
RST 5	0028H
RST 6	0030H
RST 7	0038H

The 8085 has four additional interrupts and these interrupts generate RST instructions internally and thus do not require any external hardware. These instructions and their Restart addresses are:

Interrupt	*Restart Address*
TRAP	0024H
RST 5.5	002CH
RST 6.5	0034H
RST 7.5	003CH

Logical Instructions

Opcode	Operand	Description
Compare register or memory with accumulator		

CMP	R
	M
		The contents of the operand (register or memory) are compared with the contents of the accumulator. Both contents are preserved. The result of the comparison is shown by setting the flags as follows:
		if (A) < (reg/mem): carry flag is set
		if (A) = (reg/mem): zero flag is set
		if (A) > (reg/mem): carry and zero flags are reset
		Example: CMP B or CMP M

Compare immediate with accumulator

CPI 8-bit data

The second byte (8-bit data) is compared with the contents of the accumulator. The values being compared remain unchanged. The result of the comparison is shown by setting the flags as follows:

if (A) < data: carry flag is set

if (A) = data: zero flag is set

if (A) > data: carry and zero flags are reset

Example: CPI 89H

Logical AND register or memory with accumulator

ANA R
M

The contents of the accumulator are logically ANDed with the contents of the operand (register or memory), and the result is placed in the accumulator. If the operand is a memory location, its address is specified by the contents of HL registers. S, Z, P are modified to reflect the result of the operation. CY is reset. AC is set.

Example: ANA B or ANA M

Logical AND immediate with accumulator

ANI 8-bit data

The contents of the accumulator are logically ANDed with the 8-bit data (operand) and the result is placed in the accumulator. S, Z, P are modified to reflect the result of the operation. CY is reset. AC is set.

Example: ANI 86H

Exclusive OR register or memory with accumulator

XRA R
M

The contents of the accumulator are Exclusive ORed with the contents of the operand (register or memory), and the result is place in the accumulator. If the operand is a memory location, its address is specified by the contents of HL registers. S, Z, P are modified to reflect the result of the operation. CY and AC are reset.

Example: XRA B or XRA M

Exclusive OR immediate with accumulator

XRI 8- bit data

The contents of the accumulator are Exclusive ORed with the 8-bit data (operand) and the result is placed in the accumulator. S, Z, P are modified to reflect the result of the operation. CY and AC are reset.

Example: XRI 86H

Logical OR register or memory with accumulator

ORA R
M

The contents of the accumulator are logically ORed with the contents of the operand (register or memory), and the result is placed in the accumulator. If the operand is a memory location, its address is specified by the contents of HL registers. S, Z, P are modified to reflect the result of the operation. CY and AC are reset.

Example: ORA B or ORA M

Logical OR immediate with accumulator		
ORI	8-bit data	The contents of the accumulator are logically ORed with the 8-bit data (operand) and the result is placed in the accumulator. S, Z, P are modified to reflect the result of the operation. CY and AC are reset. *Example:* ORI 86H
Rotate accumulator left		
RLC	none	Each binary bit of the accumulator is rotated left by one position. Bit D7 is placed in the position of D0 as well as in the Carry flag. CY is modified according to bit D7. S, Z, P, AC are not affected. *Example:* RLC
Rotate accumulator right		
RRC	none	Each binary bit of the accumulator is rotated right by one position. Bit D0 is placed in the position of D7 as well as in the Carry flag. CY is modified according to bit D0. S, Z, P, AC are not affected. *Example:* RRC
Rotate accumulator left through carry		
RAL	none	Each binary bit of the accumulator is rotated left by one position through the Carry flag. Bit D7 is placed in the Carry flag, and the Carry flag is placed in the least significant position D0. CY is modified according to bit D7. S, Z, P, AC are not affected. *Example:* RAL
Rotate accumulator right through carry		
RAR	none	Each binary bit of the accumulator is rotated right by one position through the Carry flag. Bit D0 is placed in the Carry flag, and the Carry flag is placed in the most significant position D7. CY is modified according to bit D0. S, Z, P, AC are not affected. *Example:* RAR
Complement accumulator		
CMA	none	The contents of the accumulator are complemented. No flags are affected. *Example:* CMA
Complement carry		
CMC	none	The Carry flag is complemented. No other flags are affected. *Example:* CMC
Set Carry		
STC	none	The Carry flag is set to 1. No other flags are affected. *Example:* STC

Control Instructions

Opcode	Operand	Description
No operation		
NOP	none	No operation is performed. The instruction is fetched and decoded. However no operation is executed. *Example:* NOP
Halt and enter wait state		
HLT	none	The CPU finishes executing the current instruction and halts any further execution. An interrupt or reset is necessary to exit from the halt state. *Example:* HLT
Disable interrupt		
DI	none	The interrupt enable flip-flop is reset and all the interrupt except the TRAP are disabled. No flags are affected. *Example:* DI
Enable interrupt		
EI	none	The interrupt enable flip-flop is set and all interrupt are enabled. No flags are affected. After a system reset or the acknowledgment of an interrupt, the interrupt enable flip-flop is reset, thus disabling the interrupt. This instruction is necessary to reenable the interrupt (except TRAP). *Example:* EI
Read interrupt mask		
RIM	none	This is a multipurpose instruction used to read the status of interrupt 7.5, 6.5, 5.5 and read serial data input bit. The instruction loads eight bits in the accumulator with the following interpretations. *Example:* RIM

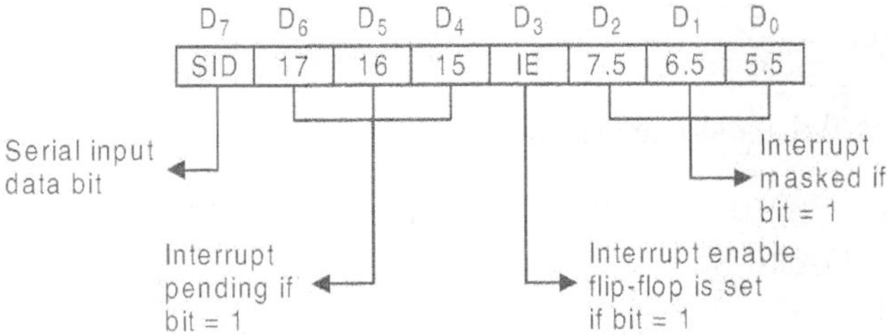

Opcode	Operand	Description
Set interrupt mask		
SIM	none	This is a multipurpose instruction and used to implement the 8085 interrupt 7.5, 6.5, 5.5, and serial data output. The instruction interprets the accumulator contents as follows. *Example:* SIM

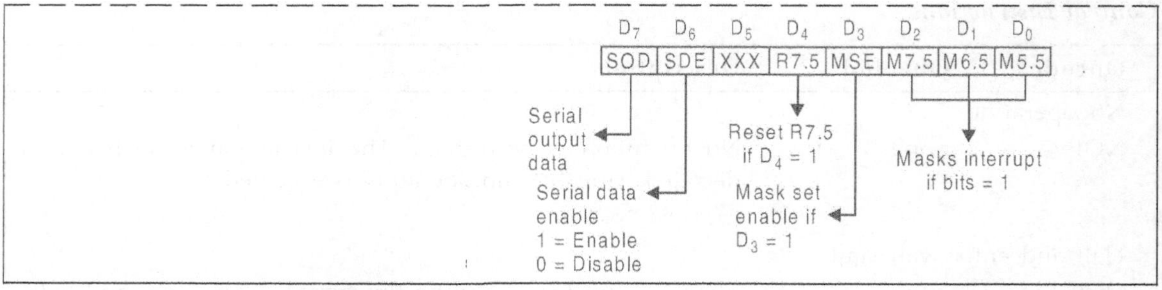

3.5 ADDRESSING MODES OF 8085

The various ways to specify a operand in a instruction are called addressing mode. In other words, it can be said that these are the various ways so that a microprocessor can access a data. In 8085, generally used addressing modes are:

(*a*) Immediate Addressing mode.
(*b*) Register Addressing mode.
(*c*) Direct Addressing mode.
(*d*) Indirect Addressing mode.

Immediate Addressing Mode

In this addressing mode, the operand is a part of instruction itself.

MVI B, 8-bit - load 8-bit data immediately to the register B.

Register Addressing Mode

In this addressing mode, the operand is kept in a register and that register is a part of instruction itself.

MOV Rd, Rs means load source register to destination register thus operand is in the source register.

Direct Addressing Mode

Here the address of operand is a part of instruction itself.

As LDA 16-bit.

Indirect Addressing Mode

Here the address of operand is the address of other location such as

ADD M Add accumulator to content of memory location and store result in accumulator.

3.6 STACK OF 8085

In 8085, the stack is defined as a set of memory locations in the R/W memory, specified by the programmer in a main program. These memory locations are used to store information (in the form of binary) temporarily during the execution of a program.

(*a*) Data bytes in the register pairs of microprocessor can be stored on the stack in reverse order by using PUSH instruction.

(*b*) As we store the dat a in reverse order so stack should be defined from maximum memory location of stack by loading address in the stack pointer by the instruction.

LXI SP, 16 Bit address.

(*c*) As we are saving register pair in the stack; so by each storing process, the content of stack pointer is decreased by two.

(*d*) Now if we want to retrieve the data from stack, use instruction POP.

(*e*) After every POP; the content of stack pointer should be increased by two locations.

3.6.1 Subroutine

A subroutine is a group of instructions written separately from the main program to perform a function that occurs repeadly in the main program.

For Example if we again and again need delay of 100ns between two successive operations; so it will be good to write a program of delay of 100ns separately and execute that when ever required. For this purpose 8085 use CALL and RET instruction.

We will also see the use of stack practically. Now,

(*i*) CALL 16 Bit address

By this instruction, we call the subroutine written at location starting from specified address unconditionally. Here the microprocessor will perform the following operations.

(*a*) Saves the contents of program counter on the stack.

(*b*) Now decrement the stack pointer by two address.

(c) Now program counter takes the value of address specified by the 16 bit address in the CALL instruction and thus jumps to the subroutine.

(*ii*) To return from subroutine, we use instruction RET.

By this instruction; microprocessor returns from the subroutine unconditionally. Here the microprocessor will perform the following operations.

(*a*) Inserts the two bytes of top of stack in to program counter thus program counter again has the address of main program instruction.

(*b*) Now increment the stack pointer by two address. Thus we return from subroutine conditionally to main programe.

(*iii*) We also have some instructions for calling subroutine on the basis of some conditions (specified completely in the instruction set of 8085) such as CC, CNC, CZ, CNZ, CP, CNP etc.

For returning from these subroutine we also have RC, RNC, RZ, RNZ etc.

3.7 INTERRUPT OF 8085

Here we will discuss some interrupt features of 8085.

Interrupt is a process by which a external device can inform to the processor that device needs the service of processor. Now according to priority of that work; processor may provide immediate service 'or' provide service after some delay.

3.7.1 Interrupt Classification of 8085

8085 has totally '5' interrupts given below. They are classified by a number of ways.

Maskable Interrupt	INTR	Unvectored Interrupt
	RST 5.5	Vector Interrupt
	RST 6.5	
	RST 7.5	
Non-Maskable Interrupt	TRAP	

Now we will discusses each one by one; firstly explain INTR.

3.7.2 General Procedure of INTR

(*i*) Set EI flag by writing EI (Enable Interrupt) in the main program if you want to accept any interrupt during the execution of main program.
But it should be clear that TRAP flag cannot be controlled by using EI flag. If you don't want to accept any interrupt, then apply disable interrupt DI.

(*ii*) Now if EI has been written, Microprocessor will check the INTR (Pin No. 10) during the execution of each instruction.

(*iii*) If the INTR is high: microprocessor completes the current instruction, disables the EI flag and it will send $\overline{\text{INTA}}$ to the device. Now it is ready to give service to that device but it will not receive any other interrupt as EI is disabled; If you want to accept another one also then after this step again Enable EI.

(*iv*) Now the content of program counter is saved in the stock; and program counter takes the starting address of the interrupt routine called service routine this time and executes that.

(*v*) At the last of interrupt service routine; we have a RET instruction that will cause to return microprocessor to the main programme.

3.7.3 RST (Restart) Instructions

The 8085 instruction set includes '8' RST (Restart) instructions as shown in instruction set.

Instruction	Restart address
RST 0	0000H
RST 1	0008H
RST 2	0010H
RST 3	0018H
RST 4	0020H
RST 5	0028H
RST 6	0030H
RST 7	0038H

(*i*) Basically RST instruction is a 1-byte call instruction that transfer the program execution to a specific location at page 00H.

(*ii*) Execution of RST is same as call instruction.

These are implemented by hardware but generated using software instruction in the main program.

Vectored Interrupt

Intercepts RST 5.5, RST 6.5, RST 7.5 & TRAP flags are called vectored interrupt because as 8085 gets such type of interrupt, it will automatically goes to specified location. These don't require any $\overline{INTA}$ signal and hardware as neccessary hardware is already implemented inside 8085.

Intercept	Location
TRAP	0024H
RST 7.5	003CH
RST 6.5	0034H
RST 5.5	002CH

These are written in the desending order of priority; Highest priority is of TRAP and lowest priority is of RST 5.5.

TRAP. Trap flag is of high priority; this can not masked 'or' demasked. It is used for emergency purpose such as power failure.

RST 7.5, 6.5 and 5.5. These first of all place EI in software program and then use SIM (set interrupt mask) instruction (explained earlier) mask any two and unmask only desired one for service.

3.8 MEMORY INTERFACING

Memory is an integrated part of microcomputer system. We will discuss the interfacing of a memory chip with the microprocessor. While executing a program, the microprocessor needs to access memory quite frequently to read instruction codes and data stored in memory; the interfacing circuit enables that access. Memory has certain signal requirements to write into and read from its register. Similarly microprocessor initiates a set of signals when it wants to read from and write into memory. The interfacing process involves desiging a circuit that will match the memory requirements with microprocessor signals.

Here we will discuss memory interfacing using memory chips such as 2732 ROM and 6116 static R/W memory. We will discuss address decoding and memory address.

3.8.1 Memory Structure and Its Requirement

Read/Write memory (R/W M) is a group of register to store binary information. Figure shows a typical R/W memory chip. It is represented by (2048 x 8) means it has total 8-bit data line input/output line. $2048 = 2^{11}$ means, it has 11 address lines, It has also one chip select $\overline{CS}$ and two control lines : Read ($\overline{RD}$) to enable the output buffer and write ($\overline{WR}$) to enable the input buffer. It also has one internal address decoder that decodes the address lines.

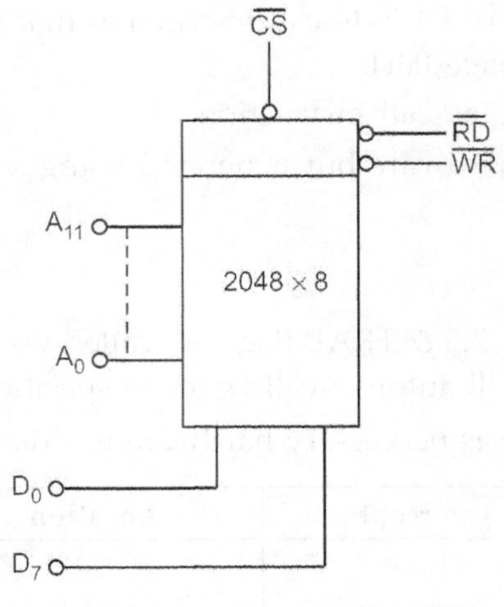

Fig. 3.8 Read/Write Memory Chip

3.8.2 Interfacing Circuit

Figure 3.9 shows an interfacing circuit using a 3 to 8 decoder to interface 2732 EPROM memory chip. It has been assumed that the chip has already been programmed. We can analysis the interfacing circuit in the following steps.

1. Connect the required address lines of the address bus to the address lines of the memory chip.
2. Decode the remaining address lines of address bus to generate the chip select signal and connect the signal to select the chip.
3. Generate control signals $\overline{\text{MEMR}}$ and $\overline{\text{MEMW}}$ by combining $\overline{\text{RD}}$ and $\overline{\text{WR}}$ signals with IO/$\overline{\text{M}}$ and to enable appropriate buffers.

3.8.3 Address Decoding and Memory Addresses

Now for the circuit diagram, follow the above '3' steps.

Step I. The 8085 address lines A11-A0 are connected to pin A11-A0 of the memory chip to address 4096 registers.

Step II. The decoder is used to decode four address lines A15-A12 . The output O0 of the decoder is connected to chip enable ($\overline{\text{CE}}$). The $\overline{\text{CE}}$ is asserted only when the address on A15-A12 is 0000; A15 (low) enables the decoder and the input 000 assets the output O0.

Step III. For this EPROM, we need one control signal: Memory Read ($\overline{\text{MEMR}}$) active low. The $\overline{\text{MEMR}}$ is connected to enable the output buffer.

We can obtain the address range of this memory chip by analyzing the possible logic levels on the 16 address lines. The logic levels on the address lines A11 - A0 can assume any combination from all 0's to all 1's. Therefore, we have

	A15	A14	A13	A12	A11	A10	A9	A8	A7	A6	A5	A4	A3	A2	A1
0000H =	0	0	0	0	0	0	0	0	0	0	0	0	0	0	0
0FFFH =	0	0	0	0	1	1	1	1	1	1	1	1	1	1	1
	Chip enable				Register select										

So we have chip ranges from 0000H to 0FFFH as described above.

Now we should see the 4096 × 8 chip by following analogy.

1. It has totally $4096 = 2^{12} = 12$ address lines requirement.
2. Each word written would be of 8-bits.
3. We have totally 16 pages with 256 lines each.

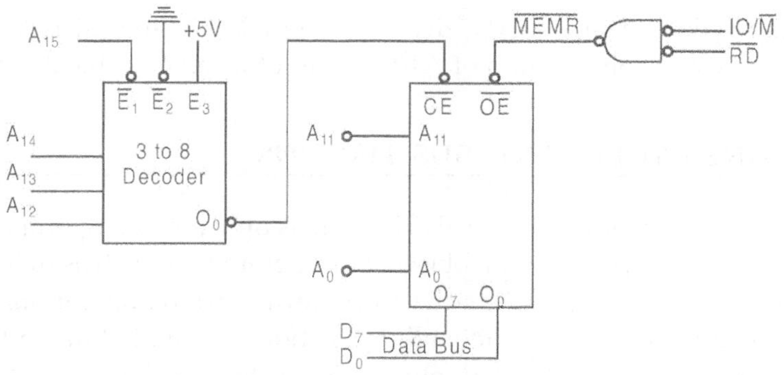

Fig. 3.9 Memory Mapping

Example : *For the figure given below, specify the range of memory.*

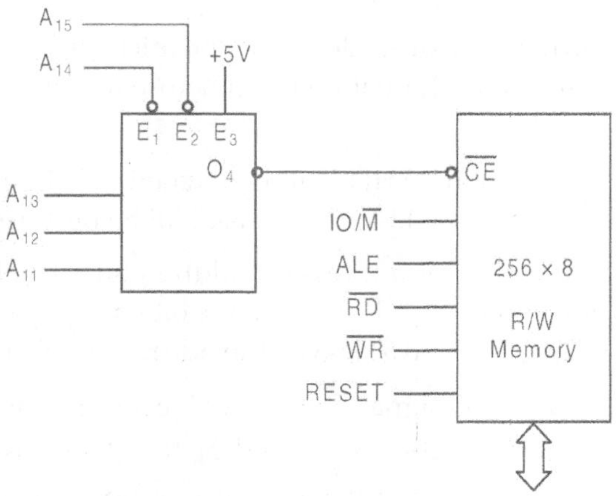

Solution: It should be clear that

(*i*) E1 and E2 should be always active low to choose this decoder so. A14 and A15 should always be 00.

(*ii*) We are using O4 of decoder to select 256 X 8 R/W memory, so.

A13 A12 A11 = 100

(*iii*) We have totally 256 required address.

So 256 = 2^8

Total address line required = 8.

So we will use A7 – A0 for that purpose now finally we have.

A15	A14	A13	A12	A11	A10	A9	A8	A7	A6	A5	A4	A3	A2	A1	A0
0	0	1	0	0	0	0	0	0	0	0	0	0	0	0	0
0	0	1	0	0	0	0	0	1	1	1	1	1	1	1	1
Used to select decoder		Chip enable			Don't care			Register select							

Thus finally we have memory range 2000H to 20FFH.

It should be clear to you that AD0 to AD7 denotes that ROM is using A0 to A7. Now A10–A8 are free so you can use any one combination of A10–A8 out of its all possible $2^3 = 8$ combinations.

3.9 8085 MACHINE CYCLES AND BUS TIMINGS

Instruction has Two Parts: Operation code also known as op-code and operand. The op-code is a command such as ADD and operand is an object to be operated on, such as byte 'or' the content of register. To execute a instruction, 8085 needs to perform various operations such as memory read/write and I/O Read/Write. Fortunately all instruction are divided into a few basic machine cycle and these machine cycles are divided into precise system clock periods. Now we will study some Basic Instructions.

3.9.1 Opcode Fetch Machine Cycle

The first operation in any instruction is op-code fetch. The microprocessor needs to get (fetch) this machine code from memory register where it is stored before microprocessor or begin to execute the instruction let us example that.

MOV C, A that is stored at location 2004H is to be fetched. We know from data sheet of 8085 that MOV C, A has code 4FH. (0100.1111) Now this task will be performed by the following steps

1. The microprocess places the 16-bit address on address bus. The higher order 8-bit address is put at high order address bus and low order 8-bits are put in the lower order address bus. As the ALE goes low; AD7-AD0 leaves the address in buffer and ready to take data.
2. Control unit sends the control signal $\overline{RD}$ to enable the memory chip.
3. The byte from the memory location is placed at the data bus.
4. The byte is placed in the instruction decoder of microprocessor, and task, is carried out according to inspection complete machine cycle has been shown.

3.9.2 Memory Read and Memory Write Cycle

It is now very easy to understand by the Fig. 3.10 and Fig. 3.11.

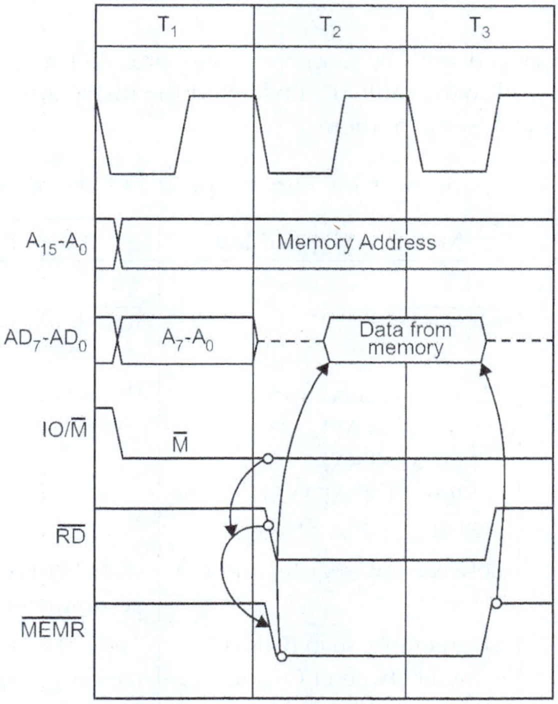

Fig. 3.10 Timing of the Memory Read Cycle

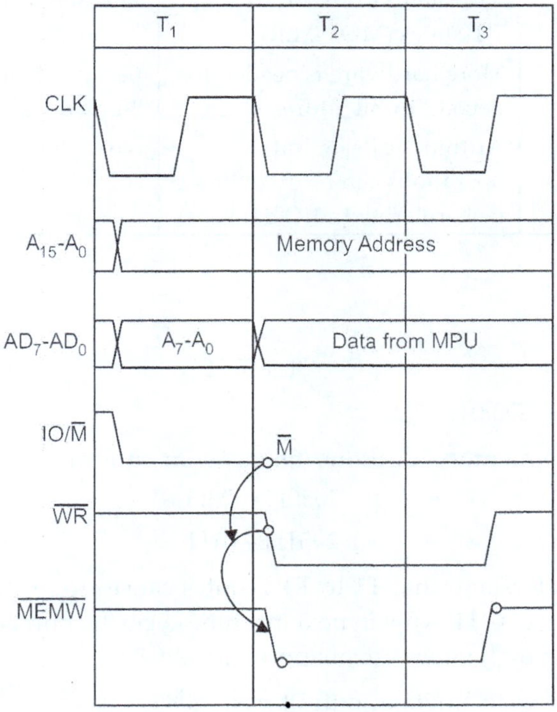

Fig. 3.11 Timing of the Memory Write Cycle

3.9.3 I/O Execution

An I/O device can be interfaced with 8085 microprocessor either as a pheripheral I/O 'or' as a memory mapped I/O. The following table 3.5 has noted the instruction for memory mapped I/O and pheripheral I/O, let us take some of them.

Table 3.5 Comparison of Memory-Mapped I/O and Peripheral I/O

Characteristics	Memory-Mapped I/O	Peripheral I/O
1. Device address	16-bit	8-bit
2. Control signals for Input/Output	$\overline{MEMR}/\overline{MEMW}$	$\overline{IOR}/\overline{IOW}$
3. Instructions available	Memory-related instructions such as STA; LDA; LDAX; STAX; MOV M,R: ADD M; SUB M; ANA M: etc.	IN and OUT
4. Data transfer	Between any register and I/O	Only between I/O and the accumulator
5. Maximum number of I/Os possible	The memory map (64K) is shared between I/Os and system memory	The I/O map is independent of the memory map; 256 input devices and 256 output devices can be connected
6. Execution speed	13 T-states (STA, LDA) 7 T-states (MOV M,R)	10 T-states
7. Hardware requirements	More hardware is needed to decode 16-bit address	Less hardware is needed to decode 8-bit address
8. Other features	Arithmetic or logical operations can be directly performed with I/O data	Not available

OUT Instruction

Let instruction to executed

OUT 07H

For OUT machine code = D3:01

So Finally we have two memory locations (any as your choice)

2050H – D3H

2051H – 01H

Now we can easily understand that T1 to T4 T-states causes to fetch D3H and Next T1 to T4 T- state causes the fetching of 01H. Now in next machine cycle the port address 07 H is taken. This is I/O execution so IO/$\overline{M}$ = high when use get port address.

This port will cause to write a data at output from 8085.

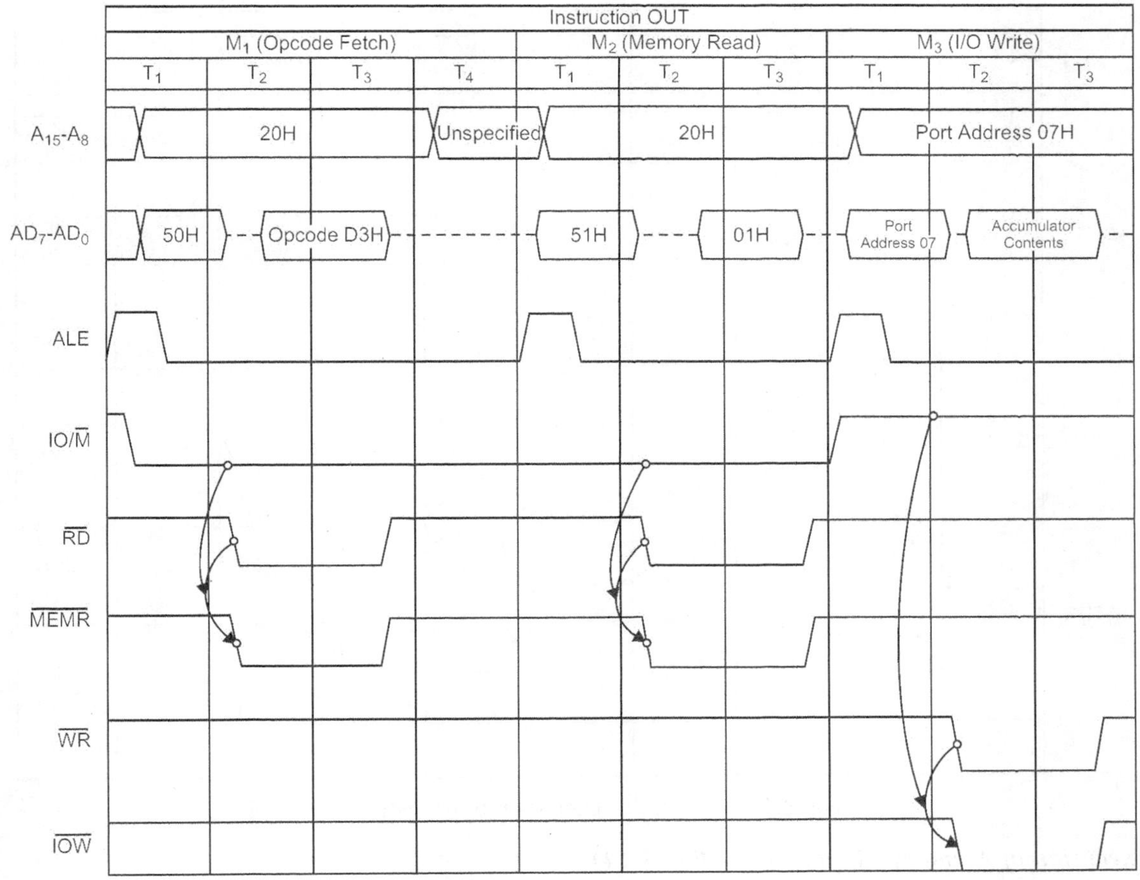

Fig. 3.12 OUT Timing Diagram

IN Instruction

The same process as OUT. Let we draw timing diagram for.

Memory address	Machine code
2065	DB 84 H
2066	64

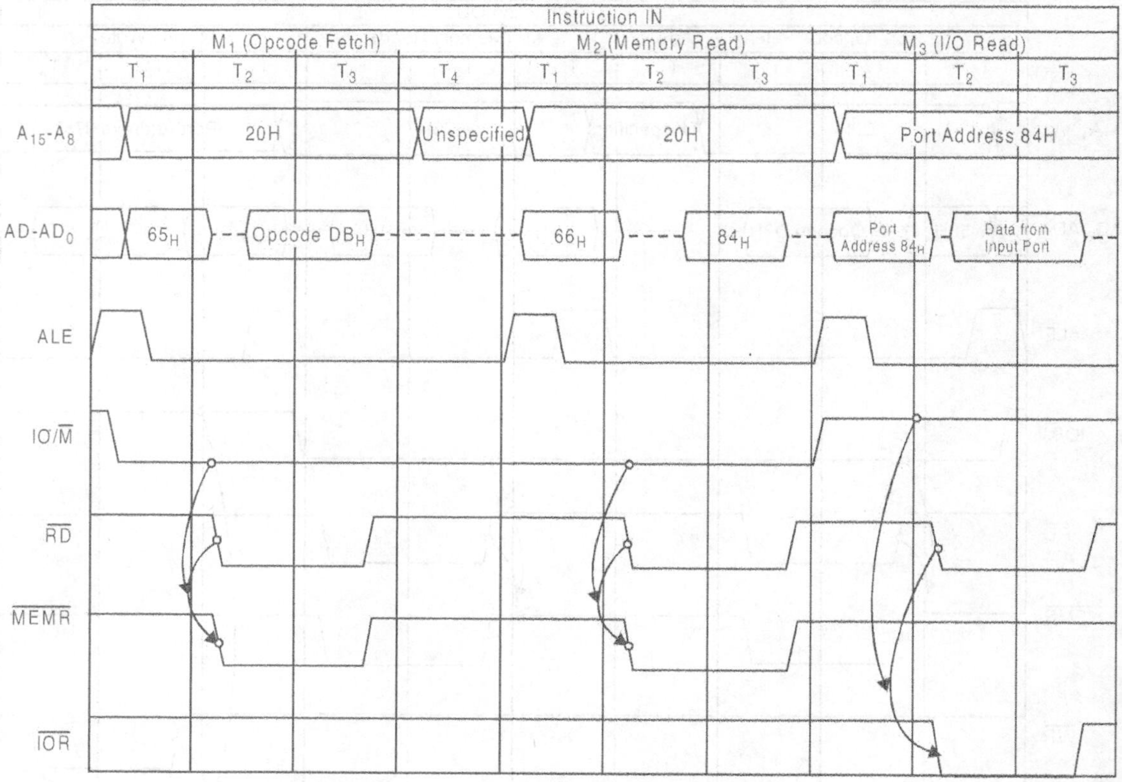

Fig. 3.13 Timing for Execution of IN Instruction

Execution of Memory Mapped I/O (Fig. 3.14)

The execution of memory related data transfer instruction is similar to the execution of IN 'or' OUT instruction, except that the memory related instruction have 16-bit addresses.

For example, to execute the instruction STA 8000H in the fourth machine cycle (M4), the microprocessor places memory address 8000H on the entire address bus (A15–A0). The accumulator content are sent on the data bus, followed by the control signal $\overline{MEMW}$.

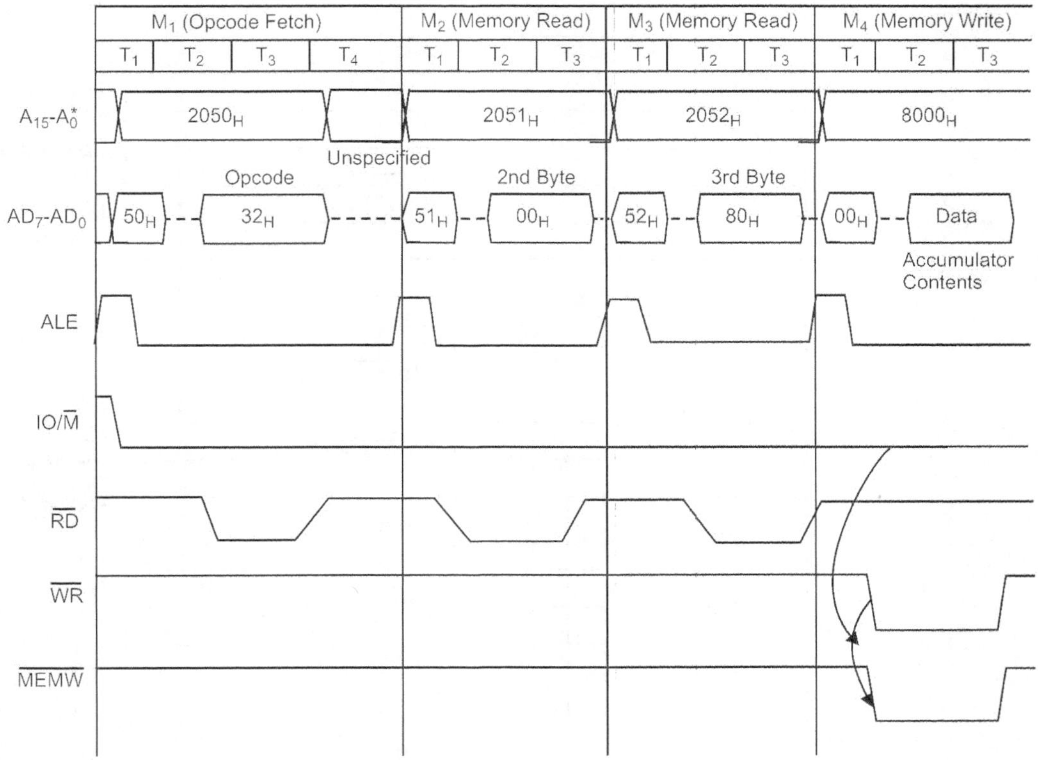

Fig. 3.14 Timing for Execution of the Instruction: STA 8000H

3.10 DEMULTIPLEXING THE BUS AD7–AD0

Just take example that we have to read address 2005H. By Fig. 3.15, it is clear that for three clock periods higher order address bus has 20H. But lower order address (05)H is lost after the first clock period. Thus we can see that after first T1. If we have complete 2005H identification; then we can perform the task.

Here we have seen the Basic Demultiplexing using latch.

During T1, ALE is high so latch allows to pass address at output but as it goes low, data buses (D7–D0) is available.

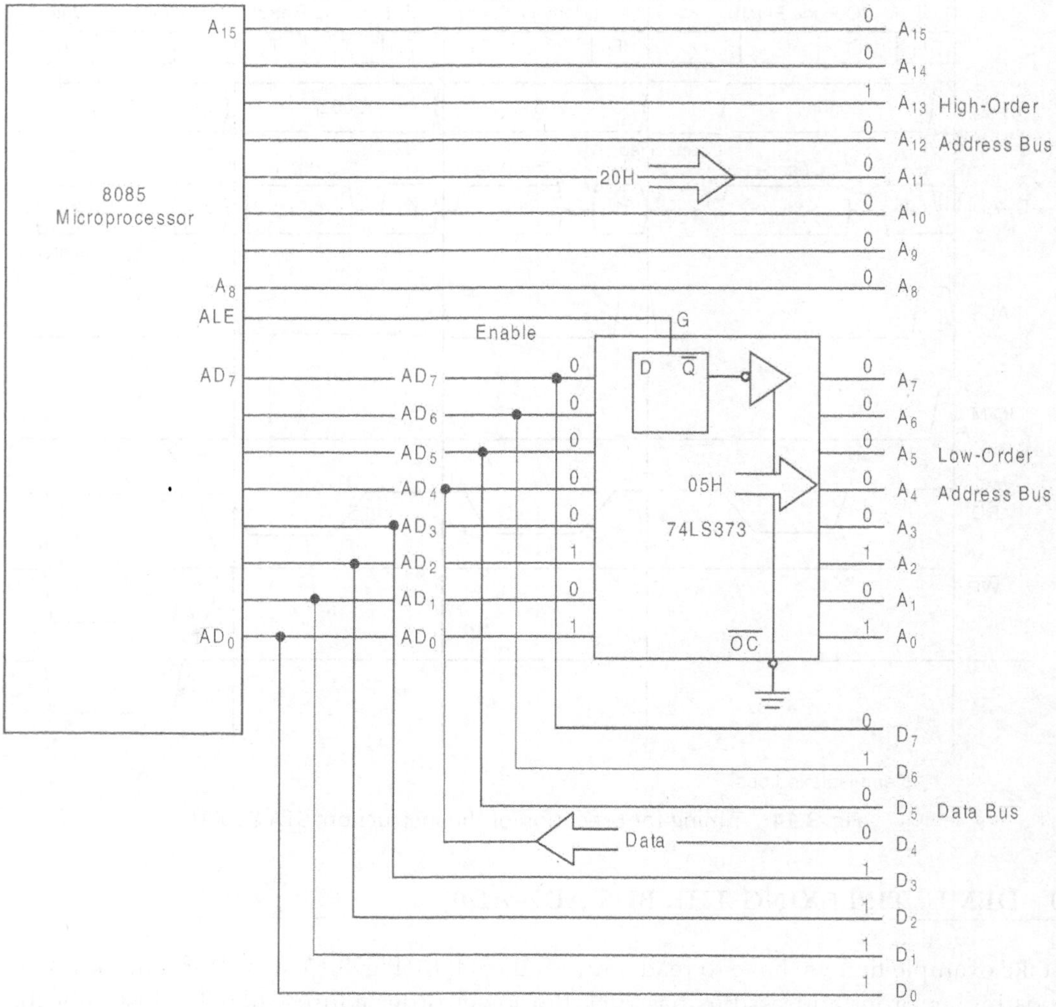

Fig. 3.15 Schematic of Latching Low-order Address Bus

REVIEW QUESTIONS

1. Draw the pin diagram of 8085. Explain the function of each pin.
2. Draw the general block diagram of 8085. Explain each block diagram.
3. What are the various addressing modes of 8085?
4. What do you understand by subroutine? Differentiate with service routine.
5. Classify the interrupts of 8085.
6. Explain the vector interrupts of 8085.
7. Explain the function of ALE, HOLD, HLDA, SFD, SOD.
8. Explain the stack and stack operation of 8085 in detail.
9. Explain the flag register of 8085 in detail.

4

ASSEMBLY LANGUAGE PROGRAMMING OF 8085

4.1 INTRODUCTION

A computer can only do what the programmer has told it to do, in the form of a program. A program is just a sequence of very simple commands that lead the computer to solve some problem. Once the program is written and debugged, the computer can execute the instructions very fast, and always do it the same, every time, without a mistake.

And herein lies the power of a computer. Even though the program consists of very simple instructions, the overall result can be very impressive, due mostly to the speed at which the computer can process the instructions. Even though each step in the program is very simple, the sequence of instructions, executing at millions of steps per second, can appear to be very complicated, when taken as a whole. The trick is not to think of it as a whole, but as a series of very simple steps, or commands.

The microprocessor itself is usually a single integrated circuit (IC). Most microprocessors, or very small computers, have much the same commands or instructions that they can perform. They vary mostly in the names used to describe each command. In a typical microprocessor, there are commands to move data around, do simple math (add, subtract, multiply, and divide), bring data into the microprocessor from the outside world, and send data out of the microprocessor to the outside world.

A typical micro has three basic parts inside. These are the Program Counter (PC), Memory, and Input/Output (I/O). The Program Counter keeps track of which command is to be executed. The Memory contains the commands to be executed. The Input/Output handles the transfer of data to and from the outside world (outside the microprocessor's physical package). There are many other actual parts inside our microprocessor however, we will learn about each and every single one, one step at a time.

4.1.1 A Simple Program

As stated before, a program is a sequence or series of very simple commands or instructions. A real world example program might be the problem of crossing a busy street.

Step 1: Walk up to the traffic lights and stop.

Step 2: Look at the traffic light.

Step 3: Is your light green?

Step 4: If the light is red, go to step 2. (Otherwise continue to step 5)

Step 5: Look to the left.

Step 6: Are there cars still passing by?

Step 7: If yes, go to step 5. (Otherwise continue to step 8).

Step 8: Look to the right.

Step 9: Are there cars still passing by? (There shouldn't be any by now, but you never know!)

Step 10: If yes, go to step 8. (Otherwise continue to step 11)

Step 11: Proceed across the street, carefully!!

Now this may seem childish at first glance, but this is exactly what you do every time you cross a busy street, that has a traffic light (at least, I hope you do). This is also exactly how you would tell a microprocessor to cross the street, if one could. This is what I mean by a sequence or series of very simple steps.

The micro would not, however, in this case, know what to do when it got to the other side, since we didn't tell it. A person, on the other hand, could decide what to do next, at a moments notice, without any apparent programming. In the case of a person, though, there has been some programming; it's called past experiences.

Another program might be to fill a glass with water from a tap.

Step 1: Turn on the water.

Step 2: Put the glass under the tap.

Step 3: Look at the glass.

Step 4: Is it full?

Step 5: If no, go to step 3. (Otherwise, continue to step 6)

Step 6: Remove the glass from under the tap

Step 7: Turn off the water.

This is a simpler program, with fewer steps, but it solves a problem, to fill a glass with water. In a microprocessor, the problems are different but the logical steps to solve the problem are similar, that is, a series of very simple steps, leading to the solution of a larger problem.

Also notice that since the steps are numbered. 1 through 7, that is the order in which they are executed. The Program Counter, in this case, you reading each line, starting with 1 and ending with 7, doing what each one says. In a micro, the Program Counter automatically advances to the next step, after doing what the current step says, unless a Branch, or jump, is encountered. A branch is an instruction that directs the Program Counter to go to a specific step, other than the next in the sequence. The branch in this example is step 5. Not only is this a branch, but also it is a conditional branch. In other words, based on whether the glass is full or not, the branch is taken, or not. A micro has both branch and conditional branch instructions. Without this ability to reuse instructions, in a sort of looping action, a solution would take many more steps, if it would be possible at all.

The point of this lesson is to show how a simple set of instructions can solve a bigger problem. Taken as a whole, the solution could appear to be more complicated than any of the separate steps it took to solve it.

The most difficult problem to be solved in programming a microprocessor is to define the problem you are trying to solve.

4.2 MICROPROCESSOR INSTRUCTIONS

We refer to a binary number like 1111 as 1111b, a decimal number like 123 as 123, and a hex number like A5 as A5h. So don't be confused by the letters following numbers, we use both caps and lowercase to denote binary, decimal, or hexadecimal so there is no doubt what base a multidigit, or multibit, number is in.

Also there is another kind of memory, called flags. Flags are single bit numbers used to indicate different conditions. They are called flags because they flag the program of events or conditions. If a flag is raised, or has a 1 in it, it is said to be SET. If it is a 0, it is said to be CLEARED or RESET.

One other thing, in an 8-bit byte, the 8-bits are referred to as bits 0 through 7, with bit 0 being the right most, or least significant (LSB), and bit 7 as the left most or most significant (MSB).

Lastly, there are various Registers inside a microprocessor or Central Processing Unit (CPU). These vary from CPU to CPU, but all contain a register called the Accumulator. It is also referred to in some as the **A** register. We will be using the accumulator in the following discussion. It is a type of memory for storing temporary results and is 8-bits wide, or a byte, as are most places that data can be put inside the CPU.

In the CPU we will be using, there are 5 different types of instructions and several variations of each, resulting in over 100 different instructions. These 4 types are ARITHMETIC, LOGICAL, BRANCHING, and DATA TRANSFER.

4.2.1 Arithmetic

The arithmetic instructions usually include addition, subtraction, division, multiplication, incrementing, and decrementing although division and multiplication were not available in most early CPU's. There are two flags used with arithmetic that tell the program what was the outcome of an instruction. One is the Carry (C) flag. The other is the Zero (Z) flag. The Z flag, if set, says that the result of the instruction left a value of 0 in the accumulator.

Addition

This is straightforward and is simply to add two numbers together and get the result. However there is one more thing. If, in the addition, the result was too big to fit into the accumulator, part of it might be lost. There is a safeguard against this. Take the case of 11111111b (255) and 11111111b (255). These are the largest numbers that can fit into an 8-bit register or memory location. You can add these as decimal numbers, since I gave you their values in decimal also, and you would get 510. The binary value for 510 is 111111110b (9-bits).

The accumulator is only 8-bits wide, it is a byte. How do you fit a 9-bit number into 8-bits of space? The answer is, you can't, and it's called an OVERFLOW condition. So how do we get around this dilemma? We do it with the CARRY (C) flag. If the result of the addition is greater than

8-bits, the CARRY (C) flag will hold the 9th-bit. In this case the accumulator would have in 11111110b (254) and the C flag would be a 1, or set. This 1 has the value of 256 because this is the 9th-bit. We haven't covered a 9-bit number, but they come up all the time as overflows in addition. Since we are using base 2, the 8th-bit (bit 7) in a byte is worth 128, then the 9th-bit is worth 2 times that, or 256. Adding 254 and 256, we get 510, the answer, and we didn't loose anything, because of the C flag. The result of the addition not caused an overflow, the C flag would be 0, or cleared.

Subtraction

In the case of subtraction, the process is more difficult to explain, and as such, I'm not going to cover it here. It involves 1's compliment and 2's compliment representation. But I will tell you this, you can subtract two numbers and if there is an under flow, the C flag will be a 1; otherwise it will be a 0. An under flow is where you subtract a larger number from a smaller number.

Multiplication and Division

In the microprocessor we will be using, the 8085, multiply and divide instructions are available so we will later talk about them. They do, however, do just what the names suggest.

Increment and Decrement

Two other instructions are included in the arithmetic group. They are increment and decrement. These instructions are used to count events or loops in a program. Each time an increment is executed, the value is incremented by 1. A decrement, decrements the value by 1. These can be used with conditional jumps to loop a section of program, a certain number of times. We will see these used later.

Logical

In microprocessors there are other mathematical instructions called logical instructions. These are OR, AND, XOR, ROTATE, COMPLEMENT and CLEAR. These commands are usually not concerned with the value of the data they work with, but, instead, the value, or state, of each bit in the data.

OR

The OR function can be demonstrate by taking two binary numbers, 1010b and 0110b. When OR'ing two numbers, it doesn't matter at which end you start, right or left. Let's start from the left. In the first bit position there is a 1 in the first number and a 0 in the second number. This would result in a 1. The next bit has a 0 in the first number and a 1 in the second number. The result would be 1. The next bit has a 1 in the first number and a 1 in the second number. The result would be a 1. The last bit has a 0 in the first number and a 0 in the second number, resulting in a 0. So the answer would be 1110b. The rule that gives this answer says that with an OR, a 1 in either number result in a 1, or said another way, any 1 in, gives a 1 out.

AND

AND'ing uses a different rule. The rule here is a 0 in either number will result in a 0, for each corresponding bit position. Using the same two numbers 1010b and 0110b the result would be 0010b. You can see that every bit position except the third has a zero in one or the other number. Another way of defining an AND is to say that a 1 AND a 1 results in a 1.

XOR (Exclusive OR)

XOR'ing is similar to OR'ing with one exception. An OR can also be called an inclusive OR. This means that a 1 in either number or both will result in a 1. An exclusive OR says that if either number has a 1 in it, but not both, a 1 will result. A seemingly small difference, but crucial. Using the same two numbers, the result would be 1100b. The first two bits have a 1 in either the first or the second number but not both. The third bit has a 1 in both numbers, which results in a 0. The fourth has no 1's at all, so the result is 0. The difference may seem small, even though the OR and XOR result in different answers. The main use of an XOR is to test two numbers against each other. If they are the same, the result will be all 0's, otherwise the answer will have 1's where there are differences.

Compliment

Complimenting a number results in the opposite state of all the 1's and 0's. Take the number 1111b. Complimenting results in 0000b. This is the simplest operator of all and the easiest to understand. Its uses are varied, but necessary, as you'll see later.

Rotate

These instructions rotate bits in a byte. The rotation can be left or right, and is done one bit each instruction. An example might be where the accumulator has a 11000011b in it. If we rotate left, the result will be 10000111b. You can see that bit 7 has now been moved into bit 0 and all the other bits have move 1 bit position in, the left direction.

Clear

This instruction clears, or zero's out the accumulator. This is the same as moving a 0 into the accumulator. This also clears the C flag and sets the Z flag.

4.2.2 Branching

There are also program flow commands. These are branches or jumps. They have several different names reflecting the way they do the jump or on what condition causes the jump, like an overflow or under flow, or the results being zero or not zero. But all stop the normal sequential execution of the program, and jump to another location, other than the next instruction in sequence.

Jump on Condition (of a Bit)

These instructions let you make a jump based on whether a certain bit is set (a 1) or cleared (a 0). This bit can be the CY (carry) flag, the Z (zero) flag, or any other bit.

Call

There is also a variation on a jump that is referred to as a CALL. A CALL does a jump, but then eventually comes back to the place where the CALL instruction was executed and continues with the next instruction after the CALL. This allows the programmer to create little sub-programs, or subroutines, that do repetitive tasks needed by the main program. This saves programming time because once the subroutine is written, it can be used by the main program whenever it needs it, a kind of way to create your own instructions.

4.2.3 Data Transfer

Moving

These instructions do exactly what you would think. They move data around between the various registers and memory.

Exchanging

Exchanging is a variation on the move instruction. Here data is exchanged between two places.

4.3 MEMORY AND ADDRESSING

There are several different types of memory in a microprocessor. One is Program memory. This is where the program is located. Another is Data memory. This is where data that might be used by the program is located. The neat (or strange) thing is that they both reside in the same memory space and can be altered by the program. That's right, a program can actually alter itself if that was necessary. Two terms are used when talking about memory. Reading (load) is getting a value from memory and Writing (store) is putting a value into memory.

There are three buses associated with the memory subsystem. One is the **address bus,** the second is the **data bus,** and the third is the **control bus.** It's important for you to know exactly how all this works, because these busses transport data and addresses everywhere. All three are connected to the memory subsystem.

In the 8085 CPU, the address bus is 16 bits wide. It acts to select one of the unique 2^{16} (64K) memory locations. The control bus determines whether this will be a read or a write. In the case of an instruction fetch, the control bus is set up for a read operation. Data is read or written through the data bus, which is 8 bits wide. This is why all registers and memory are 8 bits wide; it's the width of the data bus on the 8085 CPU. A bus is just a group of connections that all share a common function. Instead of speaking of each bit or connection in the address separately, for example, all 16 are taken together and referred to simply as the address bus. The same is true for the control and data buses.

A byte is the most used number in a micro because each memory location or register is one byte wide. Memory has to be thought of as a sort of file cabinet with each location in it being a folder in the cabinet. In a file cabinet, you go through the tabs on the folders until you find the right one. To get to each memory location, a different method is used.

Instead, a unique address is assigned to each location. In most micros this address is a word or 16 bits, or 4-digit hex. This allows for a maximum of 65536 (2^{16} or 64K) unique addresses or memory locations that can be accessed. A 4 digit hex number usually refers to these addresses. Memory usually starts at address 0000h and could go up to FFFFh (2^{16} or 64K or 65536 in total). To access these locations, a 16-bit address is presented to memory and the byte at that location is either read or written.

The Program Counter is what holds this address when the micro is executing instructions. The reason instructions are read sequentially is because the program counter automatically increments after fetching the current instruction. It does this even before the current instruction is acted upon. The sequence is that the program counter's contents are placed on the memory address bus and the instruction is fetched from memory through the data bus, and immediately the program counter is incremented by 1. Then the micro looks at the instruction and starts processing it. If the instruction is not some kind of jump or call, the instruction is completed and the program counter is presented to the memory address bus again and the next instruction is fetched, the program counter is incremented and the process starts over. This is referred to, in computer jargon, as fetch, decode, and execute.

In the case of reading or writing data, the process is a little different. Data can be read from or written to memory in similar fashion to the fetch. But data does not need the decode and execute steps.

We start looking at the microprocessor, INTEL. 8085.

4.4 ASSEMBLY LANGUAGE PROGRAMMING OF 8085

We have discussed many important 8085 instruction set previously. Now we will study the 8085 programmes. Let us start with some very short programmes.

Example 1. *Get a data to accumulator and then move it to register B.*

Solution.	MVI	A, 05	Get 05 (data) into register (accumulator)
	MOV	B, A	Transfer the data from accumulator to register B
	HLT		

Example 2. *Load the content of memory location 2000H directly to the accumulator, then transfer it to the register B.*

Solution.	LDA	2000H	Get the content of memory location 2000H into accumulator
	MOV	B, A	Now move the content of accumulator to register B.
	HLT		

Example 3. *Move the content of memory location 2000H to register C.*

Solution.	LXI	H, 2000H	Get the memory address 2000H to H-L pair.
	MOV	C, M	Move the content of memory location, whose address in H-L pair, to register C.
	HLT		

Example 4. *Place the content of memory location 2000H in register B, and that of 2001H in register C.*

Solution.	LXI	H, 2000H	Get 2000H in H-L pair.
	MOV	B, M	Move the content of 2000H to register B.
	INX	H	Increment H-L pair by one.
	MOV	C, M	Move the content of 2001H to C.
	HLT		Stop

4.5 FLOW CHART

So to perform a particular work, we first all develop a concept. To represent a concept in a easy work, we represent that in the form of flow chart. In flow chart, each step of our concept is presented by very small and easy steps.

Now next step is to write a program by using right instructions.

Programming of 8085

Addition of two 8-bit Number and Answer is in 8-bits

Problem. *Add 15H and 58H*

First Number is in memory location 2001H

Second Number is in memory location 2002H

Answer to be stored in memory location 2003H

Flow Diagram:

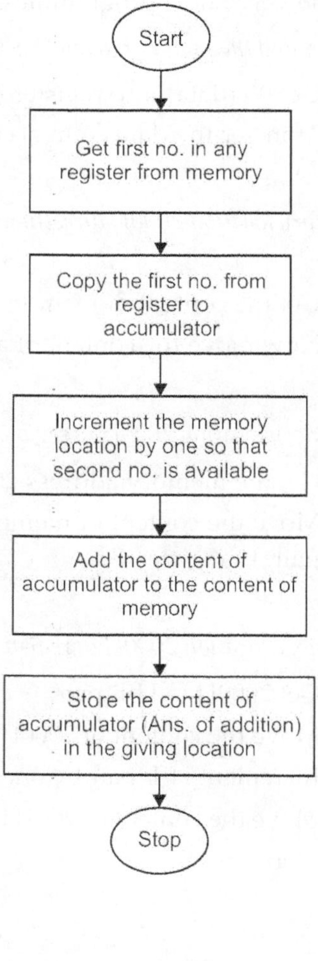

Program

```
LXI     H, 2001H    Get first number from memory to register
MOV     A, M        Copy the first no. from memory to accumulator
INX     H           Increment the register pair to get the second no.
ADD     M           Add the Content of accumulator to register pair
STA     2003H       Store the result in the location
HLT
```

Addition of two 8-bit Number and Answer is in 16-bits

Problem. *First Number is in memory location 2001H*

Second Number is in memory location 2002H

Answer to be stored in memory location 2003H and 2004H

Program

```
        LXI     H, 2001H    Get first no. from memory to register
        MVI     B, 00H      The initial value of register B is 00H (used for carry)
        MOV     A, M        Copy the first number from memory to accumulator
        INX     H           Increment the resister pair to get the second no.
        ADD     M           Add the content of accumulator to register pair
        JNC     NEW         If there is no carry jump to NEW
        INR     C           If carry increment the B by one
NEW     STA     2003H       Store the result in the location
        MOV     A, B        Copy the carry in accumulator
        STA     2004H       Store the carry in the location
        HLT
```

Flow Diagram:

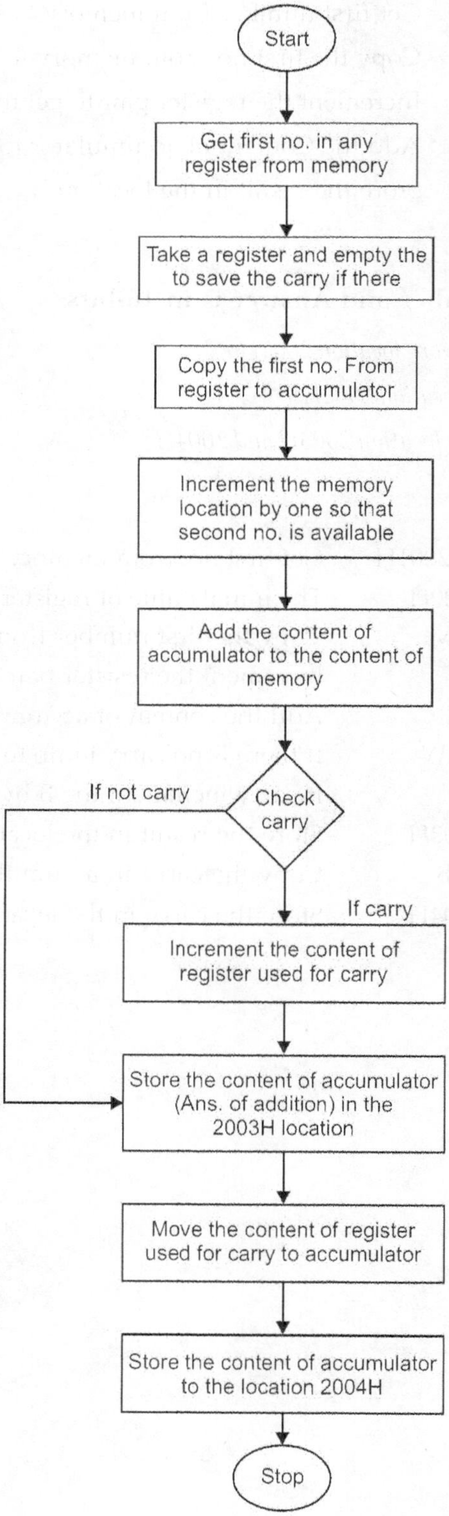

Addition of two 16-bit Number and Answer is in 16-bits (or more)

Problem. *First Number is in memory location 2001H and 2002H*

Second Number is in memory location2003H and 2004H

Answer to be stored in memory location 2005H and 2006H

Flow Diagram:

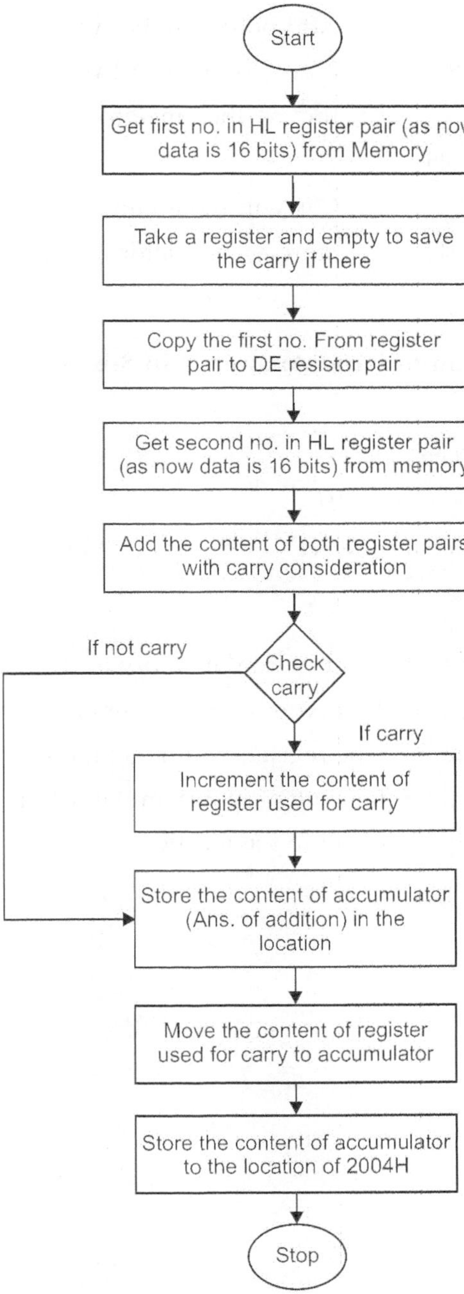

Program

	LHLD	2001H	First Number in the HL pair
	MVI	B,00	Empty register for carry
	XCHG		Content of HL pair into DE pair
	LHLD	2003H	Second Number in the HL pair
	DAD	D	Add both Number with carry
	JNC	NEW	If no carry go to NEW
	INR	B	If carry then increment B by one
NEW	SHLD	2005H	
	MOV	A,B	Carry in the accumulator
	STA	2006H	Store accumulator in location
	HLT		

Subtraction of two 8-bit Number and Answer is in 8-bits

Problem. *Subtract 58H and 12H*

First Number is in memory location 2001H

Second Number is in memory location 2002H

Answer to be stored in memory location 2003H

Program

LXI	H, 2001H	Get first Number from memory to Register
MOV	A, M	Copy the first Number from memory to accumulator
INX	H	Increment the register pair to get the second no.
SUB	M	Subtract the content of accumulator to register pair.
STA	2003H	Store the result in the location
HLT		

Flow Diagram:

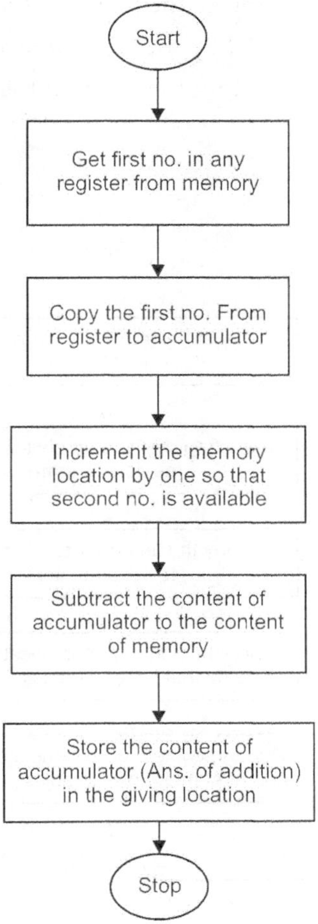

Decimal Addition of two 8-bit Number and Answer is in 16-bits

Problem. *First Number is in memory location 2001H*

Second Number is in memory location 2002H

Answer to be stored in memory location 2003H and 2004H

Program

LXI	H, 2001H	Get first Number from memory to register
MVI	B, 00	The initial value of register B is 00H (used for carry)
MOV	A, M	Copy the first Number from memory to accumulator
INX	H	Increment the register pair to get the second Number.
ADD	M	Add the content of accumulator to register pair.
DAA		Adjust in the decimal form
JNC	NEW	If there is no carry, jump to NEW

	INR	B	If carry, increment the B by one.
NEW	STA	2003H	Store the result in the location
	MOV	A, B	Copy the carry in accumulator
	STA	2004H	Store the carry in the location
	HLT		

Flow Diagram:

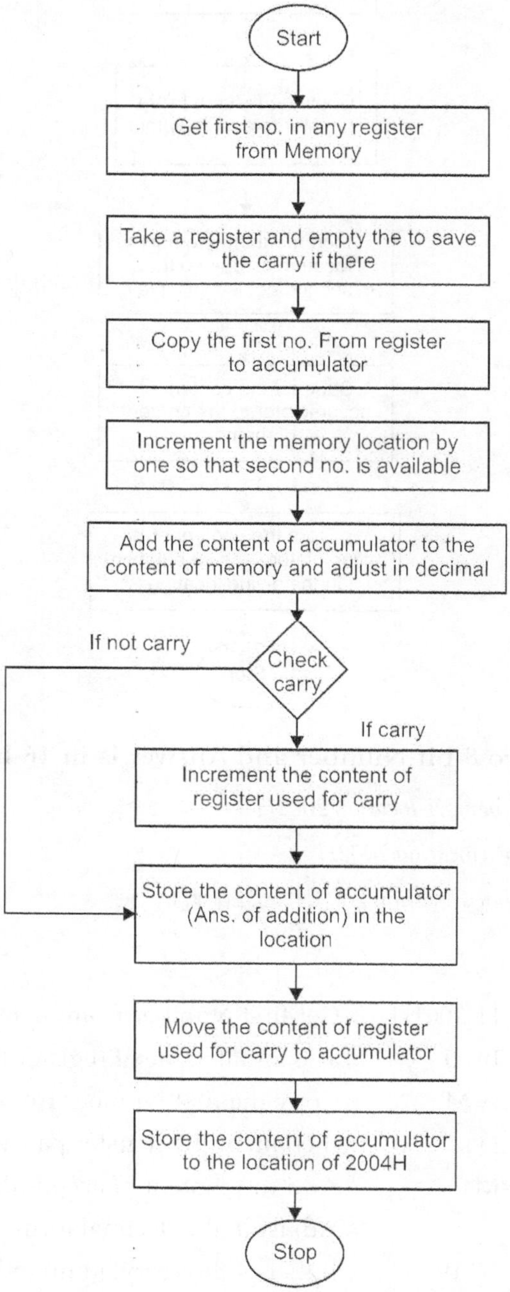

To find the one's complement of an 8-bit number

Problem. *Number Stored in the location 2001H*

Answer to be stored in the location 2002H

Flow Diagram:

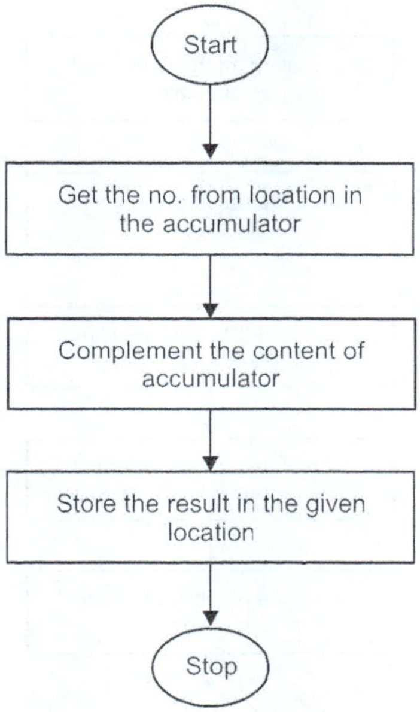

Program

LDA	2001H	Load number in accumulator
CMA		Complement the content of accumulator
STA	2002H	Store the result in the location 2002h
HLT		

To find the one's complement of a 16-bit number

Problem. *Numbers Stored in the location 2001H, 2002H*

Answer to be stored in the location 2004H, 2005H

Flow Diagram: (Simply perform complement operation on LSB 8 bits, store that result and again take MSB 8-bits and complement them and store result).

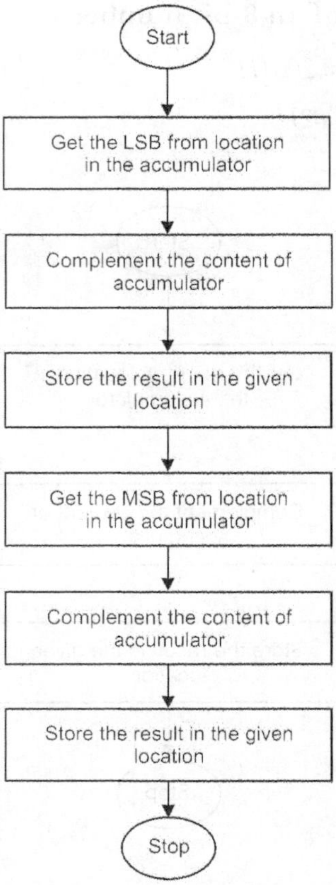

Program

LDA	2001H	Load LSB, in accumulator
CMA		Complement the content of accumulator
STA	2004H	Store the result in the location 2002h
LDA	2002H	Load MSB, in accumulator
CMA		Co mplement the content of accumulator
STA	2005H	Store the result in the location 2002h
HLT		

To find the two's complement of an 8-bit number

Problem. *Number Stored in the location 2001H*

Answer to be stored in the location 2002H

Flow Diagram:

It should be clear that

Two's complement = One's complement + 1

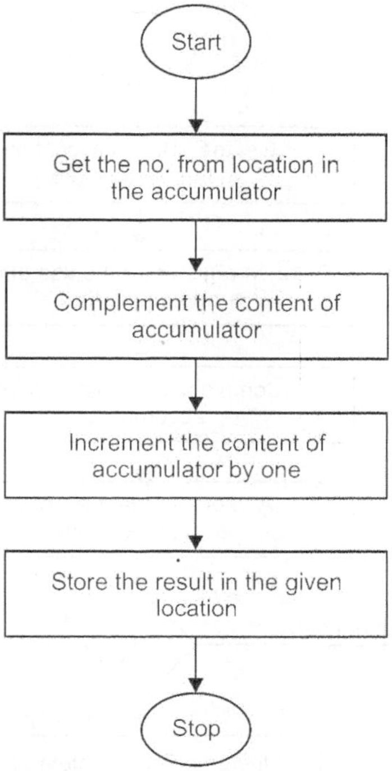

Program

LDA	2001H	Load number in accumulator
CMA		Complement accumulator
INR	A	Increment accumulator by one
STA	2002H	Store the content of accumulator
HLT		

To find the two's complement of a 16-bit number

Problem. *Number Stored in the location 2001H, 2002H*

Answer to be stored in the location 2004H, 2005H

Flow Diagram:

- Simply take one's complement of LSB 8 bits
- Add one to this result
- Store this result in the given location
- Also store the carry of this operation
- Now take MSB of Number.
- Complement this MSB.
- Add one and previous carry to this result
- Again store this result

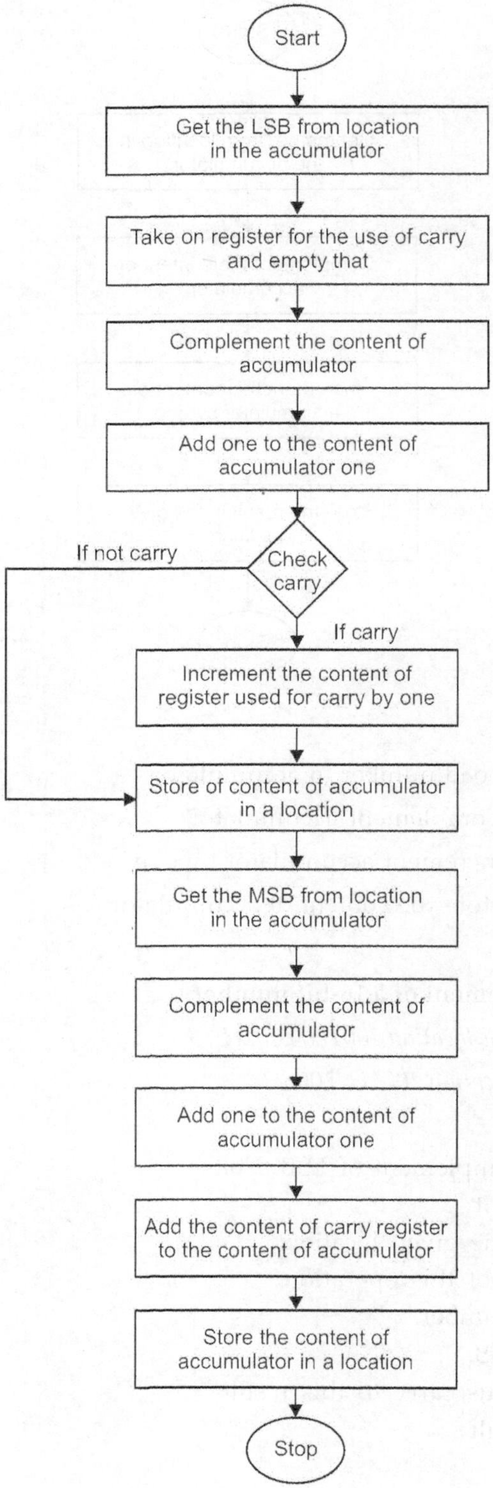
Start
Get the LSB from location in the accumulator
Take on register for the use of carry and empty that
Complement the content of accumulator
Add one to the content of accumulator one
Check carry
If not carry
If carry
Increment the content of register used for carry by one
Store of content of accumulator in a location
Get the MSB from location in the accumulator
Complement the content of accumulator
Add one to the content of accumulator one
Add the content of carry register to the content of accumulator
Store the content of accumulator in a location
Stop

Program

LDA	2001H	Load accumulator with LSB
MVI	B,00	Load the register B with zero to use it to store carry.
CMA		Complement the content of accumulator
ADI	01H	Add accumulator by one
JNC	NEW	Jump if not carry to NEW
INR	B	Increment register B by one
NEW STA	2004H	Store accumulator
LDA	2002H	Load accumulator with MSB
CMA		Complement accumulator
ADI	01H	Add accumulator by 1
ADD	B	Add carry of LSB
STA	2005 H	Store accumulator
HLT		

To mask off LSB 4-bits of 8-bit number

Problem. *Given number = C5 at the location 2001H*

Flow Diagram:

- C5 = 1100 0101, If LSB is masked, the answer should be C0.
- It is possible when it is ANDed with the Number F0
- So it is a trick that bits to be masked are always ANDed with 0.
- The bits allowed to be passed without masking are ANDed with 1.

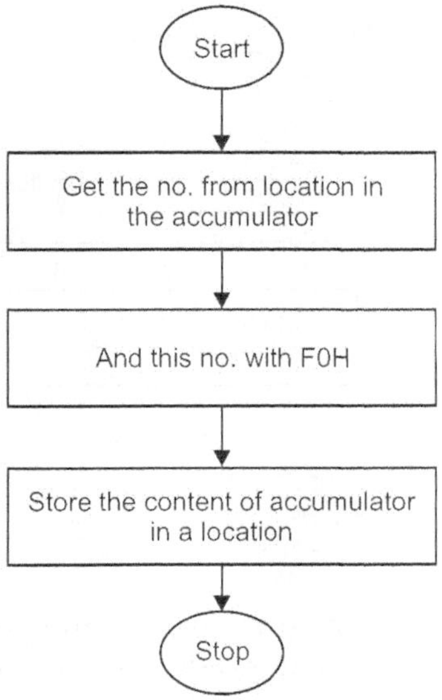

Program

LDA	2001H	Get the number in the accumulator
ANI	F0	AND the accumulator with F0
STA	2002H	Store the result in the location
HLT		

To find larger of two number

Problem. *First number is in memory location 2001H*
Second number is in memory location 2002H
Answer to be stored in memory location 2003H

Flow Diagram:

- Take First number in the accumulator
- Compare this number to the second number
  CMP B Means that A (accumulator)–B
  If Result is positive the A>B, Otherwise A<B in case of carry.
- Store the Larger one

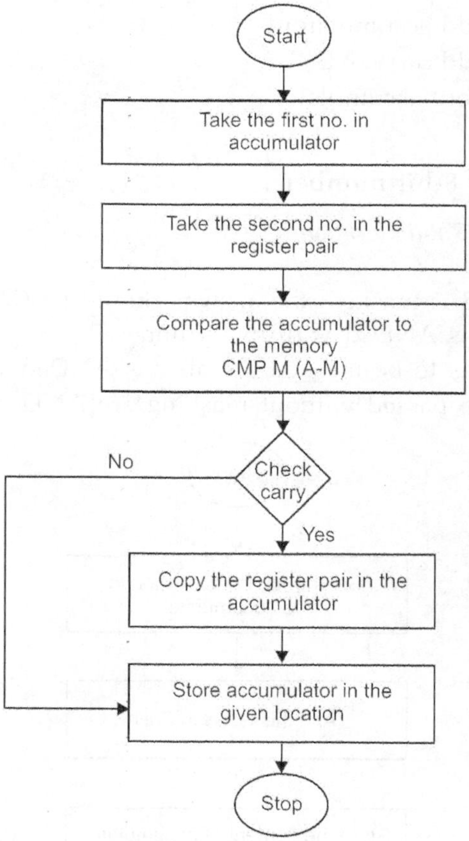

Program

	LXI	H, 2001H	Address of first number in H-L pair
	MOV	A, M	First number in the accumulator
	INX	H	Address of the second number in the H-L pair
	CMP	M	Compare the accumulator to the memory
	JNC	NEW	
	MOV	A, M	If carry get second number in accumulator
NEW	STA	2003H	Store the larger number in given location
	HLT		

To find larger of in a data array

Problem. *Total Numbers. Stored in the location starting from 2000H = 10*

Flow Diagram:

- Perform the work as previous program but now for all numbers.
- First of all compare the two numbers, find the larger one of them.
- Now take the third number and compare that with the result of previous comparison.

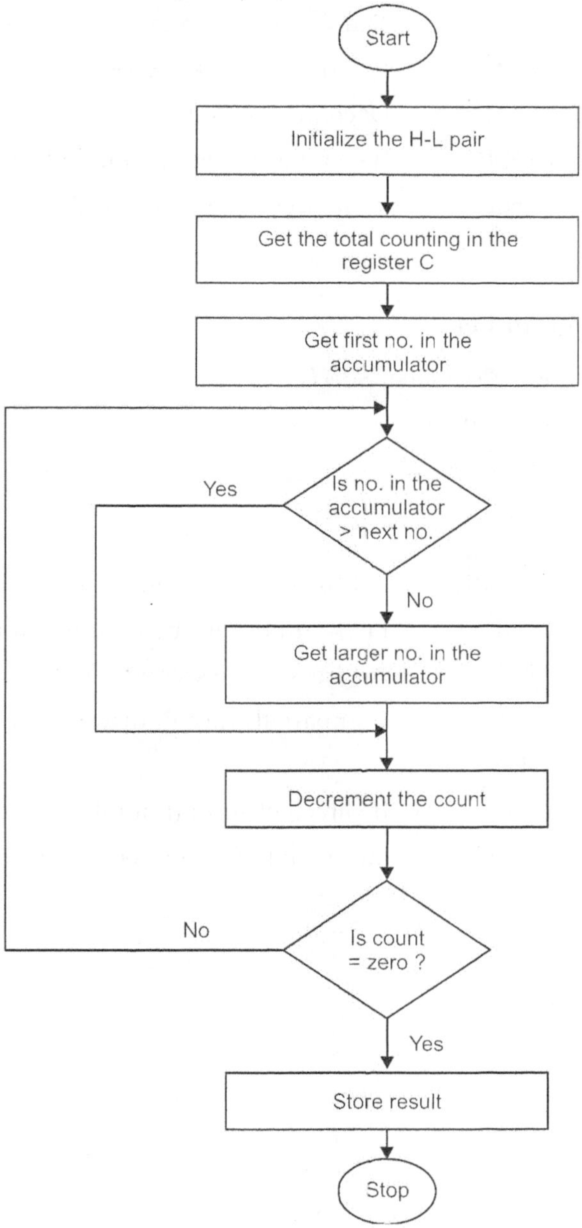

```
          LXI    H, 2001H   Initialized the H-L pair
          MVI    C, 10H     Count in the register C
          MOV    A, M       Take first number in the accumulator
          DCR    C          Decrement count
LOOP      INX    H          Takes the second number in the H-L pair
          CMP    M          Compare next number with previous maximum
          JNC    NEW        Number larger in the accumulator, goes to NEW
          MOV    A, M       If carry, then take second number in the accumulator
NEW       DCR    C          Decrement count
          JNZ    LOOP       If count is not zero, jump to LOOP
          STA    2050H      Store accumulator in the 2050H
          HLT
```

To find smaller of two number

Problem. *First number is in memory location 2001H*

Second number is in memory location 2002H

Answer to be stored in memory location 2003H

Program

```
          LXI    H, 2001H   Address of first number in H-L pair
          MOV    A, M       First number in the accumulator
          INX    H          Address of the second number in the H-L pair
          CMP    M          Compare the accumulator to the memory
          JC     NEW
          MOV    A, M       If carry get second number in accumulator
NEW       STA    2003H      Store the larger number in given location
          HLT
```

Flow Diagram:

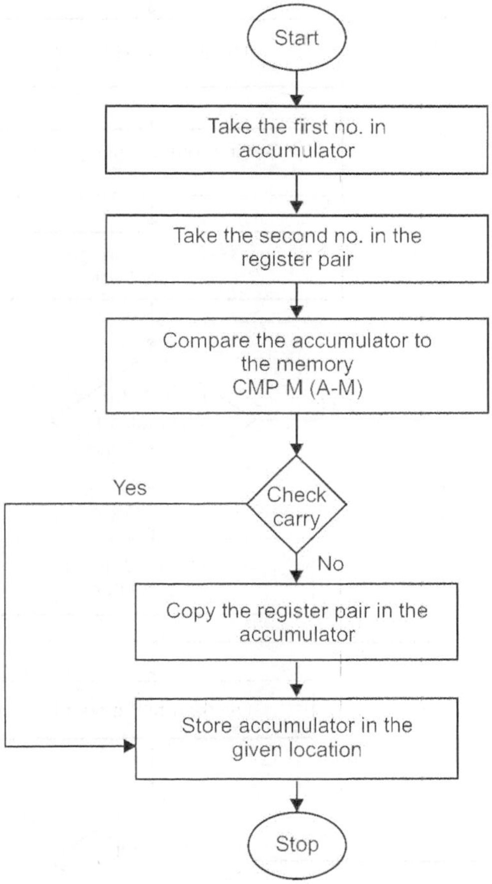

To find smaller of in a data array

Problem. *Total numbers stored in the location starting from 2000H = 10*

Flow Diagram:

- Perform the work as above program but now for all numbers.
- First of all compare the two numbers, find the larger one of them.
- Now take the third number and compare that with the result of previous comparison.

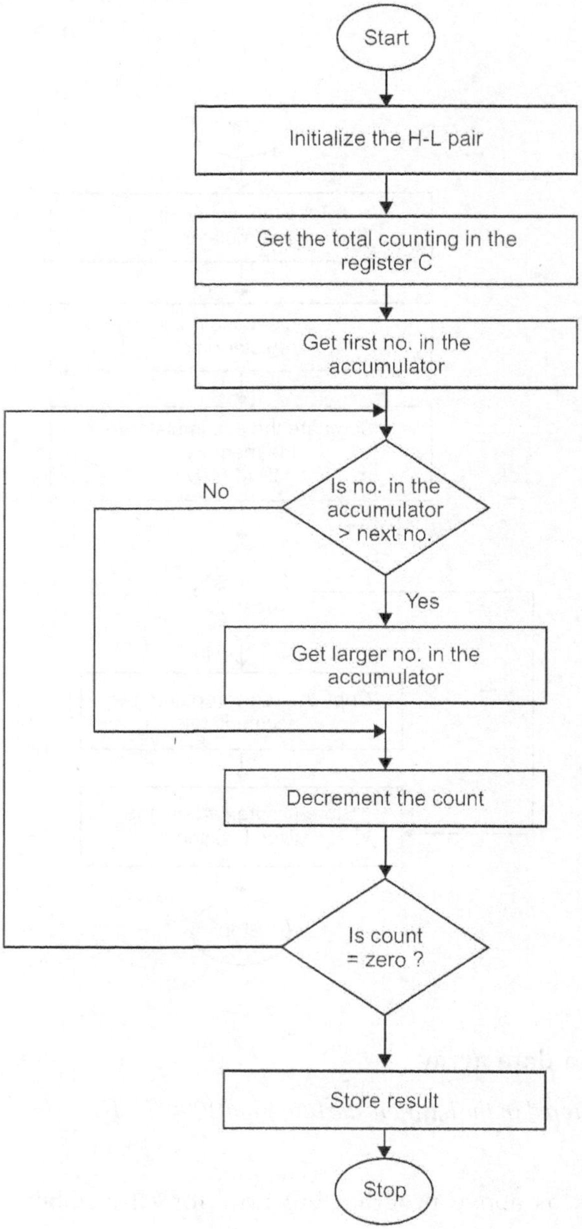

Program

LXI	H, 2001H	Initialized the H-L pair
MVI	C, 10H	Count in the register C
MOV	A, M	Take first number in the accumulator
DCR	C	Decrement count

LOOP	INX	H	Takes the second number in the H-L pair
	CMP	M	Compare next number with previous minimum
	JC	NEW	Number smaller in the accumulator, goes to NEW
	MOV	A, M	If carry, then take second number in the accumulator
NEW	DCR	C	Decrement count
	JNZ	LOOP	If count is not zero, jump to loop
	STA	2050H	Store accumulator in the 2050H
	HLT		

Sum of a series of 8-bit numbers; sum: 8-bits

Problem.

- *Addition of five numbers*
- *Numbers are placed in the location 2001H to 2005H*
- *Store result in the location 2050H*

Flow Diagram:

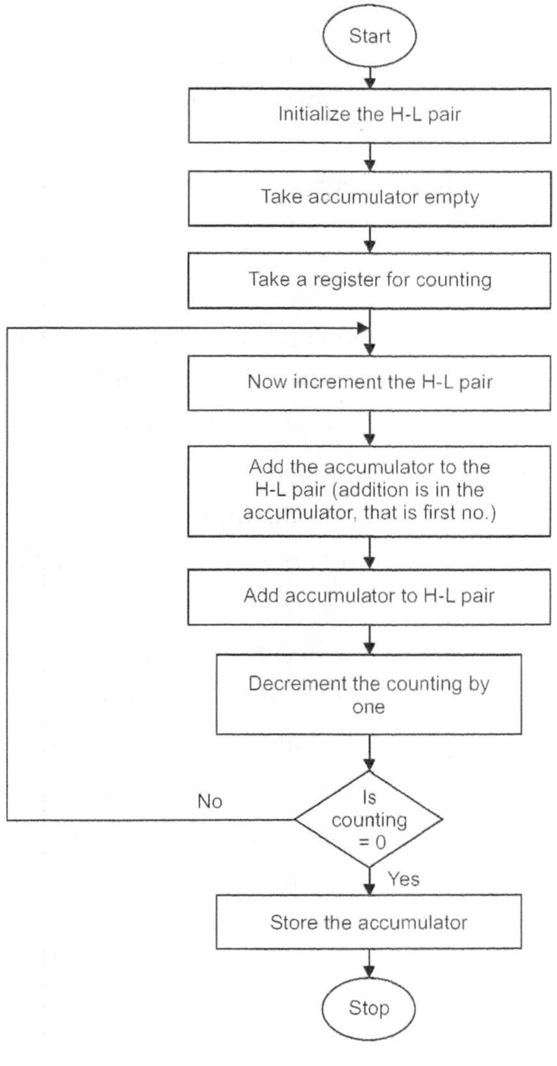

Program

	LXI	H, 2001H	Initialized The H-L register pair
	MVI	C, 05	Counter register for total numbers added
	MVI	A, 00	Initially value of sum = 0
LOOP	INX	H	Take the next data in the H-L pair
	ADD	M	Previous sum + next number
	DCR	C	Decrement count
	JNZ	LOOP	If not zero, go to loop
	STA	2050H	If count zero store result at the location
	HLT		

Sum of a series of 8-bit numbers; sum: 16-bits

Problem.

- *Addition of five numbers.*
- *Numbers are placed in the location 2001H to 2005H*
- *Store result in the location 2050H and 2051H*

Flow Diagram:

Perform the same operation as in previous Case but now also count the carry and display the at Last.

Program

	LXI	H, 2001H	Initialized the H-L register pair
	MVI	C, 05	Register set for counting
	MVI	A, 00	Take accumulator empty
	MVI	B, 00	Take empty register for carry (MSB)
LOOP	INX	H	Takes first number.
	ADD	M	Add accumulator to the first number.
	JNC	NEW	Check carry
	INR	B	Increment the carry register by one in case of carry
NEW	DCR	C	Decrease the counting by one
	JNZ	LOOP	If counting is not = 0, go to LOOP
	STA	2050H	if counting is = 0, Store the accumulator
	MOV	A, B	Take the final carry in the accumulator
	STA	2051H	Store the register B
	HLT		

Flow Diagram:

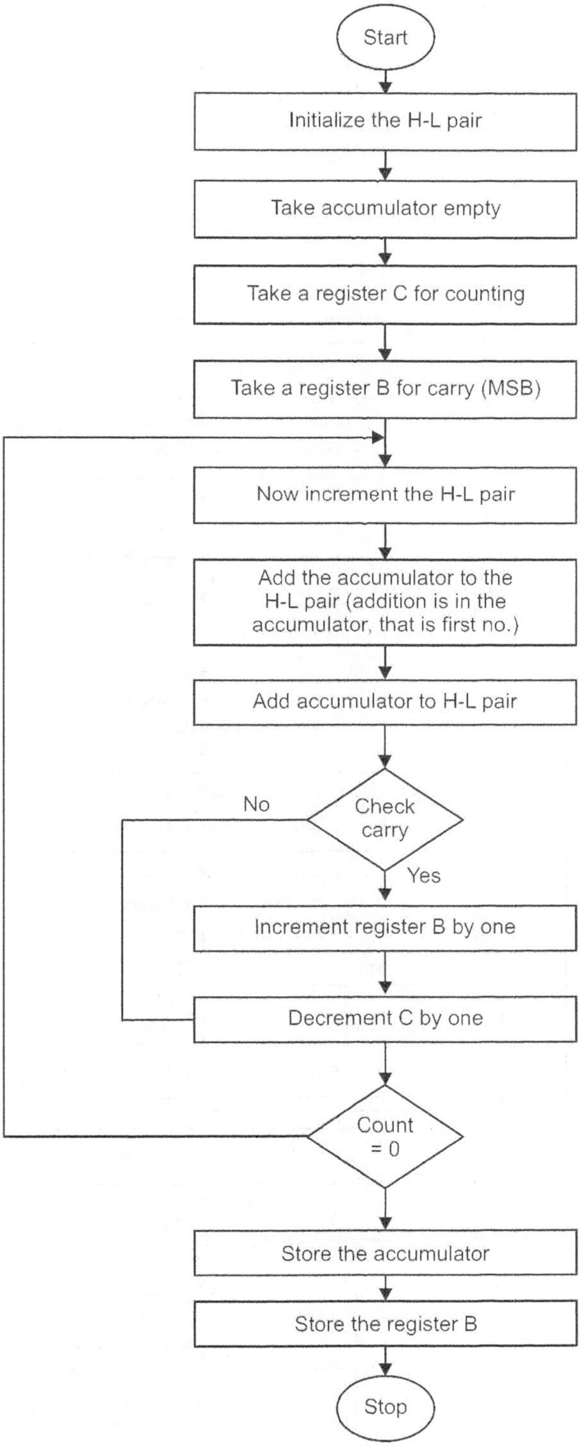

Sum of a series of 8-bit numbers; sum: 16-bits (answer should be in the decimal)

Problem.

- *Addition of five numbers.*
- *Numbers are placed in the location 2001H to 2005H*
- *Store result in the location 2050H and 2051H*

Flow Diagram:

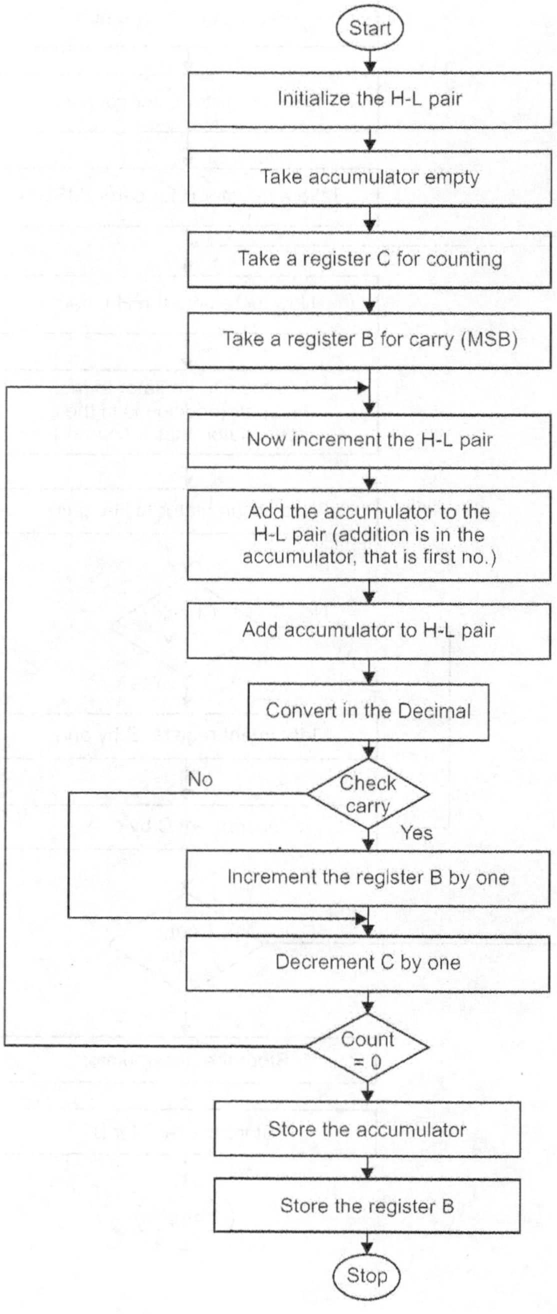

Program

```
LOOP    LXI   H, 2001H   Initialized the H-L register pair
        MVI   C, 05      Register set for counting
        MVI   A, 00      Take accumulator empty
        MVI   B, 00      Take empty register for carry (MSB)
LOOP    INX   H          Takes first number.
        ADD   M          Add accumulator to the first number.
        DAA              Decimal adjust
        JNC   NEW        Check carry
        INR   B          Increment the carry register by one in case of carry
NEW     DCR   C          Decrease the counting by one
        JNZ   LOOP       If counting is not = 0, go to LOOP
        STA   2050H      If counting is = 0, Store the accumulator
        MOV   A, B       Take the final carry in the accumulator
        STA   2051H      Store the register B
        HLT
```

4.6 DESIGNING OF A COUNTER

We may see that designing of a counter is a important programming application. Counters are basically designed to keep track of event time delays using software instructions is far more flexible and less time consuming than the designing process using hardware.

Just start with simple program. But if we want to calculate 'or' set a predefined time delay, we have to know about the total T-states during the execution of instruction. And the frequency of clock used in microprocessor.

Example 1. Here we have used a register that is loaded with acounting and the loop is executed until the loaded counting is zero.

Flow Chart of Counter

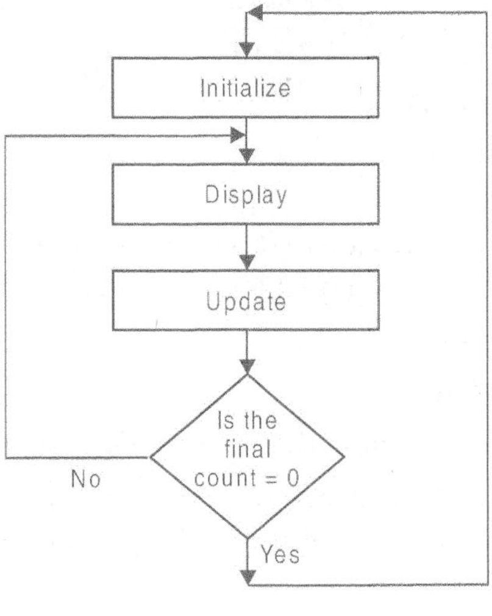

Flow Chart of Counter with Delay

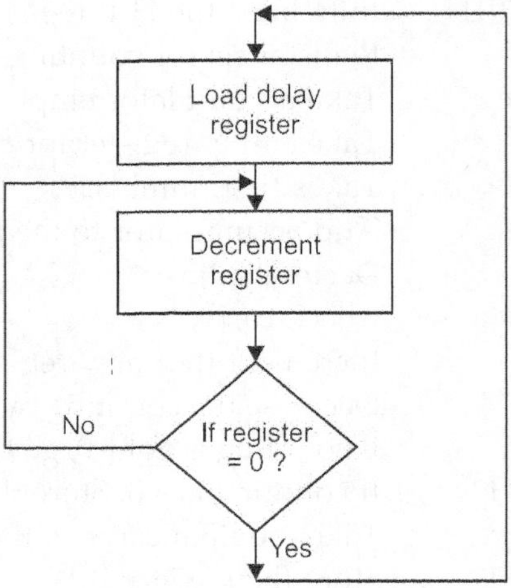

Program

	MVI	CFFH	Load counting at register C
LOOP	DCR	C	
	JNZ	LOOP	

Now we can see that

- Register C is loaded with counting FFH = 255_{10} that is executed one and requires total of T states.
- Now left two instructions DCR and JNZ requires 4 and 10 T states respectively; thus totally required states '14' for one time execution.

  But these two instructions are repeated till counting 255 is zero. So they are executed 255 times.

  If we are using 2 MHz clock frequency then total time delay

$$= 14 \times 255 \times \frac{1}{2 \times 10^6}$$

$$= 1785 \ \mu s$$

$$T_L \simeq 1.8 \text{ ms}$$

- Thus total time delay

$$T_D = \text{Delay due to (LXI) Instruction} + 1.8 \text{ ms}$$

$$= (7 \times 0.5 \ \mu s) + 1.8 \text{ ms}$$

$$T_D \simeq 1.8 \text{ ms}$$

Thus we can adjust the time delay.

Example 2. *Write a programe for counting and delay above then 1.8 ms and calculate total time delay.*

Solution. For time delay above then 1.8 ms, we have to use register pair instead of register. But this can easily be performed using register. But with loop under loop for extra long delay we will check them only be one.

(*a*) Using register pair

Op-code	Operand	Comments	T-state
LXI	B, 2384H	Load B register pair with 10-bit count	10
LOOP	B	Decrement B by	6
DCX		one	
MOV	A, C	Place C to A	4
ORA	B	OR (B) with (C) to set zero flag	4
JNZ	Loop	If zero flag not set go to loop	10/7

Note: We should remember that DCX B will not set zero flag; to check the real condition; we have to affect the zero flag at right time so we used ORA.

Now we will calculate total delay = 6 + 4 + 4 + 10 = 24 T-states.

The loop will be repeated until count 2384_H is set zero.

$$2384_H = (3092)_{10}.$$

So it will be repeated by 9092 times. So total T-states.

$$= 24 \times 9092 \text{ T-states}$$

For clock 2MHz; delay $= 24 \times 9092 \times \dfrac{1}{2 \times 10^6}$

$= 109$ ms.

Thus total delay $T_D = T_L + \text{Delay due to (LXI)}$

$$\boxed{T_D \simeq 109 \text{ ms}}$$

By Using Loop Under Loop

Let us see the following programmes

			T states	
	MVI	B, 38N	7T	
Loop 2	MVI	C, FFN	7T	
Loop 1	DCR	C	4T	Loop 1 dealy
	JNZ	Loop 1	10/7T	
	DCR	B	4T	Loop 2 dealy
	JNZ	Loop 2	10/7T	

The delay for loop 1 $T_{L1} = 1.8$ ms (as calculated in Ex.)

The delay for loop 2 $= (21 \times 0.5\ \mu s) + T_{L1}) \times 38\text{ H}$

$= (21 \times 0.5\ \mu s + 1.8 \text{ ms}) \times 56_{10}.$

$$\boxed{T_D = 100.46 \text{ ms}}$$

Example 3. *Write a program to count from '0' to 'g' with a one-second delay between each count. At the count of g, the count should reset itself to '0' and repeat the sequence continuously. Assume the clock frequency of system is 1 MHz.*

Solution. The program can be easily seen into two parts.

(*a*) The count is set up by loading a register with the appropriate No. (0) and increment it until it reaches to digit '*g*'. When count register reaches '*g*', it again start from '0'.

(*b*) Between each count, you have to provide delay of 1 second.

Program

```
                  START     MVI B, 00H
                  DISPLAY   OUT PORT # (Address)
       ┌                    LXI H, 16-bit
Delay  │          Loop      DCX H        6
= 24 T ┤                    MOV A, L     4
state  │                    ORA H        4
       └                    JNZ loop     10/7
                            INR B
                            MOV A, B
                  CPI       OAH          Compare every count with 10₁₀
                  JNZ       DISPLAY
                  JZ        START        If we get 10₁₀ at any state again start from 0₁₀
```

Thus total delay $T_L = 24 \times \text{Counting No.} \times \dfrac{1}{1 \times 10^6}$

Required = 1 second

$$1 \text{ second} = 24 \times \text{Counting No.} \times \frac{1}{1 \times 10^6}$$

$$\text{Counting No.} = \frac{10^6}{24} = 41666_{10}$$

$$= \text{A2C2H}$$

Thus this loop should be repeated for A2C2 times so.

LXIH, 16-bit should be LXIH, A2C2H.

REVIEW QUESTIONS

1. The following block of data is stored in the memory location from XX55H to XX5AH. Transfer the data to location XX80H to XX85H in the reverse direction.
2. Write a program to add the following fine data byte stored in memory locations starting from 2060H and display the sum. (The sum doesn't generate a carry).
   Data (H) 1A, 32, 4A, 12, 37
3. Write a program to add the data bytes stored in memory location starting at 2060H and display. The sum at the output port if the sum does not generate a carry. If a result generates a carry, stop the addition and display 02H at the output port.

4. A set of eight data bytes are stored in memory location starting from 2000H. Write a program to substract two bytes at a time and store the result in a sequential order in memory location from 2070H.

   Data (H)—Fg, 38, A7, B2, C8, A2, F4, 67
5. A set of eight readings is stored in memory starting at location 2000H. Write a program to check whether a byte 60H exists in the set. If it does, stop checking and display; otherwise output FFH.
6. Design a counter that start from 0 to g_{10} and it should again restart the counting when it reaches at $0g_{10}$.
7. Write a program to count from 0 to 20H with a delay of 100 ms between each count. After the count 20H, the counter should reset itself and repeat the sequence. Take the clock cycle of 1 MHz.

5

80C86 CMOS 16-Bit Microprocessor

Features

- Compatible with NMOS 8086
- Completely Static CMOS Design
  - DC 5MHz (80C86)
  - DC 8MHz (80C86-2)
- Low Power Operation
  - ICCSB .... 500 μA Max
  - ICCOP .... 10 mA/MHz Typ
- 1M Byte of Direct Memory Addressing Capability
- 24 Operand Addressing Modes
- Bit, Byte, Word, and Block Move Operations
- 8-Bit and 16-Bit Signed/Unsigned Arithmetic
  - Binary, or Decimal
  - Multiply and Divide
- Wide Operating Temperature Range
  - C80C86 .... 0°C to +70°C

The 80C86 high performance 16-bit CMOS CPU is manufactured using a self-aligned silicon gate CMOS process. Two modes of operation, minimum for small systems and maximum for larger applications such as multiprocessing, allow user configuration to achieve the highest performance level. Full TTL compatibility (with the exception of CLOCK) and industry standard operation allow use of existing NMOS 8086 hardware and software designs.

Pinouts

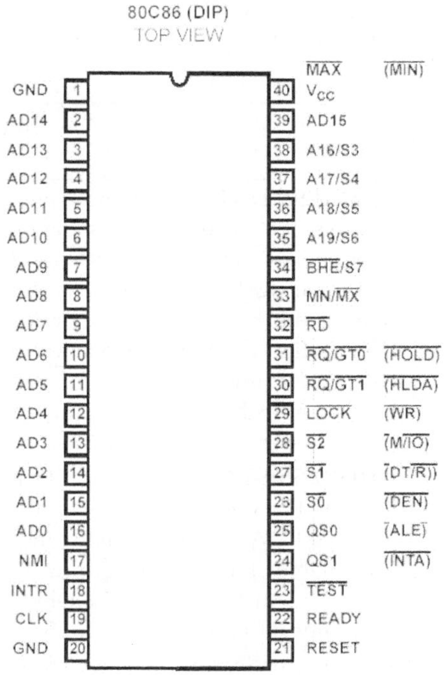

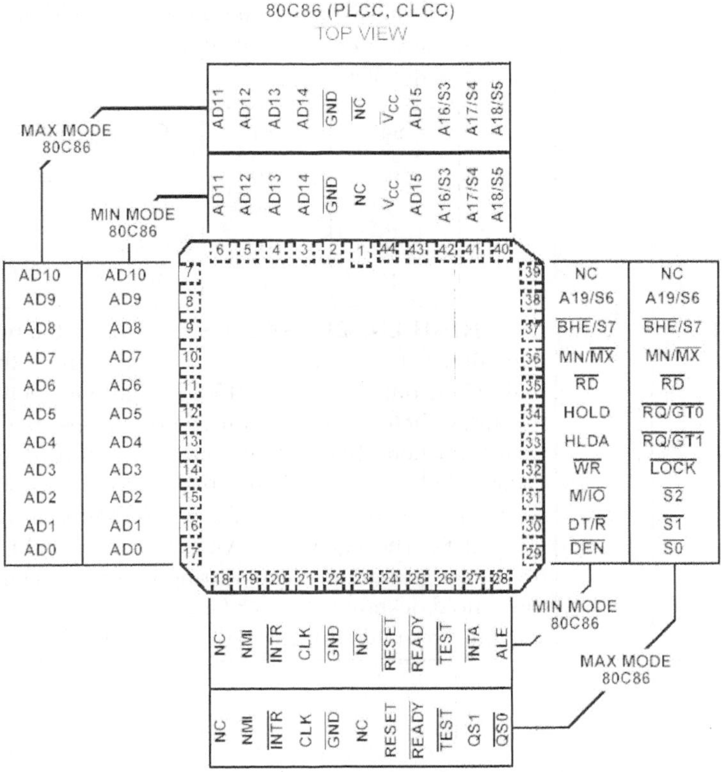

Fig. 5.1 Pinouts

5.1 PIN DESCRIPTION

The following pin function descriptions are for 80C86 systems in either minimum or maximum mode.

Symbol	Pin Number	Type	Description
AD15–AD0	2–16, 39	I/O	**ADDRESS DATA BUS:** These lines constitute the time multiplexed memory/IO address (T1) and data (T2, T3, TW, T4) bus. A0 is analogous to $\overline{BHE}$ for the lower byte of the data bus, pins D7-D0. It is LOW during Ti when a byte is to be transferred on the lower portion of the bus in memory or I/O operations. Eight-bit oriented devices tied to the lower half would normally use A0 to condition chip select functions (See $\overline{BHE}$). These lines are active HIGH and are held at high impedance to the last valid logic level during interrupt acknowledge and local bus "hold acknowledge' or "grant sequence".
A19/S6 A18/S5 A17/S4 A16/S3	35-38	O	**ADDRESS/STATUS:** During T1, these are the four most significant address lines for memory operations. During I/O operations these lines are LOW. During memory and I/O operations, status information is available on these lines during T2, T3, TW, T4. S6 is always LOW. The status of the interrupt enable FLAG bit (S5) is updated at the beginning of each clock cycle. S4 and S3 are encoded as shown. This information indicates which segment register is presently being used for data accessing. These lines are held at high impedance to the last valid logic level during local bus "hold acknowledge" or "grant sequence.

S4	S3	Characteristics
0	0	Alternate Data
0	1	Stack
1	0	Code or None
1	1	Data

Symbol	Pin Number	Type	Description
$\overline{BHE}$/S7	34	O	**BUS HIGH ENABLE/STATUS:** During T1 the bus high enable signal ($\overline{BHE}$) should be used to enable data onto the most significant half of the data bus, pins D15-D8. Eight bit oriented devices tied to the upper half of the bus would normally use $\overline{BHE}$ to condition chip select functions. BHE is LOW during T1 for read, write, and interrupt acknowledge cycles when a byte is to be transferred on the high portion of the bus. The S7 status information is available during T2, T3 and T4. The signal is active LOW, and is held at high impedance to the last valid logic level during interrupt acknowledge and local bus "hold acknowledge" or "grant sequence", it is LOW during T1 for the first interrupt acknowledge cycle.

BHE	A0	Characteristics
0	0	Whole Word
0	1	Upper Byte From/to Odd Address
1	0	Lower Byte From/to Even address
1	1	None

$\overline{RD}$	32	O	**READ:** Read strobe indicates that the processor is performing a memory or I/O read cycle, depending on the state of the M/$\overline{IO}$ or S2 pin. This signal is used to read devices which reside on the 80C86 local bus. $\overline{RD}$ is active LOW during T2. T3 and TW of any read cycle, and is guaranteed to remain HIGH in T2 until the 80C86 local bus has floated. This line is held at a high impedance logic one state during "hold acknowledge" or "grand sequence".
READY	22	I	**READY:** is the acknowledgment from the addressed memory or I/O device that will complete the data transfer. The RDY signal from memory or I/O is synchronized by the 82C84A Clock Generator to form READY. This signal is active HIGH. The 80C86 READY input is not synchronized. Correct operation is not guaranteed if the Setup and Hold Times are not met.
INTR	18	I	**INTERRUPT REQUEST:** is a level triggered input which is sampled during the last dock cycle of each instruction to determine if the processor should enter into an interrupt acknowledge operation. A subroutine is vectored to via an interrupt vector lookup table located in system memory. It can be internally masked by software resetting the interrupt enable bit. INTR is internally synchronized. This signal is active HIGH.
$\overline{TEST}$	23	I	**TEST:** input is examined by the "Wait" instruction. If the $\overline{TEST}$ input is LOW execution continues, otherwise the processor waits in an "Idle" state. This input is synchronized internally during each clock cycle on the leading edge of CLK.
NMI	17	I	**NON-MASKABLE INTERRUPT:** is an edge triggered input which causes a type 2 interrupt. A subroutine is vectored to via an interrupt vector lookup table located in system memory. NMI is not maskable internally by software. A transition from LOW to HIGH initiates the interrupt at the end of the current instruction. This input is internally synchronized.
RESET	21	I	**RESET:** causes the processor to immediately terminate its present activity. The signal must transition LOW to HIGH and remain active HIGH for at least four clock cycles. It restarts execution, as described in the Instruction Set description, when RESET returns LOW. RESET is internally synchronized.
CLK	19	I	**CLOCK:** provides the basic timing for the processor and bus controller. It is asymmetric with a 33% duty cycle to provide optimized internal timing.
VCC	40		**VCC:** +5V power supply pin. A 0.1u,F capacitor between pins 20 and 40 is recommended for decoupling.
GND	1,20		**GND:** Ground. Note: both must be connected. A 0.1uF capacitor between pins 1 and 20 is recommended for decoupling.
MN/$\overline{MX}$	33	I	**MINIMUM/MAXIMUM:** Indicates what mode the processor is to operate in. The two modes are discussed in the following sections.

Minimum Mode System

The following pin function descriptions are for the 80C86 in minimum mode (*i.e.*, $MN/\overline{MX} = V_{CC}$). Only the pin functions which are unique to minimum mode are described;

Symbol	Pin Number	Type	Description
$M/\overline{IO}$	28	O	**STATUS LINE:** logically equivalent to $\overline{S2}$ in the maximum mode. It is used to distinguish a memory access from an I/O access. $M/\overline{IO}$ becomes valid in the T4 preceding a bus cyde and remains valid until the final T4 of the cyde (M = HIGH, I/O = LOW). $M/\overline{IO}$ is held to a high impedance logic one during local bus "hold acknowledge".
$\overline{WR}$	29	O	**WRITE:** indicates that the processor is performing a write memory or write I/O cyde, depending on the state of the $M/\overline{IO}$ signal. $\overline{WR}$ is active for T2, T3 and TW of any write cycle. It is active LOW, and is held to high impedance logic one during local bus "hold acknowledge".
$\overline{INTA}$	24	O	**INTERRUPT ACKNOWLEDGE:** is used as a read strobe for interrupt acknowledge cycles. It is active LOW during T2, T3 and TW of each interrupt acknowledge cycle. Note that $\overline{INTA}$ is never floated.
ALE	25	O	**ADDRESS LATCH ENABLE:** is provided by the processor to latch the address into the 82C82/82C83 address latch. It is a HIGH pulse active during clock LOW of T1 of any bus cycle. Note that ALE is never floated.
$DT/\overline{R}$			**DATA TRANSMIT/RECEIVE:** is needed in a minimum system that desires to use a data bus transceiver. It is used to control the direction of data flow through the transceiver. Logically, $DT/\overline{R}$ is equivalent to $\overline{S1}$ in maximum mode, and its timing is the same as for $M/\overline{IO}$ (T = HIGH, R = LOW). $DT/\overline{R}$ is held to a high impedance logic one during local bus "hold acknowledge".
$\overline{DEN}$	26	O	**DATA ENABLE:** provided as an output enable for a bus transceiver in a minimum system which uses the transceiver. $\overline{DEN}$ is active LOW during each memory and I/O access and for $\overline{INTA}$ cycles. For a read or $\overline{INTA}$ cycle it is active from the middle of T2 until the middle of T4, while for a write cycle it is active from the beginning of T2 until the middle of T4. $\overline{DEN}$ is held to a high impedance logic one during local bus "hold acknowledge".
HOLD HLDA	31, 30	I O	**HOLD:** indicates that another master is requesting a local bus "hold". To be an acknowledged, HOLD must be active HIGH. The processor receiving the "hold" will issue a "hold acknowledge" (HLDA) in the middle of a T4 or Tl clock cycle. Simultaneously with the issuance of HLDA, the processor will float the local bus and control lines. After HOLD is detected as being LOW, the processor will lower HLDA, and when the processor needs to run another cycle, it will again drive the local bus and control lines. HOLD is not an asynchronous input. External synchronization should be provided if the system cannot otherwise guarantee the setup time.

Maximum Mode System

The following pin function descriptions are for the 80C86 system in maximum mode (*i.e.*, MN/$\overline{MX}$ = GND). Only the pin functions which are unique to maximum mode are described below.

<table>
<tr><th>Symbol</th><th>Pin Number</th><th>Type</th><th>Description</th></tr>
<tr><td>$\overline{S0}$
$\overline{S1}$
$\overline{S2}$</td><td>26
27
28</td><td>O
O
O</td><td>STATUS: is active during T4, T1 and T2 and is returned to the passive state (1, 1, 1) during T3 or during TW when READY is HIGH. This status is used by the 82C88 Bus controller to generate all memory and I/O access control signals. Any change $\overline{S2}$, $\overline{S1}$ or $\overline{S0}$ during T4 is used to indicate the beginning of a bus cycle, and the return to the passive state in T3 or TW is used indicate the end of a bus cycle.
These signals are held at a high impedance logic one state during "grant sequence".
<table>
<tr><th>$\overline{S2}$</th><th>$\overline{S1}$</th><th>$\overline{S0}$</th><th>Characteristics</th></tr>
<tr><td>0</td><td>0</td><td>0</td><td>Interrupt Acknowledge</td></tr>
<tr><td>0</td><td>0</td><td>1</td><td>Read I/O Port</td></tr>
<tr><td>0</td><td>1</td><td>0</td><td>Write I/O Port</td></tr>
<tr><td>0</td><td>1</td><td>1</td><td>Halt</td></tr>
<tr><td>1</td><td>0</td><td>0</td><td>Code Access</td></tr>
<tr><td>1</td><td>0</td><td>1</td><td>Read Memory</td></tr>
<tr><td>1</td><td>1</td><td>0</td><td>Write Memory</td></tr>
<tr><td>1</td><td>1</td><td>1</td><td>Passive</td></tr>
</table></td></tr>
<tr><td>$\overline{RQ}/\overline{GT0}$
$\overline{RQ}/\overline{GT1}$</td><td>31, 30</td><td>I/O</td><td>REQUEST/GRANT: pins are used by other local bus masters to force the processor to release the local bus at the end of the processor's current bus cycle. Each pin is bidirectional with $\overline{RQ}/\overline{GT0}$ having higher priority than $\overline{RQ}/\overline{GT1}$. $\overline{RQ}/\overline{GT}$ has an internal pull-up bus hold device so it may be left unconnected. The request/grant sequence is as follows (see $\overline{RQ}/\overline{GT}$ Sequence Timing)
1. A pulse of 1 CLK wide from another local bus master indicates a local bus request ("hold") to the 80C86 (pulse 1).
2. During a T4 or T1 clock cycle, a pulse 1 CLK wide from the 80C86 to the requesting master (pulse 2) indicates that the 80C86 has allowed the local bus to float and that it will enter the "grant sequence" state at the next CLK. The CPU's bus interface unit is disconnected logically from the local bus during 'grant sequence'.
3. A pulse 1 CLK wide from the requesting master indicates to the 80C86 (pulse 3) that the "hold" request is about to end and that the 80C86 can reclaim the local bus at the next CLK. The CPU then enters T4 (or Tl if no bus cycles pending).
Each Master-Master exchange of the local bus is a sequence of 3 pulses. There must be one idle CLK cycle after each bus exchange. Pulses are active low.</td></tr>
</table>

			If the request is made while the CPU is performing a memory cycle, it will release the local bus during T4 of the cycle when all the following conditions are met: 1. Request occurs on or before T2. 2. Current cycle is not the low byte of a word (on an odd address). 3. Current cycle is not the first acknowledge of an interrupt acknowledge sequence. 4. A locked instruction is not currently executing. If the local bus is idle when the request is made the two possible events will follow: 1. Local bus will be released during the next cycle. 2. A memory cycle will start within three clocks.
$\overline{LOCK}$	29	O	**LOCK:** output indicates that other system bus masters are not to gain control of the system bus while $\overline{LOCK}$ is active LOW. The $\overline{LOCK}$ signal is activated by the 'LOCK' prefix instruction and remains active until the completion of the next instruction. This signal is active LOW, and is held at a high impedance logic one state during 'grant sequence'. In MAX mode, $\overline{LOCK}$ is automatically generated during T2 of the first $\overline{INTA}$ cycle and removed during T2 of the second $\overline{INTA}$ cycle.
QS1, QS0	24,25	O	**QUEUE STATUS:** The queue status is valid during the CLK cycle after which the queue operation is performed. QS1 and QS0 provide status to allow external tracking of the internal 80C86 instruction queue. Note that QS1, QS0 never become high impedance.

QS1	QS	Characteristics
0	0	No Operation
0	1	First byte of op code from queue
1	0	Empty the queue
1	1	Subsequent byte from queue

5.2 FUNCTIONAL DESCRIPTION

5.2.1 Static Operation

All 80C86 circuitry is of static design. Internal registers, counters and latches are static and require no refresh as with dynamic circuit design. This eliminates the minimum operating frequency restriction placed on other microprocessors. The CMOS 80C86 can operate from DC to the specified upper frequency limit. The processor clock may be stopped in either state (HIGH/LOW) and held there indefinitely. This type of operation is especially useful for system debug or power critical applications.

The 80C86 can be single stepped using only the CPU clock. This state can be maintained as long as is necessary. Single step clock operation allows simple interface circuitry to provide critical information for bringing up your system.

Static design also allows very low frequency operation (down to DC). In a power critical situation, this can provide extremely low power operation since 80C86 power dissipation is directly related to operating frequency. As the system frequency is reduced, so is the operating power until, ultimately, at a DC input frequency, the 80C86 power requirement is the standby current, (500μA maximum).

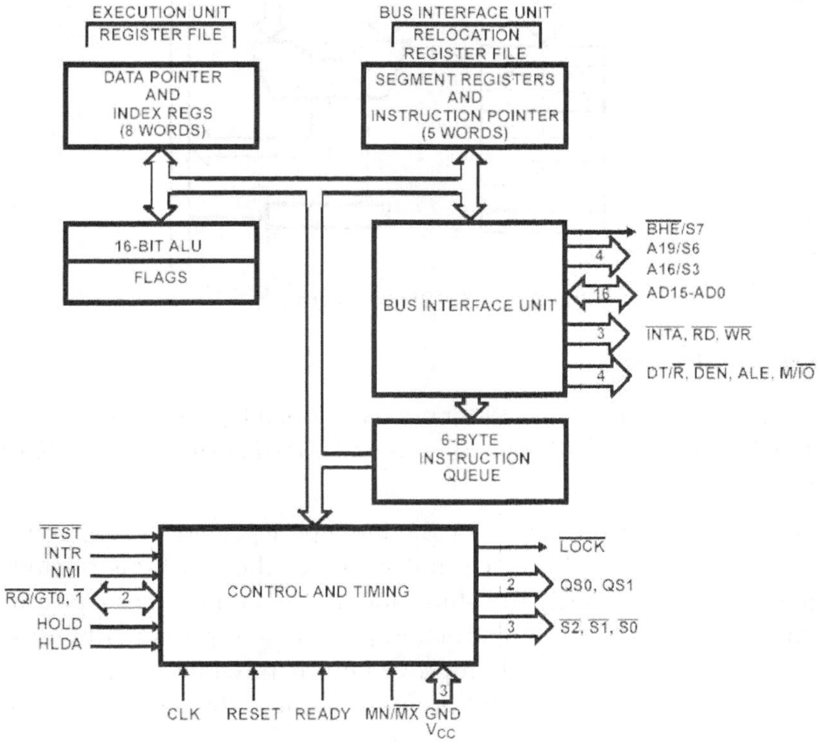

Fig. 5.2 Functional Diagram

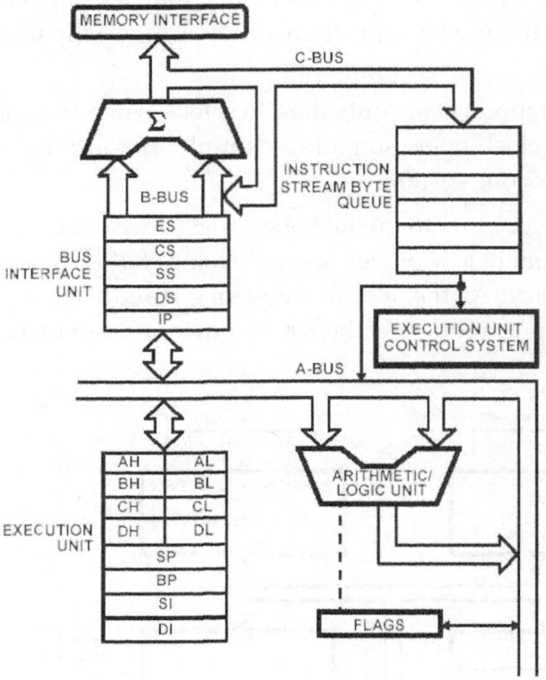

Fig. 5.3 Block Diagram

5.2.2 Internal Architecture

The internal functions of the 80C86 processor are partitioned logically into two processing units. The first is the **bus Interface Unit (BIU)** and the second is the **Execution Unit (EU)** as shown in the CPU functional diagram.

These units can interact directly, but for the most part perform as separate asynchronous operational processors. The bus interface unit provides the functions related to instruction fetching and queuing, cperand fetch and store, and address relocation. This unit also provides the basic bus control. The overlap of instruction pre-fetching provided by this unit serves to increase processor performance through improved bus bandwidth utilization. Up to 6 bytes of the instruction stream can be queued while waiting for decoding and execution.

The instruction stream queuing mechanism allows the BIU to keep the memory utilized very efficiently. Whenever there is space for at least 2 bytes in the queue, the BIU will attempt a word fetch memory cycle. This greatly reduces "dead-time" on the memory bus. The queue acts as a First-In-First-Out (FIFO) buffer, from which the EU extracts instruction bytes as required. If the queue is empty (following a branch instruction, for example), the first byte into the queue immediately becomes available to the EU.

The execution unit receives pre-fetched instructions from the BIU queue and provides un-relocated operand addresses to the BIU. Memory operands are passed through the BIU for processing by the EU, which passes results to the BIU for storage.

5.2.3 Memory Organization

The processor provides a 20-bit address to memory, which locates the byte being referenced. The memory is organized as a linear array of up to 1 million bytes, addressed as 00000(H) to FFFFF(H). The memory is logically divided into code, data, extra and stack segments of up to 64K bytes each, with each segment falling on 16-byte boundaries. (See Fig. 5.4).

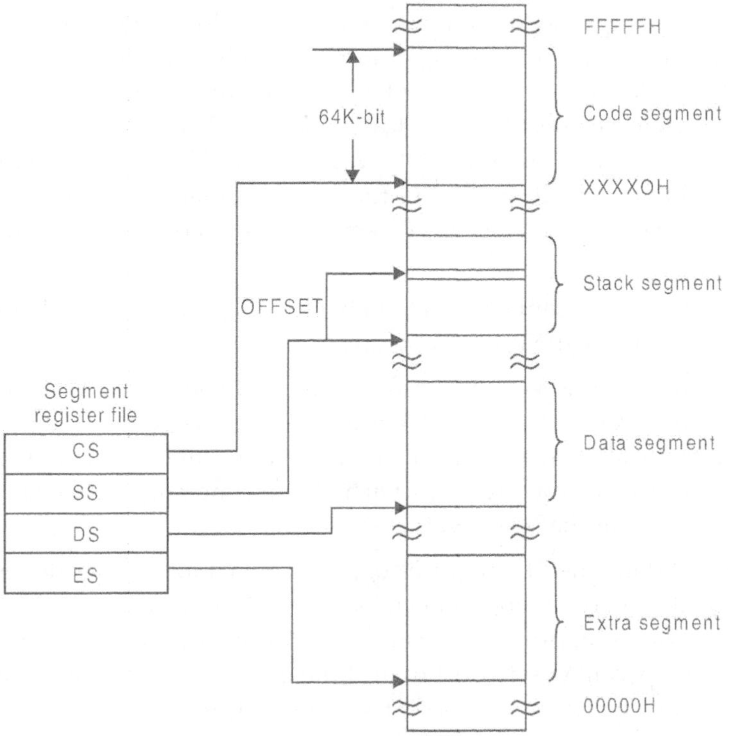

Fig. 5.4 80C86 Memory Organization

Table 5.1

Type of Memory Reference	Default Segment Base	Alternate Segment Base	Offset
Instruction Fetch	CS	None	IP
Stack Operation	SS	None	SP
Variable (except following)	DS	CS, ES, SS	Effective Address
String Source	DS	CS, ES, SS	SI
String Destination	ES	None	DI
BP Used As Base Register	SS	CS, DS, ES	Effective Address

All memory references are made relative to base addresses contained in high speed segment registers. The segment types were chosen based on the addressing needs of programs. The segment register to be selected is automatically chosen according to the specific rules of Table 5.1. All information in one segment type share the same logical attributes (*e.g.* code or data). By structuring memory into re-locatable areas of similar characteristics and by automatically selecting segment registers, programs are shorter, faster and more structured. (See Table 5.1).

Word (16-bit) operands can be located on even or odd address boundaries and are thus, not constrained to even boundaries as is the case in many 16-bit computers. For address and data operands, the least significant byte of the word is stored in the lower valued address location and the most significant byte in the next higher address location. The BIU automatically performs the proper number of memory accesses; one, if the word operand is on an even byte boundary and two, if it is on an odd byte boundary. Except for the performance penalty, this double access is transparent to the software. The performance penalty does not occur for instruction fetches; only word operands.

Physically, the memory is organized as a high bank (D15-D8) and a low bank (D7-D0) of 512K bytes addressed in parallel by the processor's address lines.

Byte data with even addresses is transferred on the D7-D0 bus lines, while odd addressed byte data (A0 HIGH) is transferred on the D15-D8 bus lines. The processor provides two enable signals, BHE and A0, to selectively allow reading from or writing into either an odd byte location, even byte location, or both. The instruction stream is fetched from memory as words and is addressed internally by the processor at the byte level as necessary.

In referencing word data, the BIU requires one or two memory cycles depending on whether the starting byte of the word is on an even or odd address, respectively. Consequently, in referencing word operands performance can be optimized by locating data on even address boundaries. This is an especially useful technique for using the stack, since odd address references to the stack may adversely affect the context switching time for interrupt processing or task multiplexing.

Certain locations in memory are reserved for specific CPU operations (See Fig. 5.5). Locations from address FFFF0H through FFFFFH are reserved for operations including a jump to the initial program loading routine. Following RESET, the CPU will always begin execution at location FFFF0H where the jump must be located. Locations 00000H through 003FFH are reserved for interrupt operations. Each of the 256 possible interrupt service routines is accessed thru its own pair of 16-bit pointers (segment address pointer and offset address pointer). The first pointer, used as the offset address, is loaded into the IP and the second pointer, which designates the base address is loaded into the CS. At this point program control is transferred to the interrupt routine. The pointer elements are assumed to have been stored at the respective places in reserved memory prior to occurrence of interrupts.

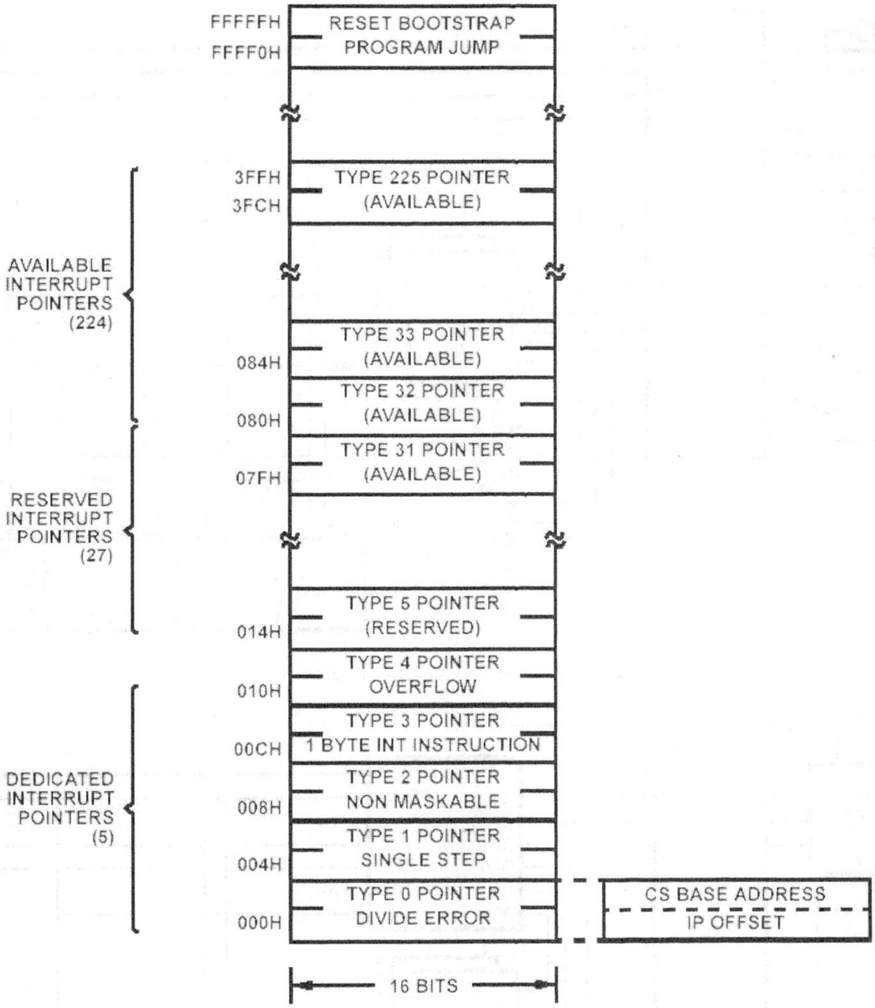

Fig. 5.5 Reserved Memory Locations

5.2.4 Minimum and Maximum Operation Modes

The requirements for supporting minimum and maximum 80C86 systems are sufficiently different that they cannot be met efficiently using 40 uniquely defined pins. Consequently, the 80C86 is equipped with a strap pin (MN/$\overline{MX}$) which defines the system configuration. The definition of a certain subset of the pins changes, dependent on the condition of the strap pin. When the MN/$\overline{MX}$ pin is strapped to GND, the 80C86 defines pins 24 through 31 and 34 in maximum mode. When the MN/$\overline{MX}$ pin is strapped to V_{CC}, the 80C86 generates bus control signals itself on pins 24 through 31 and 34.

The minimum mode 80C86 can be used with either a multiplexed or demultiplexed bus. This architecture provides the 80C86 processing power in a highly integrated form.

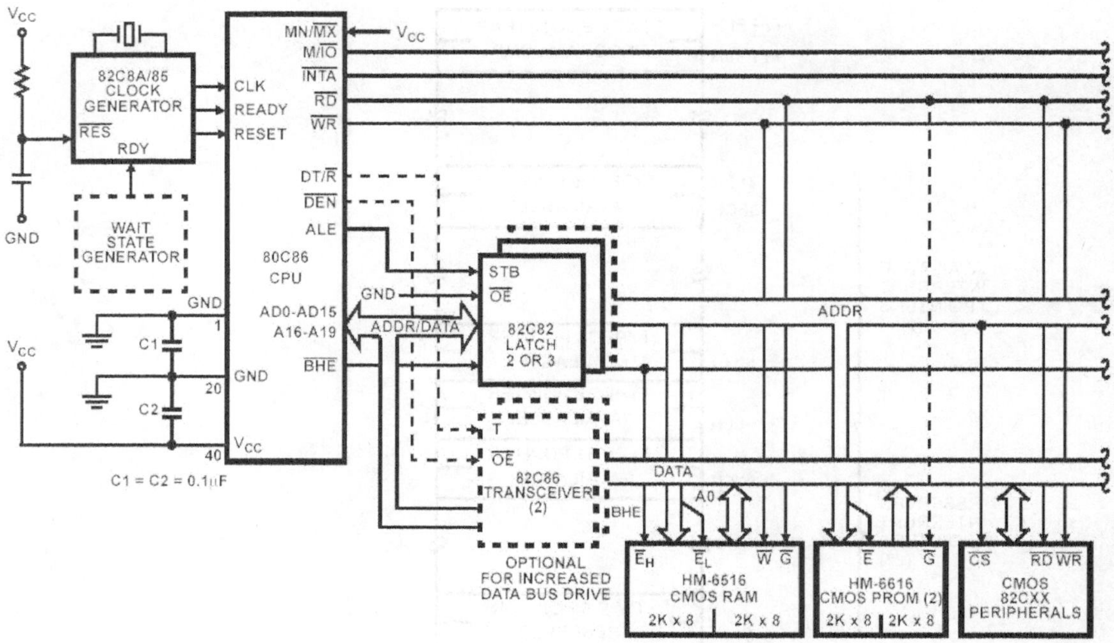

Fig. 5.6 Minimum Mode 80C86 Typical Configuration

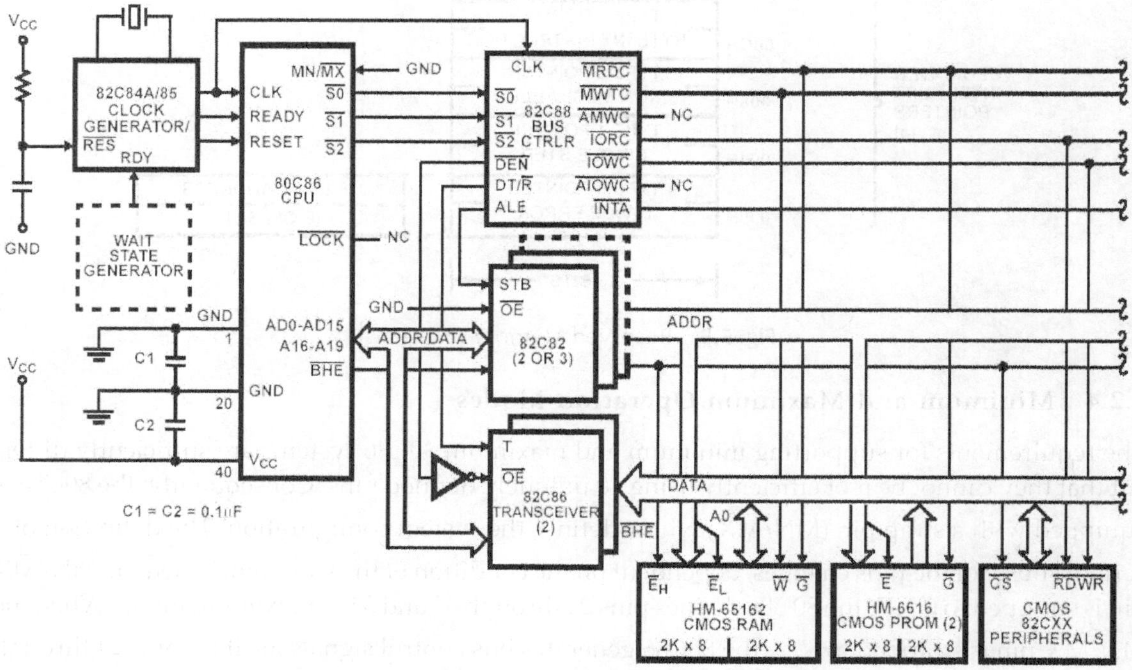

Fig. 5.7 Maximum Mode 80C86 Typical Configuration

The demultiplexed mode requires two 82C82 latches (for 64K addressability) or three 82C82 latches (for a full megabyte of addressing). An 82C86 or 82C87 transceiver can also be used if data bus buffering is required. (See Fig. 5.6) The 80C86 provides $\overline{DEN}$ and $DT/\overline{R}$ to control the transceiver, and ALE to latch the addresses. This configuration of the minimum mode provides the standard demultiplexed bus structure with heavy bus buffering and relaxed bus timing requirements.

The maximum mode employs the 82C88 bus controller (See Fig. 5.7). The 82C88 decodes status lines $\overline{S0}$, $\overline{S1}$ and $\overline{S2}$, and provides the system with all bus control signals.

Moving the bus control to the 82C88 provides better source and sink current capability to the control lines, and frees the 80C86 pins for extended large system features. Hardware lock, queue status, and two request/grant interfaces are provided by the 80C86 in maximum mode. These features allow coprocessors in local bus and remote bus configurations.

5.2.5 Bus Operation

The 80C86 has a combined address and date bus commonly referred to as a time multiplexed bus. This technique provides the most efficient use of pins on the processor while permitting the use of a standard 40 lead package. This "local bus" can be buffered directly and used throughout the system with address latching provided on memory and I/O modules. In addition, the bus can also be demultiplexed at the processor with a single set of 82C82 address latches if a standard non-multiplexed bus is desired for the system.

Each processor bus cycle consists of at least four CLK cycles. These are referred to as T1, T2, T3 and T4 (see Fig. 5.8). The address is emitted from the processor during T1 and data transfer occurs on the bus during T3 and T4. T2 is used primarily for changing the direction of the bus during read operations. In the event that a "NOT READY" indication is given by the addressed device, "Wait" states (TW) are inserted between T3 and T4. Each inserted wait state is the same duration as a CLK cycle. Periods can occur between 80C86 driven bus cycles. These are referred to as idle" states (T_1) or inactive CLK cycles. The processor uses these cycles for internal housekeeping and processing.

During T1 of any bus cycle, the ALE (Address Latch Enable) signal is emitted (by either the processor or the 82C88 bus controller, depending on the MN/$\overline{MX}$ strap). At the trailing edge of this pulse, a valid address and certain status information for the cycle may be latched.

Status bits $\overline{S0}$, $\overline{S1}$ and $\overline{S2}$ are used by the bus controller, in maximum mode, to identify the type of bus transaction according to Table 5.2

Table 5.2

$\overline{S2}$	$\overline{S1}$	$\overline{S0}$	Characteristics
0	0	0	Interrupt
0	0	1	Read I/O
0	1	0	Write I/O
0	1	1	Halt
1	0	0	Instruction Fetch
1	0	1	Read Data from Memory
1	1	0	Write Data to Memory
1	1	1	Passive (No Bus Cycle)

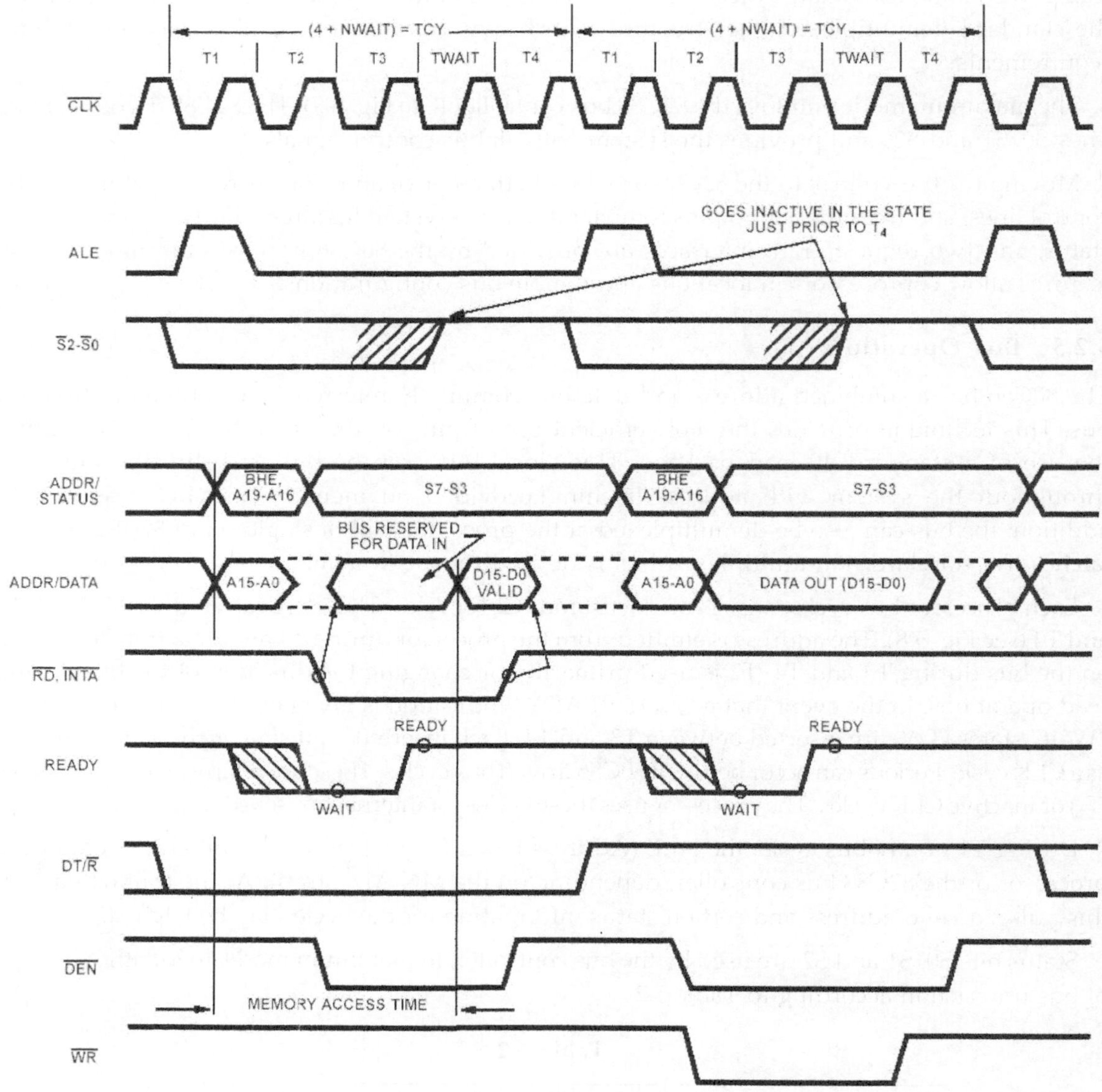

Fig. 5.8 Basic System Timing

Status bits S3 through S7 are time multiplexed with high order address bits and the $\overline{BHE}$ signal, and are therefore valid during T2 through T4. S3 and S4 indicate which segment register (see Instruction Set Description) was used for this bus cycle in forming the address, according to Table 5.3.

S5 is a reflection of the interrupt enable bit in the flag register. S3 is always zero and S7 is a spare status bit.

Table 5.3

S4	S3	Characteristics
0	0	Alternate Data (Extra Segment)
0	1	Stack
1	0	Code or None
1	1	Data

5.2.6 I/O Addressing

In the 80C86, I/O operations can address up to a maximum of 64K I/O byte registers or 32K I/O word registers. The I/O address appears in the same format as the memory address on bus lines A15-A0. The address lines A19-A16 are zero in I/O operations. The variable I/O instructions which use register DX as a pointer have full address capability while the direct I/O instructions directly address one or two of the 256 I/O byte locations in page 0 of the I/O address space.

I/O ports are addressed in the same manner as memory locations. Even addressed bytes are transferred on the D7-D0 bus lines and odd addressed bytes on D15-D8. Care must be taken to ensure that each register within an 8-bit peripheral located on the lower portion of the bus be addressed as even.

5.2.7 Flag Register

A flag register indicates the condition produced by the execution of an instruction and also controls certain operation of exceution unit.

In 8086, we have totaly 16-bit flag register but only 9 flags are active. Here we have 6 flags called conditional flags; they are called conditional flag because they reflect the conditions due to execution of an instruction. These flags include:

1. Carry flag (CF)
2. Parity flag (PF)
3. Auxiliary carry flag (AF)
4. Zero flag (ZF)
5. Sign flag (SF)
6. Over flow flag (OF)

These are same as used in 8085.

The remaining three flags are used for control certain operations of the processor. Control flags are set/reset by software instructions; these three flags are:

1. Trap flag (TF)—used for single stepping through a programme.
2. Interrupt flag (IF)—that allows 'or' not allow to interrupt a programme.
3. Direction flag (DF)—that its used for string operation. So in summary:

Table 5.4 Flags of 8086

Total active flags = 9	
Conditional Flag (6)	**Control Flag (3)**
1. Used to reflect the condition due to execution of an instruction.	1. Used to control a operation performed by execution unit.
2. These are CF, PF, AF, ZF, SF and OF	2. These are TF, IF, DF.
3. These are set/reset by microprocessor according to condition generated by execution of an instruction.	3. These are set/reset by software instruction.

Table 5.5 80C86 Register

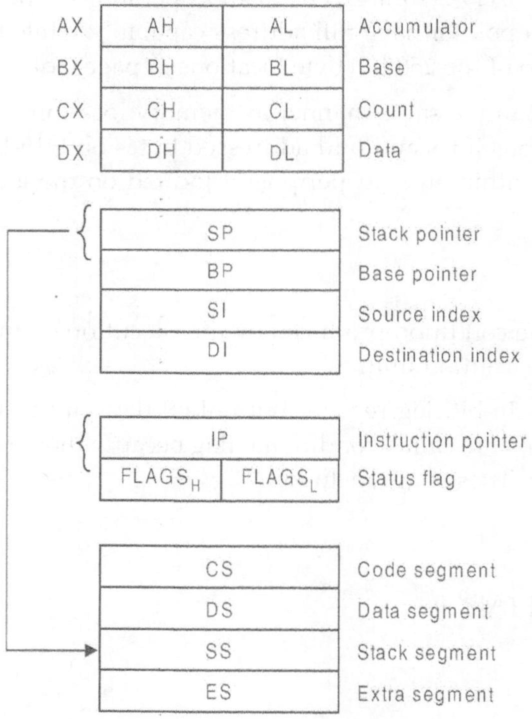

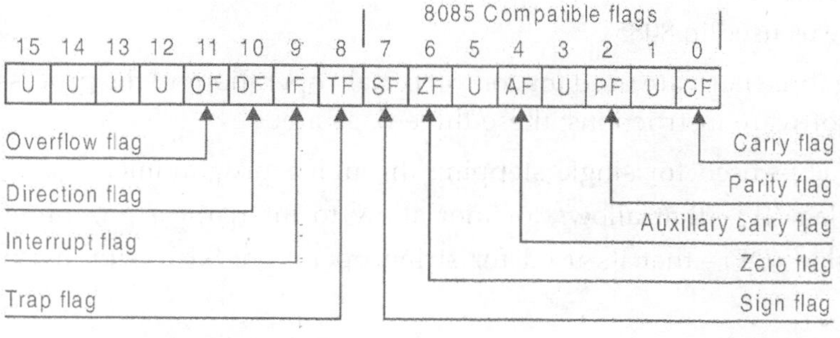

*U = Undefined

Fig. 5.9 8086 Flag Register

5.2.8 Dedicated Interrupts

As we have seen in Fig. 5.5 that totaly we have 256 interrupts called vector interrupt because each have a particular memory address.

The lowest 5 types of a interrupt are deciated for particular tasks and remaining 5 to 31 interrupts location has been reserved for future by intel co-operation. Now we will explain one by one in detail.

Type '0' Divide Zero Error Interrupt

The divide zero error occures when a division inspection is executed.

If the quotient that results from a DIV instruction 'or' an IDIV instruction is larger than the specified destination, a divide zero error has occurred.

This condition causes automatically initiation of type '0' interrupt and access of service routine starting by address 0000_{16} to 0002_{16}. Generally NMI is assigned to the hardware events that require immediate operation such as power failure.

Due to this type of interrupt, the service routine of location 0008_{16} to $000A_{16}$ has been executed.

Type 1 Single Step Interrupt

If the trap flag (TF) bit is set, the single mode function is enabled. This TF flag as described easier is set by software instruction.

The single step interrupt feature is used to monitor 'or' debug a program. Due to this type of function; the debugging is done by instruction to instruction. If any error caused to a particular instruction, that will be find and removed.

Due to this interrupt the service routing of address 0004_{16} to 0006_{16} is executed.

Type 2 Non Markable Interrupt

It is initiated by external hardware; but its different to the external hardware interrupt as it can not be marked by using interrupt flag.

Type-3 Break Point Interrupt

It is initiated by INT3 instruction. It is also a debugging tool; but now in stead of each instruction; it will check a part of program where INT3 has been inserted; if after '5' instructions it is inserted it will check 5 instructions and show errors if any.

During executions of this instruction; the service routine of location $000E_{16}$ to $000C_{H.}$

Type-4 Overflow Error Interrupt

It is set if the operation between two operand is giving result bigger than memory location. It is same as type '0' but now it can be used for any arithmetic operation. When a overflow occurs; 'OF' flag is set then if we have entered 'INTO', Type 4 interrupt will cause to execute service routine occurs between 0010_{16} to 0012_{16}.

The advantage of such service is that; we do not require any program to check, whether a overhead occurred 'or' not. The process will go without any problem and process will be interrupted only when such error occurs.

5.3 EXTERNAL INTERFACE

5.3.1 Processor RESET and Initialization

Processor initialization or start up is accomplished with activation (HIGH) of the RESET pin. The 80C86 RESET is required to be HIGH for greater than 4 CLK cycles. The 80C86 will terminate operations on the high-going edge of RESET and will remain dormant as long as RESET is HIGH. The low-going transition of RESET triggers an internal reset sequence for approximately 7 clock cycles. After this interval, the 80C86 operates normally beginning with the instruction in absolute location FFFF0H. (See Fig. 5.5). The RESET input is internally synchronized to the processor clock. At initialization, the HIGH-to-LOW transition of RESET must occur no sooner than 50μs (or 4 CLK cycles, whichever is greater) after power-up, to allow complete initialization of the 80C86.

NMI will not be recognized prior to the second CLK cycle following the end of RESET. If NMI is asserted sooner than nine clock cycles after the end of RESET, the processor may execute one instruction before responding to the interrupt.

5.3.2 Bus Hold Circuitry

To avoid high current conditions caused by floating inputs to CMOS devices and to eliminate need for pull-up/down resistors, "bus-hold" circuitry has been used on the 80C86 pins 2-16,26-32 and 34-39. (See Fig. 5.10 and Fig. 5.11). These circuits will maintain the last valid logic state if no driving source is present (*i.e.*, an unconnected pin or a driving source which goes to a high impedance state). To overdrive the "bus hold" circuits, an external driver must be capable of supplying approximately 400μA minimum sink or source current at valid input voltage levels. Since this "bus hold" circuitry is active and not a "resistive" type element the associated power supply current is negligible and power dissipation is significantly reduced when compared to the use of passive pull-up resistors.

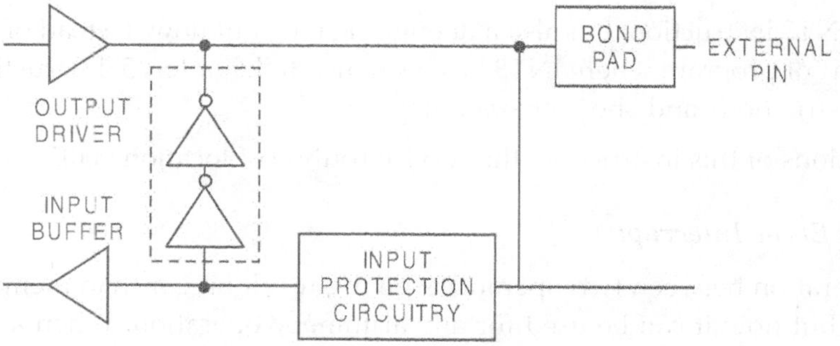

Fig. 5.10 Bus Hold Circuitry Pin 2-16, 34-39

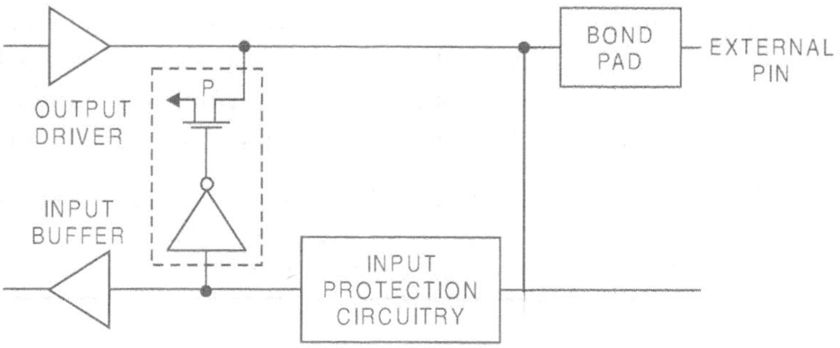

Fig. 5.11 Bus Hold Circuitry Pin 26-32

5.3.3 Interrupt Operations

Interrupt operations fall into two classes: software or hardware initiated. The software initiated interrupts and software aspects of hardware interrupts are specified in the Instruction Set Description. Hardware interrupts can be classified as non-maskable or maskable.

Interrupts result in a transfer of control to a new program location. A 256-element table containing address pointers to the interrupt service program locations resides in absolute locations 0 through 3FFH, which are reserved for this purpose. Each element in the table is 4 bytes in size and corresponds to an interrupt "type". An interrupting device supplies an 8-bit type number during the interrupt acknowledge sequence, which is used to "vector" through the appropriate element to the new interrupt service program location. All flags and both the Code Segment and Instruction Pointer register are saved as part of the $\overline{\text{INTA}}$ sequence. These are restored upon execution of an Interrupt Return (IRET) instruction.

(*a*) **Non-Maskable Interrupt (NMI) :** The processor provides a single non-maskable interrupt pin (NMI) which has higher priority than the maskable interrupt request pin (INTR). A typical use would be to activate a power failure routine. The NMI is edge-triggered on a LOW-to-HIGH transition. The activation of this pin causes a type 2 interrupt.

NMI is required to have a duration in the HIGH state of greater than two CLK cycles, but is not required to be synchronized to the clock. Any positive transition of NMI is latched on-chip and will be serviced at the end of the current instruction or between whole moves of a block-type instruction. Worst case response to NMI would be for multiply, divide, and variable shift instructions. There is no specification on the occurrence of the low-going edge; it may occur before, during or after the servicing of NMI. Another positive edge triggers another response if it occurs after the start of the NMI procedure. The signal must be free of logical spikes in general and be free of bounces on the low-going edge to avoid triggering extraneous responses.

(*b*) **Maskable Interrupt (INTR) :** The 80C86 provides a single interrupt request input (INTR) which can be masked internally by software with the resetting of the interrupt enable flag (IF) status bit. The interrupt request signal is level triggered. It is internally synchronized during each clock cycle on the high-going edge of CLK. To be responded to, INTR must be present (HIGH) during the clock period preceding the end of the current instruction or the end of a whole move for a block type instruction. INTR may be removed anytime after the falling edge of the first $\overline{\text{INTA}}$ signal. During the interrupt response sequence further interrupts are disabled. The enable bit is reset as part of the response to any interrupt (INTR, NMI, software interrupt or single-step), although the FLAGS register which is automatically pushed onto the stack reflects the state of the processor prior to the interrupt. Until the old FLAGS register is restored, the enable bit will be zero unless specifically set by an instruction.

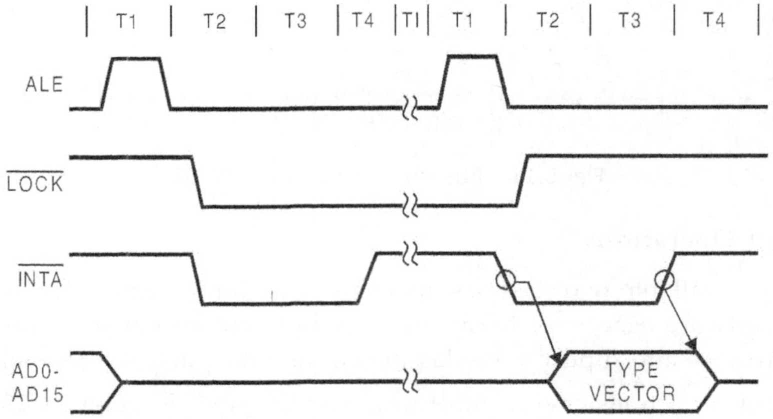

Fig. 5.12 Interrupt Acknowledge Sequence

During the response sequence (Fig. 5.12), the processor executes two successive (back-to-back) interrupt acknowledge cycles. The 80C86 emits the $\overline{\text{LOCK}}$ signal (Max mode only) from T2 of the first bus cycle until T2 of the second. A local bus "hold" request will not be honored until the end of the second bus cycle. In the second bus cycle, a byte is supplied to the 80C86 by the 82C59A Interrupt Controller, which identifies the source (type) of the interrupt. This byte is multiplied by four and used as a pointer into the interrupt vector lookup table. An INTR signal left HIGH will be continually responded to within the limitations of the enable bit and sample period. The INTERRUPT RETURN instruction includes a FLAGS pop which returns the status of the original interrupt enable bit when it restores the FLAGS.

(*c*) **Halt :** When a software "HALT" instruction is executed the processor indicates that it is entering the "HALT" state in one of two ways depending upon which mode is strapped. In minimum mode, the processor issues one ALE with no qualifying bus control signals. In maximum mode, the processor issues appropriate HALT status on $\overline{S2}, \overline{S1}, \overline{S0}$ and the 82C88 bus controller issues one ALE. The 80C86 will not leave the "HALT" state when a local bus "hold" is entered while in "HALT". In this case, the processor reissues the HALT indicator at the end of the local bus hold. An NMI or interrupt request (when interrupts enabled) or RESET will force the 80C86 out of the "HALT" state.

5.4 READ/MODIFY/WRITE

5.4.1 Operations via Lock

The LOCK status information is provided by the processor when consecutive bus cycles are required during the execution of an instruction. This gives the processor the capability of performing read/modify/write operations on memory (via the Exchange Register With Memory instruction, for example) without another system bus master receiving intervening memory cycles. This is useful in multiprocessor system configurations to accomplish "test and set lock" operations. The LOCK signal is activated (forced LOW) in the clock cycle following decoding of the software "LOCK" prefix instruction. It is deactivated at the end of the last bus cycle of the instruction following the "LOCK" prefix instruction. While LOCK is active a request on a RQ/GT pin will be recorded and then honored at the end of the LOCK.

5.4.2 External Synchronization via TEST

As an alternative to interrupts, the 80C86 provides a single software-testable input pin ($\overline{\text{TEST}}$). This input is utilized by executing a WAIT instruction. The single WAIT instruction is repeatedly executed until the $\overline{\text{TEST}}$ input goes active (LOW). The execution of WAIT does not consume bus cycles once the queue is full.

If a local bus request occurs during WAIT execution, the 80C86 three-states all output drivers white inputs and I/O pins are held at valid logic levels by internal bus-hold circuits. If interrupts are enabled, the 80C86 will recognize interrupts and process them when it regains control of the bus. The WAIT instruction is then refetched, and re-executed.

5.5 BASIC SYSTEM TIMING

Typical system configurations for the processor operating in minimum mode and in maximum mode are shown in Figs. 5.6 and 5.7, respectively. In minimum mode, the MN/MX pin is strapped to V_{CC} and the processor emits bus control signals *e.g.*, $\overline{\text{RD}}$, $\overline{\text{WR}}$, etc.) directly. In maximum mode, the $\overline{\text{MN}}$/MX pin is strapped to GND and the processor emits coded status information which the 82C88 bus controller uses to generate MULTIBUS compatible bus control signals. Fig. 5.8 shows the signal timing relationships.

5.5.1 System Timing—Minimum System

The read cycle begins in T1 with the assertion of the Address Latch Enable (ALE) signal. The trailing (low-going) edge of this signal is used to latch the address information, which is valid on the address/data bus (AD0-AD15) at this time, into the 82C82/82C83 latch. The $\overline{\text{BHE}}$ and A0 signals address the low, high or both bytes. From T1 to T4 the M/$\overline{\text{IO}}$ signal indicates a memory or I/O operation. At T2, the address is removed from the address/data bus and the bus is held at the last valid logic state by internal bus hold devices. The read control signal is also asserted at T2. The read (RD) signal causes the addressed device to enable its data bus drivers to the local bus. Some time later, valid data will be available on the bus and the addressed device will drive the READY line HIGH. When the processor returns the read signal to a HIGH level, the addressed device will again three-state its bus drivers. If a transceiver (82C86/82C87) is required to buffer the 80C86 local bus, signals DT/$\overline{\text{R}}$ and $\overline{\text{DEN}}$ are provided by the 80C86.

A write cycle also begins with the assertion of ALE and the emission of the address. The M/IO signal is again asserted to indicate a memory or I/O write operation. In T2, immediately following the address emission, the processor emits the data to be written into the addressed location. This data remains valid until at least the middle of T4. During T2, T3 and TW, the processor asserts the write control signal. The write ($\overline{WR}$) signal becomes active at the beginning of T2 as opposed to the read which is delayed somewhat into T2 to provide time for output drivers to become inactive.

The $\overline{BHE}$ and A0 signals are used to select the proper byte(s) of the memory/IO word to be read or written according to Table 5.6.

Table 5.6

$\overline{BHE}$	A0	Characteristics
0	0	Whole word
0	1	Upper Byte From/To Odd Address
1	0	Lower Byte From/To Even Address
1	1	None

I/O ports are addressed in the same manner as memory location. Even addressed bytes are transferred on the D7-D0 bus lines and odd address bytes on D15-D8.

The basic difference between the interrupt acknowledge cycle and a read cycle is that the interrupt acknowledge signal ($\overline{INTA}$) is asserted in place of the read ($\overline{RD}$) signal and the address bus is held at the last valid logic state by internal bus hold devices. (See Figs. 5.10 and 5.11). In the second of two successive $\overline{INTA}$ cycles a byte of information is read from the data bus (D7-D0) as supplied by the interrupt system logic (*i.e.*, 82C59A Priority Interrupt Controller). This byte identifies the source (type) of the interrupt. It is multiplied by four and used as a pointer into an interrupt vector lookup table.

5.5.2 Bus Timing—Medium Size Systems

For medium complexity systems, the MN/$\overline{MX}$ pin is connected to GND and the 82C88 Bus Controller is added to the system as well as an 82C82/82C83 latch for latching the system address, and an 82C86/82C87 transceiver to allow for bus loading greater than the 80C86 is capable of handling. Signals ALE, $\overline{DEN}$, and DT/$\overline{R}$ are generated by the 82C88 instead of the processor in this configuration, although their timing remains relatively the same. The 80C86 status outputs ($\overline{S2}$, $\overline{S1}$ and $\overline{S0}$) provide type-of-cycle information and become 82C88 inputs. This bus cycle information specifies read (code, data or I/O), write (data or I/O), interrupt acknowledge, or software halt. The 82C88 issues control signals specifying memory read or write, I/O read or write, or interrupt acknowledge. The 82C88 provides two types of write strobes, normal and advanced, to be applied as required. The normal write strobes have data valid at the leading edge of write. The advanced write strobes have the same timing as read strobes, and hence, data is not valid at the leading edge of write. The 82C86/82C87 transceiver receives the usual T and $\overline{OE}$ inputs from the 82C88 DT/$\overline{R}$ and DEN signals.

The pointer into the interrupt vector table, which is passed during the second $\overline{INTA}$ cycle, can be derived from an 82C59A located on either the local bus or the system bus. If the master 82C59A Priority Interrupt Controller is positioned on the local bus, the 82C86/82C87 transceiver must be disabled when reading from the master 82C59A during the interrupt acknowledge sequence and software "poll".

5.6 INSTRUCTION SET AND ADDRESSING MODE

5.6.1 Instruction Set Summary

Mnemonic and Description	Instruction Code			
	76543210	76543210	76543210	76543210
DATA TRANSFER				
MOV = MOVE:				
Register/Memory to/from	100010 dw	mod reg r/m		
Register Immediate to Register/Memory	1100011 w	mod 000 r/m	data	data if w 1
Immediate to Register	1011 w reg	data	data if w 1	
Memory to Accumulator	1010000 w	addr-low	addr-high	
Accumulator to Memory	1010001 w	addr-low	addr-high	
Register/Memory to Segment Register ††	10001110	mod 0 reg r/m		
Segment Register to Register/Memory	10001100	mod 0 reg r/m		
PUSH = Push:				
Register/Memory	11111111	mod 110 r/m		
Register	01010 reg			
Segment Register	000 reg 110			
POP = Pop:				
Register/Memory	10001111	mod 000 r/m		
Register	01011 reg			
Segment Register	000 reg 111			
XCHG = Exchange:				
Register/Memory with Register	1000011 w	mod reg r/m		
Register with Accumulator	10010 reg			
IN = Input from:				
Fixed Port	1110010 w	port		
Variable Port	1110110 w			
OUT = Output to:				
Fixed Port	1110011 w	port		
Variable Port	1110111 w			
XLAT = Translate Byte to AL	11010111			
LEA = Load EA to Register	10001101	mod reg r/m		
LDS = Load Pointer to DS	11000101	mod reg r/m		
LES = Load Pointer to ES	11000100	mod reg r/m		
LAHF = Load AH with Flags	10011111			
SAHF = Store AH into Flags	10011110			
PUSHF = Push Flags	10011100			
POPF = Pop Flags	10011101			

ARITHMETIC				
ADD = Add:				
Register/Memory with Register to Either	000000 dw	mod reg r/m		
Immediate to Register/Memory	100000 sw	mod 000 r/m	data	data if s:w = 01
Immediate to Accumulator	0000010 w	data	data if w = 1	
ADC = Add with Carry:				
Register/Memory with Register to Either	000100 dw	mod reg r/m		
Immediate to Register/Memory	100000 sw	mod 010 r/m	data	data if s:w = 01
Immediate to Accumulator	0001010 w	data	data if w = 1	
INC = Increment:				
Register/Memory	1111111 w	mod 000 r/m		
Register	01000 reg			
AAA = ASCII Adjust for Add	00110111			
DAA = Decimal Adjust for Add	00100111			
SUB = Subtract:				
Register/Memory and Register to Either	001010 dw	mod reg r/m		
Immediate from Register/Memory	100000 sw	mod 101 r/m	data	data if s:w = 01
Immediate from Accumulator	0010110 w	data	data if w = 1	
SBB = Subtract with Borrow				
Register/Memory and Register to Either	000110 dw	mod reg r/m		
Immediate from Register/Memory	100000 sw	mod 011 r/m	data	data if s:w = 01
Immediate from Accumulator	0001110 w	data	data if w = 1	
DEC = Decrement:				
Register/Memory	1111111 w	mod 001 r/m		
Register	01001 reg			
NEG = Change Sign	1111011 w	mod 011 r/m		
CMP = Compare:				
Register/Memory and Register	001110 dw	mod reg r/m		
Immediate with Register/Memory	100000 sw	mod 111 r/m	data	data if s:w = 01
Immediate with Accumulator	0011110 w	data	data if w = 1	
AAS = ASCII Adjust for Subtract	00111111			
DAS = Decimal Adjust for Subtract	00101111			
MUL = Multiply (Unsigned)	1111011 w	mod 100 r/m		
IMUL = Integer Multiply (Signed)	1111011 w	mod 101 r/m		

AAM = ASCII Adjust for Multiply	11010100	00001010		
DIV = Divide (Unsigned)	1111011 w	mod 110 r/m		
IDIV = Integer Divide (Signed)	1111011 w	mod 111 r/m		
AAD = ASCII Adjust for Divide	11010101	00001010		
CBW = Convert Byte to Word	10011000			
CWD = Convert Word to Double Word	10011001			
LOGIC				
NOT = Invert	1111011 w	mod 010 r/m		
SHL/SAL = Shift Logical/ Arithmetic Left	110100 vw	mod 100 r/m		
SHR = Shift Logical Right	110100 vw	mod 101 r/m		
SAR = Shift Arithmetic Right	110100 vw	mod 111 r/m		
ROL = Rotate Left	110100 vw	mod 000 r/m		
ROR = Rotate Right	110100 vw	mod 001 r/m		
RCL = Rotate Through Carry Flag Left	110100 vw	mod 010 r/m		
RCR = Rotate Through Carry Right	110100 vw	mod 011 r/m		
AND = And:				
Reg./Memory and Register to Either	0010000 dw	mod reg r/m		
Immediate to Register/Memory	1000000 w	mod 100 r/m	data	data if w = 1
Immediate to Accumulator	0010010 w	data	data if w = 1	
TEST = And Function to Flags, No Result:				
Register/Memory and Register	1000010 w	mod reg r/m		
Immediate Data and Register/Memory	1111011 w	mod 000 r/m	data	data if w = 1
Immediate Data and Accumulator	1010100 w	data	data if w = 1	
OR = Or:				
Register/Memory and Register to Either	000010 dw	mod reg r/m		
Immediate to Register/Memory	1000000 w	mod 101 r/m	data	data if w = 1
Immediate to Accumulator	0000110 w	data	data if w = 1	
XOR = Exclusive or:				
Register/Memory and Register to Either	001100 dw	mod reg r/m		
Immediate to Register/Memory	1000000 w	mod 110 r/m	data	data if w = 1
Immediate to Accumulator	0011010 w	data	data if w = 1	

STRING MANIPULATION				
REP = Repeat	1111001 z			
MOVS = Move Byte/Word	1010010 w			
CMPS = Compare Byte/Word	1010011 w			
SCAS = Scan Byte/Word	1010111 w			
LODS = Load Byte/Word to AL/AX	1010110 w			
STOS = Stor Byte/Word from AL/A	1010101 w			
CONTROL TRANSFER				
CALL = Call:				
Direct Within Segment	11101000	disp-low	disp-high	
Indirect Within Segment	11111111	mod 010 r/m		
Direct Intersegment	10011010	offset-low seg-low	offset-high seg-high	
Indirect Intersegment	11111111	mod 011 r/m		
JMP = Unconditional Jump:				
Direct Within Segment	11101001	disp-low	disp-high	
Direct Within Segment-Short	11101011	disp		
Indirect Within Segment	11111111	mod 100 r/m		
Direct Intersegment	11101010	offset-low seg-low	offset-high seg-high	
Indirect Intersegment	11111111	mod 101 r/m		
RET = Return from CALL:				
Within Segment	11000011			
Within Seg Adding Immed to SP	11000010	data-low	data-high	
Intersegment	11001011			
Intersegment Adding Immediate to SP	11001010	data-low	data-high	
JE/JZ = Jump on Equal/Zero	01110100	disp		
JL/JNGE = Jump on Less/Not Greater or Equal	01111100	disp		
JLE/JNG = Jump on Less or Equal/ Not Greater	01111110	disp		
JB/JNAE = Jump on Below/ Not Above or Equal	01110010	disp		
JBE/JNA = Jump on Below or Equal/Not Above	01110110	disp		
JP/JPE = Jump on Parity/ Parity Even	01111010	disp		
JO = Jump on Overflow	01110000	disp		

JS = Jump on Sign	01111000	disp		
JNE/JNZ = Jump on Not Equal/Not Zero	01110101	disp		
JNL/JGE = Jump on Not Less/Greater or Equal	01111101	disp		
JNLE/JG = Jump on Not Less or Equal/Greater	01111111	disp		
JNB/JAE = Jump on Not Below/Above or Equal	01110011	disp		
JNBE/JA = Jump on Not Below or Equal/Above	01110111	disp		
JNP/JPO = Jump on Not Par/Par Odd	01111011	disp		
JNO = Jump on Not Overflow	01110001	disp		
JNS = Jump on Not Sign	01111001	disp		
LOOP = Loop CX Times	11100010	disp		
LOOPZ/LOOPE = Loop While Zero/Equal	11100001	disp		
LOOPNZ/LOOPNE = Loop While Not Zero/Equal	11100000	disp		
JCXZ = Jump on CX Zero	11100011	disp		
INT = Interrupt				
Type Specified	11001101	type		
Type 3	11001100	disp		
INTO = Interrupt on Overflow	11001110	disp		
IRET = Interrupt Return	11001111	disp		
PROCESSOR CONTROL				
CLC = Clear Carry	11111000			
CMC = Complement Carry	11110101			
STC = Set Carry	11111001			
CLD = Clear Direction	11111100			
STD = Set Direction	11111101			
CLI = Clear Interrupt	11111010			
ST = Set Interrupt	11111011			
HLT = Halt	11110100			
WAIT = Wait	10011011			
ESC = Escape (to External Device)	11011xxx	mod xxx r/m		
LOCK = Bus Lock Prefix	11110000			

NOTES:

AL = 8-bit accumulator

AX = 16-bit accumulator

CX = Count register

DS= Data segment

ES = Extra segment

Above/below refers to unsigned value.

Greater = more positive;

Less = less positive (more negative) signed values

if d = 1 then "to" reg; if d = 0 then "from" reg

if w = 1 then word instruction; if w = 0 then byte instruction

if mod = 11 then r/m is treated as a REG field

if mod = 00 then DISP = O†, disp-low and disp-high are absent

if mod = 01 then DISP = disp-low sign-extended 16-bit, disp-high is absent

if mod = 10 then DISP = disp-high:disp-low

if r/m = 000 then EA = (BX) + (SI) + DISP

if r/m = 001 then EA = (BX) + (DI) + DISP

if r/m = 010 then EA = (BP) + (SI) + DISP

if r/m = 011 then EA = (BP) + (DI) + DISP

if r/m = 100 then EA = (SI) + DISP

if r/m = 101 then EA = (DI) + DISP

if r/m = 110 then EA = (BP) + DISP †

if r/m = 111 then EA = (BX) + DISP

DISP follows 2nd byte of instruction (before data if required)

† except if mod = 00 and r/m = 110 then EA = disp-high: disp-low.

†† MOV CS, REG/MEMORY not allowed.

if s:w = 01 then 16-bit of immediate data form the operand.

if s:w. = 11 then an immediate data byte is sign extended to form the 16-bit operand.

if v = 0 then "count" = 1; if v = 1 then "count" in (C_L)

x = don't care

z is used for string primitives for comparison with ZF FLAG.

SEGMENT OVERRIDE PREFIX

001 reg 11 0

REG is assigned according to the following table:

16-BIT (w = 1)	8-BIT (w = 0)	Segment
000 AX	000 AL	OOES
001 CX	001 CL	01 CS
010 DX	010 DL	10 SS
011 BX	011 BL	11 DS
100 SP	100 AH	OOES
101 BP	101 CH	OOES
110 SI	110 DH	OOES
111 DI	111 BH	OOES

Instructions which reference the flag register file as a 16-bit object use the symbol FLAGS to represent the file:

FLAGS=

X:X:X:X:(OF):(DF):(IF):(TF):(SF):(ZF):X:(AF):X:(PF):X:(CF)

5.6.2 Addressing Modes of 8086

The way in which a processor can access a data are refused as addressing mode. In 8086; there are a number of addressing modes. Some of them used generally are given below:

(*a*) Immediate addressing mode

(*b*) Direct addressing mode

(*c*) Register addressing mode

(*d*) Register relative addressing

(*e*) Indexed addressing

(*f*) Implied addressing

Immediate Addressing Mode

In this type of addressing mode; the operand is specified with in instruction itself.

For example: MOV CX, 253AH.

That means copy the data 253A immediately to the register CX.

Direct Addressing Mode

In this mode of addressing, the effective address of the data is directly written in the instruction itself.

This is called direct addressing mode; because the effective address of data is put directly in the instruction itself.

For example: MOV BL, [437A, H]

It means copy the data of location 437A H to the BL register.

Register Addressing Mode

In this mode; the operand is in a particular register; so to get a operand, we address a register.

For example: MOV CX, AX; mean copy the operand written in AX to register pair CX.

Register Indirect Mode

Here the effective address of the operand is the base register BX 'or' an index register that is specified by the instruction.

$$EA = \begin{Bmatrix} (BX) \\ (DI) \\ (SI) \end{Bmatrix}$$

Register Relative Addressing

The effective address of data is the sum of an 8-bit 'or' 16-bit displacement and the contents of a base register 'or' an index register. This is also called relative addressing.

$$EA = \begin{Bmatrix} (BX) \\ (BP) \\ (SI) \\ (DI) \end{Bmatrix} + \begin{Bmatrix} \text{8-bit displacement} \\ \text{or} \\ \text{16-bit displacement} \end{Bmatrix}$$

Based Indexed

The effective address of data is the sum of a base register and an index register, both of them is specified by instruction; *i.e.*,

$$EA = \begin{Bmatrix} (BX) \\ (BP) \end{Bmatrix} + \begin{Bmatrix} (SI) \\ (DI) \end{Bmatrix}$$

Relative Based Indexed

The effective address of data is sum of a base register, a index register, 8 'or' 16-bit displacement

$$EA = \begin{Bmatrix} (BX) \\ (BP) \end{Bmatrix} + \begin{Bmatrix} (SI) \\ (DI) \end{Bmatrix} + \begin{Bmatrix} \text{8-bit address} \\ \text{or} \\ \text{16-bit displacement} \end{Bmatrix}$$

Implied Addressing

In this mode; it is assumed that operand is in the accumulator itself and therefore require no address specification.

5.7 8086 IN MINIMUM MODE

8086 is in minimum mode, when its MN/$\overline{MX}$ pin is strapped to +5V. The working of pins 24 through 31 for the minimum mode are given below:

Table 5.7 Pin Description for 8086

Pin	Symbol	Type	Description
24	$\overline{INTA}$	Output	Interrupt acknowledgement
25	ALE	O	Address latch enable
26	$\overline{DEN}$	O	To inform transceiver that CPU is ready to receive 'or' send data
27	DT/$\overline{R}$	O	To show that transceiver is sending 'or' receiving data
28	M/$\overline{IO}$	O	Distinguish a memory transfer to an I/O transfer
29	$\overline{WR}$	O	To indicate the write operation
30	HLDA	O	That grant the bus control for a external requesting master
31	HOLD	I	Receive bus request from an external requesting master

(This part of pin description has been completely described in pin description of 8086. Here it is reproduced to make you clear about 8086 Minimum Mode). Now we can conclude with the following results.

(*a*) **As we know that low order address buses should be demultiplxed; so we have to use 8288 address latch;**

This latch is 8-bit latch and we have total 16-bits to be multiplexed so we have to use two 8288 address latch.

(*b*) **Now to read the data from address/data buses (AD_{15}-AD_0), we may use a transceiver 8296A; it has 8-bit input, so we have to use two 8286A for this purpose.**

(*c*) **The third component of the circuit is 8284 A clock generator.**

(*d*) **Now the remaining part of signal; transferring of data takes place as follows:**

M/IO	$\overline{RD}$	$\overline{WR}$	
0	0	1	I/O Read
0	1	0	I/O Write
1	0	1	Memory Read
1	1	0	Memory Write

5.8 8086 IN MAXIMUM MODE

A 8086 processor is in maximum mode, when its MN/$\overline{MX}$ pin is grounded. Now the descriptions of pins 24 through 31 are given below:

Table 5.8 Pin Description for Max Mode

Pin	Symbol	Type	Description
24-25	QS_1-QS_0	O	Reflects the status of instructions queue.
26, 27, 28	$\overline{S}_0, \overline{S}_1, \overline{S}_2$	O	Indicates the type of transfer take place.
29	$\overline{LOCK}$	O	Indicates the bus is not to be relinquished to other bus master.
30	$\overline{RQ/GT1}$	I/O	For inputting bus requests and outputting bus grants.
31	$\overline{RQ/GT0}$	I/O	Same as $\overline{RQ/GT1}$ except that request on $\overline{RQ/GT0}$ has higher priority.

(*a*) **Now we do not has signals such as IO/$\overline{M}$, $\overline{WR}$, $\overline{INTA}$, ALE, $\overline{DEN}$ so we generate these signal with the help of $\overline{S}0, \overline{S}1, \overline{S}2$ using a bus controller 8288.**

The 82C88 is a high performance CMOS Bus Controller manufactured using a self-aligned silicon gate CMOS process. The 82C88 provides the control and command timing signals for 80C86, 80C88, 8086, 8088, 8089, 80186, and 80188 based systems. The high output drive capability of the 82C88 eliminates the need for additional bus drivers.

5.9 8088 8-BIT MICROPROCESSOR

- 8-Bit Data Bus Interface
- 16-Bit Internal Architecture
- Direct Addressing Capability to 1 Mbyte of Memory
- Direct Software Compatibility with 8086 CPU
- 14-Word by 16-Bit Register Set with Symmetrical Operations
- 24 Operand Addressing Modes
- Byte, Word, and Block Operations
- 8-Bit and 16-Bit Signed and Unsigned Arithmetic in Binary or Decimal, Including Multiply and Divide
- Two Clock Rates:

— 5 MHz for 8088
— 8 MHz for 8088-2

The 8088 is a high performance microprocessor implemented in N-channel, depletion load, silicon gate technology (HMOS-II), and packaged in a 40-pin CERDIP package. The processor has attributes of both 8-and 16-bit microprocessors. It is directly compatible with 8086 software and 8080/8085 hardware and peripherals.

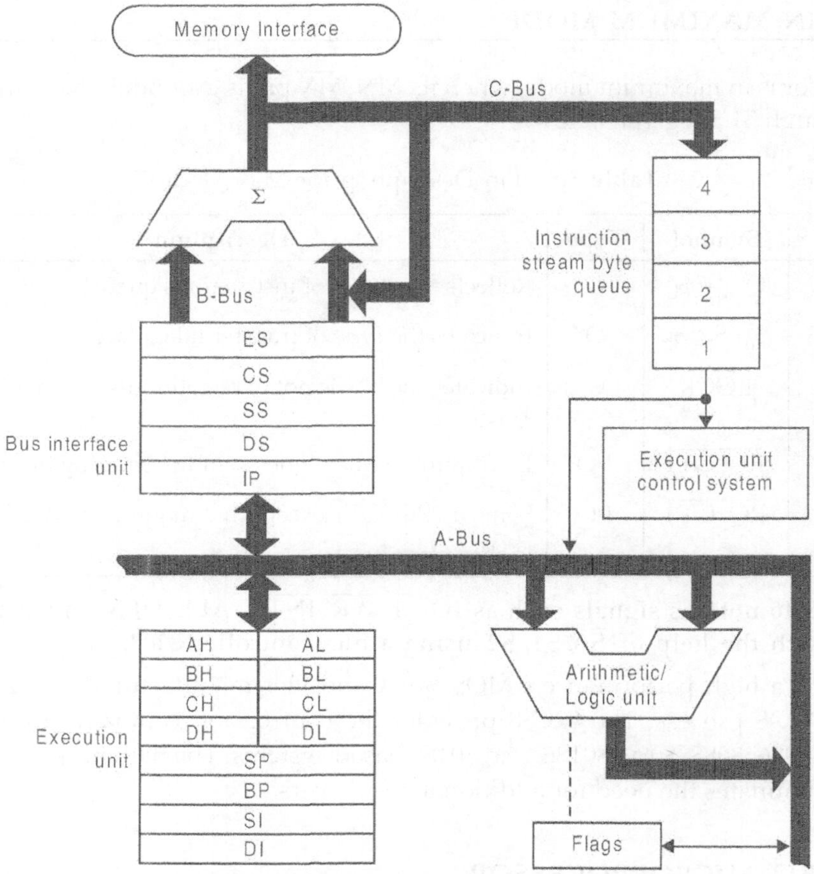

Fig. 5.13 8088 CPU Functional Block Diagram.

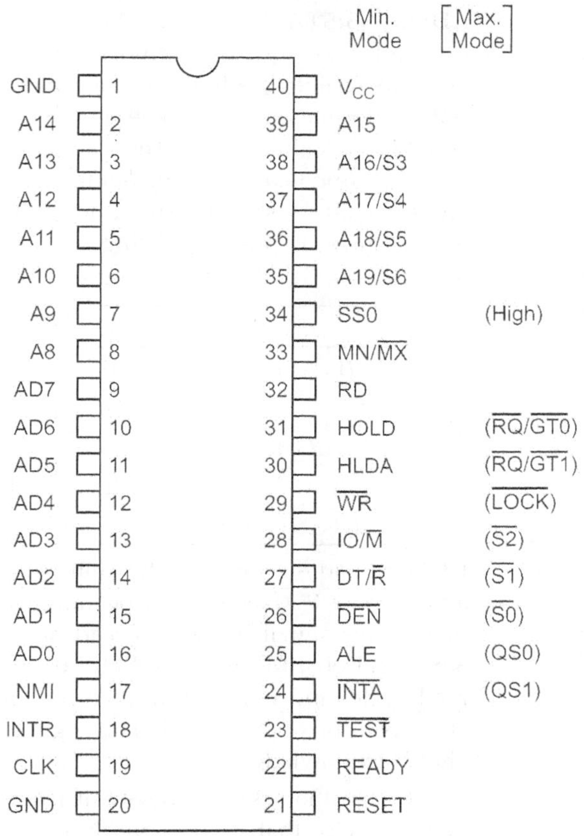

Fig. 5.14 8088 Pin Configuration

5.9.1 Pin Description

The following pin function descriptions are for 8088 systems in either minimum or maximum mode. The "local bus" in these descriptions is the direct multiplexed bus interface connection to the 8088 (without regard to additional bus buffers).

Table 5.9 Pin Description

Symbol	Pin No.	Type	Name and Function
AD7-AD0	9-16	I/O	**ADDRESS DATA BUS:** These lines constitute the time multiplexed memory/IO address (T1) and data (T2, T3, Tw, T4) bus. These lines are active HIGH and float to 3-state OFF during interrupt acknowledge and local bus "hold acknowledge".
A15-A8	2-8, 39	O	**ADDRESS BUS:** These lines provide address bits 8 through 15 for the entire bus cycle (T1-T4). These lines do not have to be latched by ALE to remain valid. A15-A8 are active HIGH and float to 3-state OFF during interrupt acknowledge and local bus "hold acknowledge".

A19/S6, A18/S5, A17/S4, A16/S3	35-38	O	**ADDRESS/STATUS:** During T1 , these are the four most significant address lines for memory operations. During I/O operations, these lines are LOW. During memory and I/O operations, status information is available on these lines during T2, T3, Tw, and T4. S6 is always low. The status of the interrupt enable flag bit (S5) is updated at the beginning of each clock cycle. S4 and S3 are encoded as shown. This information indicates which segment register is presently being used for data accessing. These lines float to 3-state OFF during local bus "hold acknowledge".

S4	S3	Characteristics
0 (LOW)	0	Alternate Data
0	1	Stack
1 (HIGH)	0	Code or None
1	1	Data
S6 is 0 (LOW)		

$\overline{RD}$	32	o	**READ:** Read strobe indicates that the processor is performing a memory or I/O read cycle, depending on the state of the IO/$\overline{M}$ pin or S2. This signal is used to read devices which reside on the 8088 local bus. $\overline{RD}$ is active LOW during T2, T3 and Tw of any read cycle, and is guaranteed to remain HIGH in T2 until the 8088 local bus has floated. This signal floats to 3-state OFF in "hold acknowledge".
READY	22	I	**READY:** is the acknowledgement from the addressed memory or I/O device that it will complete the data transfer. The RDY signal from memory or I/O is synchronized by the 8284 clock generator to form READY. This signal is active HIGH. The 8088 READY input is not synchronized. Correct operation is not guaranteed if the set up and hold times are not met.
INTR	18	I	**INTERRUPT REQUEST:** is a level triggered input which is sampled during the last clock cycle of each instruction to determine if the processor should enter into an interrupt acknowledge operation. A subroutine is vectored to via an interrupt vector lookup table located in system memory. It can be internally masked by software resetting the interrupt enable bit. INTR is internally synchronized. This signal is active HIGH.
$\overline{TEST}$	23	I	$\overline{TEST}$ input is examined by the "wait for test" instruction. If the $\overline{TEST}$ input is LOW, execution continues, otherwise the processor waits in an "idle" state. This input is synchronized internally during each clock cycle on the leading edge of CLK.
NMI	17	I	**NON-MASKABLE INTERRUPT:** is an edge triggered input which causes a type 2 interrupt. A subroutine is vectored to via an interrupt vector lookup table located in system memory. NMI is not maskable internally by software. A transition from a LOW to HIGH initiates the interrupt at the end of the current instruction. This input is internally synchronized.

RESET	21	I	**RESET:** causes the processor to immediately terminate its present activity. The signal must be active HIGH for at least four clock cycles.
CLK	19	I	**CLOCK:** provides the basic timing for the processor and bus controller. It is asymmetric with a 33% duty cycle to provide optimized internal timing.
V_{CC}	40		V_{CC}: is the + 5V ±10% power supply pin.
GND	1,20		**GND:** are the ground pins.
MN/$\overline{MX}$	33	I	**MINIMUM/MAXIMUM:** indicates what mode the processor is to operate in. The two modes are discussed in the following sections.

The following pin function descriptions are for the 8088 minimum mode (*i.e.*, MN/$\overline{MX}$ = Vcc). Only the pin functions which are unique to minimum mode are described; all other pin functions are as described above.

Symbol	Pin No.	Type	Name and Function
IO/$\overline{M}$	28	O	**STATUS LINE:** is an inverted maximum mode $\overline{S2}$. It is used to distinguish a memory access from an I/O access. IO/$\overline{M}$ becomes valid in the T4 preceding a bus cycle and remains valid until the final T4 of the cycle (I/O = HIGH, M = LOW). IO/$\overline{M}$ floats to 3-state OFF in local bus "hold acknowledge".
$\overline{WR}$	29	O	**WRITE:** strobe indicates that the processor is performing a write memory or write I/O cycle, depending on the state of the IO/$\overline{M}$ signal. WR is active for T2, T3, and Tw of any write cycle. It is active LOW, and floats to 3-state OFF in local bus "hold acknowledge".
$\overline{INTA}$	24	O	**INTA:** is used as a read strobe for interrupt acknowledge cycles. It is active LOW during T2, T3, and Tw of each interrupt acknowledge cycle.
ALE	25	O	**ADDRESS LATCH ENABLE:** is provided by the processor to latch the address into an address latch. It is a HIGH pulse active during clock low of T1 of any bus cycle. Note that ALE is never floated.
DT/$\overline{R}$	27	O	**DATA TRANSMIT/RECEIVE:** is needed in a minimum system that desires to use a data bus transceiver. It is used to control the direction of data flow through the transceiver. Logically, DT/$\overline{R}$ is equivalent to $\overline{S1}$ in the maximum mode, and its timing is the same as for IO/$\overline{M}$ (T = HIGH, R = LOW). This signal floats to 3-state OFF in local "hold acknowledge".
$\overline{DEN}$	26	O	**DATA ENABLE:** is provided as an output enable for the data bus transceiver in a minimum system which uses the transceiver. $\overline{DEN}$ is active LOW during each memory and I/O access, and for $\overline{INTA}$ cycles. For a read or $\overline{INTA}$ cycle, it is active from the middle of T2 until the middle of T4, while for a write cycle, it is active from the beginning of T2 until the middle of T4. $\overline{DEN}$ floats to 3-state OFF during local bus "hold acknowledge".

HOLD, HLDA	31,30	I, O	**HOLD:** indicates that another master is requesting a local bus "hold". To be acknowledged, HOLD must be active HIGH. The processor receiving the "hold" request will issue HLDA (HIGH) as an acknowledgement, in the middle of a T4 or Ti clock cycle. Simultaneous with the issuance of HLDA the processor will float the local bus and control lines. After HOLD is detected as being LOW, the processor lowers HLDA, and when the processor needs to run another cycle, it will again drive the local bus and control lines. HOLD and HLDA have internal pull-up resistors. Hold is not an asynchronous input External synchronization should be provided if the system cannot otherwise guarantee the set up time.
$\overline{SSO}$	34	O	**STATUS LINE:** is logically equivalent to $\overline{SO}$ in the maximum mode. The combination of $\overline{SSO}$, IO/$\overline{M}$ and DT/$\overline{R}$ allows the system to completely decode the current bus cycle status.

IO/$\overline{M}$	DT/$\overline{R}$	$\overline{SSO}$	Characteristics
1 (HIGH)	0	0	Interrupt Acknowledge
1	0	1	Read I/O Port
1	1	0	Write I/O Port
1	1	1	Halt
0 (LOW)	0	0	Code Access
0	0	1	Read Memory
0	1	0	Write Memory
0	1	1	Passive

The following pin function descriptions are for the 8088/8288 system in maximum mode (*i.e.*, MN/$\overline{MX}$ = GND). Only the pin functions which are unique to maximum mode are described; all other pin functions are as described above.

Symbol	Pin No.	Type	Name and Function
$\overline{S2}$, $\overline{S1}$, $\overline{S0}$	26-28	O	**STATUS:** is active during clock high of T4, T1 , and T2, and is returned to the passive state (1, 1, 1) during T3 or during Tw when READY is HIGH. This status is used by the 8288 bus controller to generate all memory and I/O access control signals. Any change by $\overline{S2}$, $\overline{S1}$, or $\overline{S0}$ during T4 is used to indicate the beginning of a bus cycle, and the return to the passive state in T3 and Tw is used to indicate the end of a bus cycle. These signals float to 3-state OFF during "hold acknowledge". During the first clock cycle after RESET becomes active, these signals are active HIGH. After this first clock, they float to 3-state OFF.

$\overline{S2}$	$\overline{S1}$	$\overline{S0}$	Characteristics
0 (LOW)	0	0	Interrupt Acknowledge
0	0	1	Read I/O Port
0	1	0	Write I/O Port
0	1	1	Halt

1 (HIGH)	0	0	Code Access
1	0	1	Read Memory
1	1	0	Write Memory
1	1	1	Passive

$\overline{RQ}/\overline{GT0}$ $\overline{RQ}/\overline{GT1}$	30, 31	I/O	**REQUEST/GRANT:** pins are used by other local bus masters to force the processor to release the local bus at the end of the processor's current bus cycle. Each pin is bidirectional with $\overline{RQ}/\overline{GT0}$ having higher priority than $\overline{RQ}/\overline{GT1}$. $\overline{RQ}/\overline{GT}$ has an internal pull-up resistor, so may be left unconnected. The request/grant sequence is as follows: 1. A pulse of one CLK wide from another local bus master indicates a local bus request ("hold") to the 8088 (pulse 1). 2. During a T4 or TI clock cycle, a pulse one clock wide from the 8088 to the requesting master (pulse 2), indicates that the 8088 has allowed the local bus to float and that it will enter the "hold acknowledge" state at the next CLK. The CPU's bus interface unit is disconnected logically from the local bus during "hold acknowledge". The same rules as for HOLD/HLDA apply as for when the bus is released. 3. A pulse one CLK wide from the requesting master indicates to the 8088 (pulse 3) that the "hold" request is about to end and that the 8088 can reclaim the local bus at the next CLK. The CPU then enters T4. Each master-master exchange of the local bus is a sequence of three pulses. There must be one idle CLK cycle after each bus exchange. Pulses are active LOW. If the request is made while the CPU is performing a memory cycle, it will release the local bus during T4 of the cycle when all the following conditions are met: 1. Request occurs on or before T2. 2. Current cycle is not the low bit of a word. 3. Current cycle is not the first acknowledge of an interrupt acknowledge sequence. 4. A locked instruction is not currently executing. If the local bus is idle when the request is made the two possible events will follow: 1. Local bus will be released during the next clock. 2. A memory cycle will start within 3 clocks. Now the four rules for a currently active memory cycle apply with condition number 1 already satisfied.
$\overline{LOCK}$	29	O	**$\overline{LOCK}$:** indicates that other system bus masters are not to gain control of the system bus while $\overline{LOCK}$ is active (LOW). The $\overline{LOCK}$ signal is activated by the "LOCK" prefix instruction and remains active until the completion of the next instruction. This signal is active LOW, and floats to 3-state off in "hold acknowledge".

QS1/QS0	24, 25	O	**QUEUE STATUS:** provide status to allow external tracking of the internal 8088 instruction queue. The queue status is valid during the CLK cycle after which the queue operation is performed.		
			QS1	**QS0**	**Characteristics**
			0 (LOW)	0	No Operation
			0	1	First Byte of Opcode from Queue
			1 (HIGH)	0	Empty the Queue
			1	1	Subsequent Byte from Queue
—	34	O	Pin 34 is always high in the maximum mode.		

5.9.2 Functional Description

Memory Organization

The processor provides a 20-bit address to memory which locates the byte being referenced. The memory is organized as a linear array of up to 1 million bytes, addressed as 00000(H) to FFFFF(H). The memory is logically divided into code, data, extra data, and stack segments of up to 64K bytes each, with each segment falling on 16-byte boundaries (See Fig. 5.15).

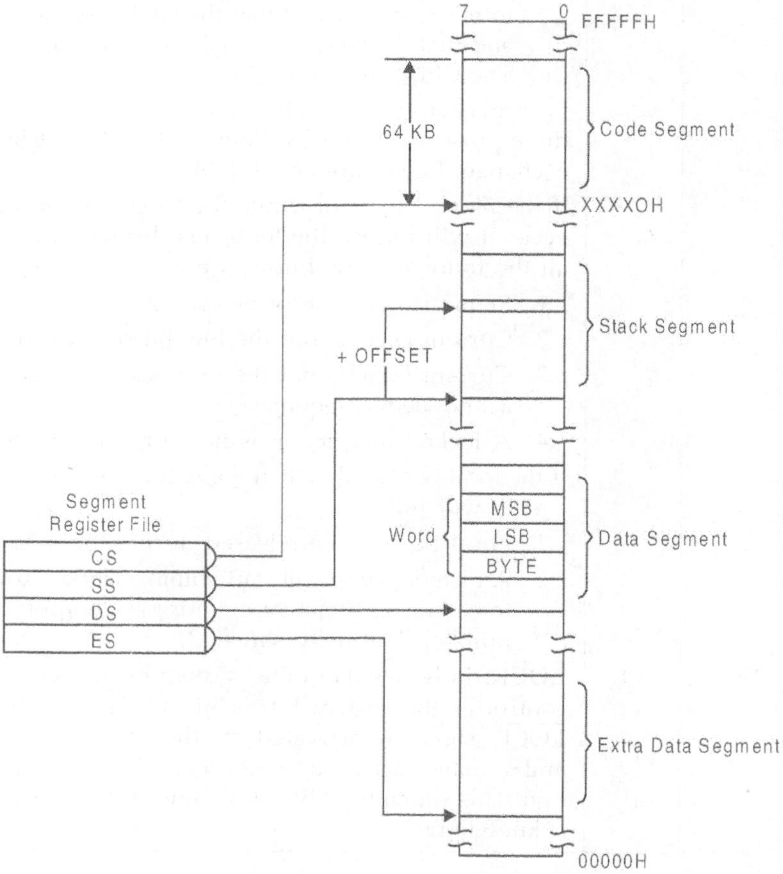

Fig. 5.15 Memory Organization

All memory references are made relative to base addresses contained in high speed segment registers. The segment types were chosen based on the addressing needs of programs. The segment register to be selected is automatically chosen according to the rules of the following table. All information in one segment type share the same logical attributes (*e.g.*, code or data). By structuring memory into relocatable areas of similar characteristics and by automatically selecting segment registers, programs are shorter, faster, and more structured.

Word (16-bit) operands can be located on even or odd address boundaries. For address and data operands, the least significant byte of the word is stored in the lower valued address location and the most significant byte in the next higher address location. The BIU will automatically execute two fetch or write cycles for 16-bit operands.

Memory Reference Used	Segment Register Used	Segment Selection Rule
Instructions	CODE(CS)	Automatic with all instruction prefetch.
Stack	STACK (SS)	All stack pushes and pops. Memory references relative to BP base register except data references.
Local Data	DATA(DS)	Data references when: relative to stack, destination of string operation, or explicity overridden.
External (Global) Data	EXTRA (ES)	Destination of string operations: Explicitly selected using a segment override.

Certain locations in memory are reserved for specific CPU operations (See Fig. 5.16). Locations from addresses FFFF0H through FFFFFH are reserved for operations including a jump to the initial system initialization routine. Following RESET, the CPU will always begin execution at location FFFF0H where the jump must be located. Locations 00000H through 003FFH are reserved for interrupt operations. Four-byte pointers consisting of a 16-bit segment address and a 16-bit offset address direct program flow to one of the 256 possible interrupt service routines. The pointer elements are assumed to have been stored at their respective places in reserved memory prior to the occurrence of interrupts.

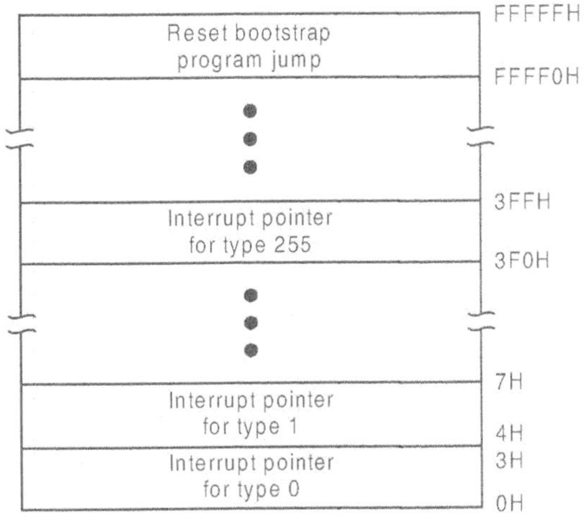

Fig. 5.16 Reserved Memory Locations

Minimum and Maximum Modes

The requirements for supporting minimum and maximum 8088 systems are sufficiently different that they cannot be done efficiently with 40 uniquely defined pins. Consequently, the 8088 is equipped with a strap pin (MN/$\overline{MX}$) which defines the system configuration.

The definition of a certain subset of the pins changes, dependent on the condition of the strap pin. When the MN/$\overline{MX}$ pin is strapped to GND, the 8088 defines pins 24 through 31 and 34 in maximum mode. When the MN/$\overline{MX}$ pin is strapped to V_{CC}, the 8088 generates bus control signals itself on pins 24 through 31 and 34.

The minimum mode 8088 can be used with either a multiplexed or demultiplexed bus. The multiplexed bus configuration is compatible with the MCS-85 multiplexed bus peripherals. This configuration (See Fig. 5.17) provides the user with a minimum chip count system. This architecture provides the 8088 processing power in a highly integrated form.

The demultiplexed mode requires one latch (for 64K addressability) or two latches (for a full megabyte of addressing). A third latch can be used for buffering if the address bus loading requires it. A transceiver can also be used if data bus buffering is required (See Fig. 5.18). The 8088 provides $\overline{DEN}$ and DT/$\overline{R}$ to control the transceiver, and ALE to latch the addresses. This configuration of the minimum mode provides the standard demultiplexed bus structure with heavy bus buffering and relaxed bus timing requirements.

The maximum mode employs the 8288 bus controller (See Fig. 5.19). The 8288 decodes status lines $\overline{S0}$, $\overline{S1}$, and $\overline{S2}$, and provides the system with all bus control signals. Moving the bus control to the 8288 provides better source and sink current capability to the control lines, and frees the 8088 pins for extended large system features. Hardware lock, queue status, and two request/grant interfaces are provided by the 8088 in maximum mode. These features allow co-processors in local bus and remote bus configurations.

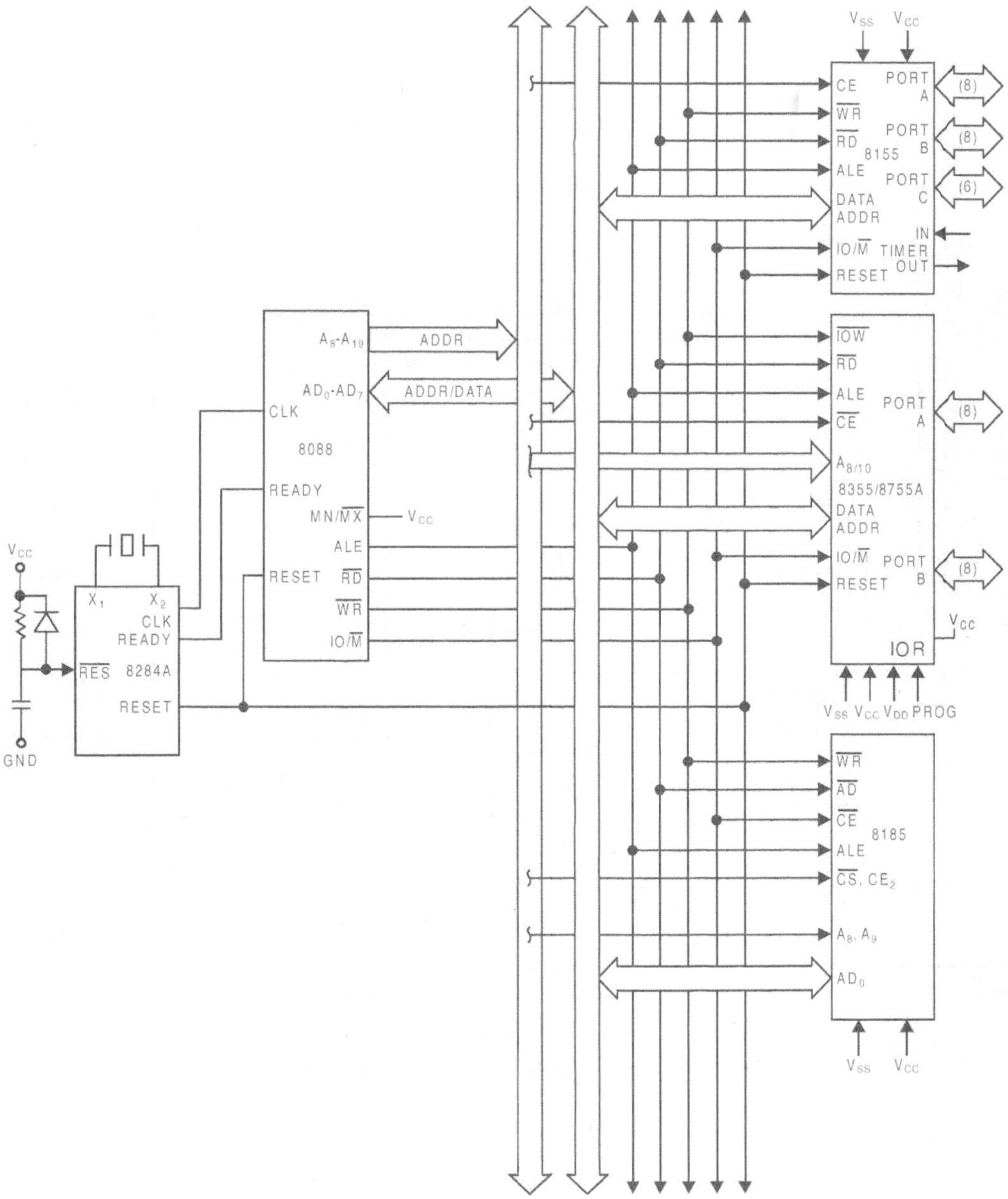

Fig. 5.17 Multiplexed Bus Configuration

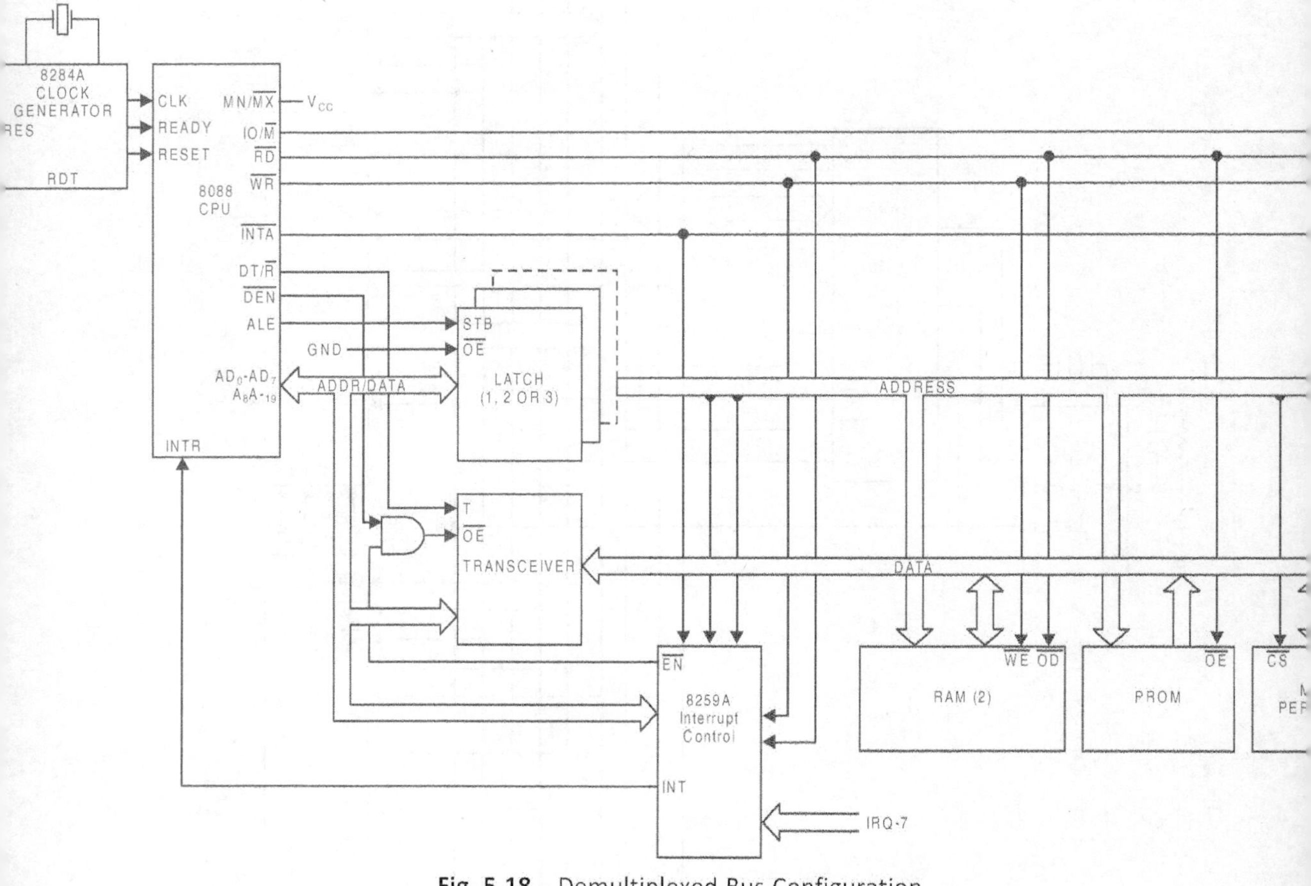

Fig. 5.18 Demultiplexed Bus Configuration

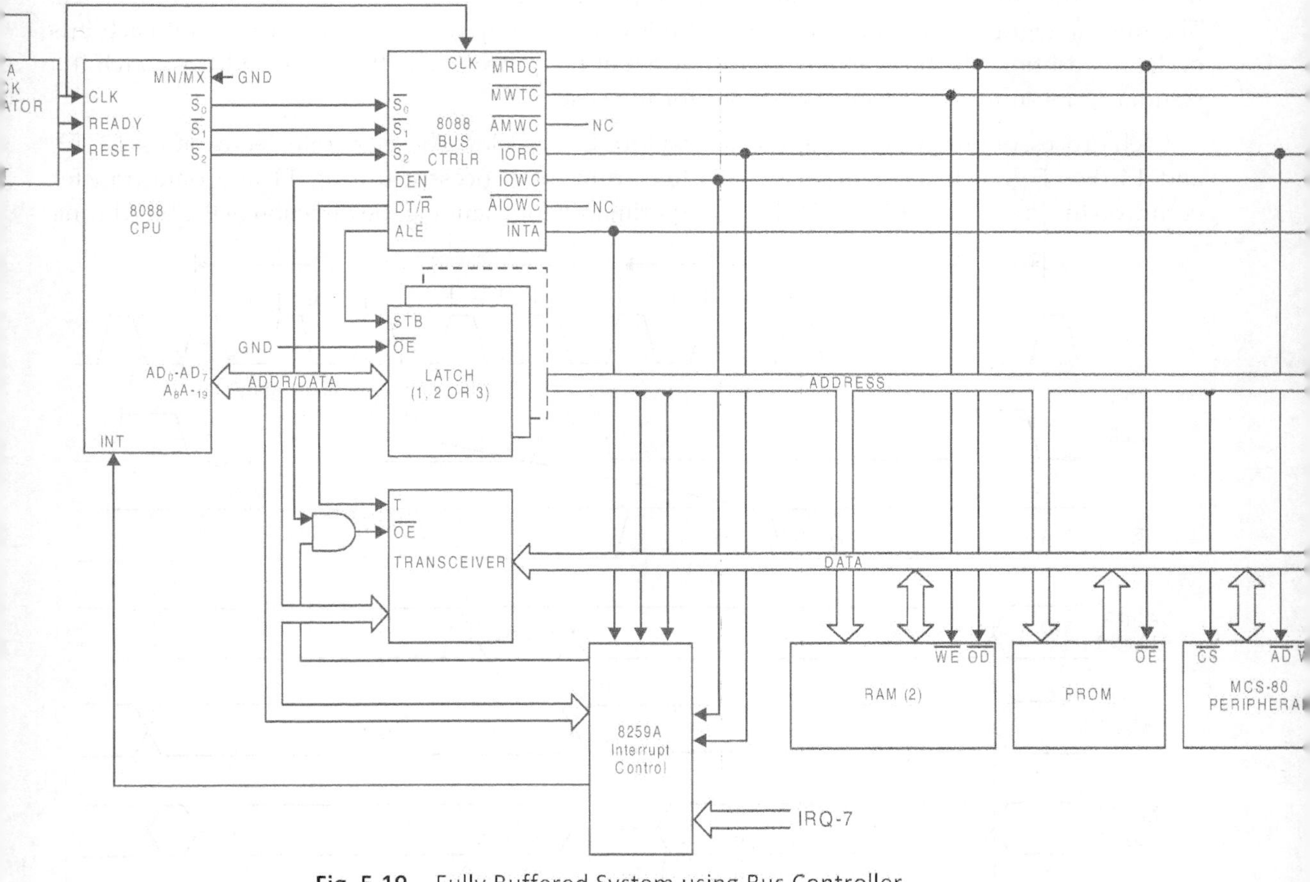

Fig. 5.19 Fully Buffered System using Bus Controller.

Bus Operation

The 8088 address/data bus is broken into three parts—the lower eight address/data bits (AD0-AD7), the middle eight address bits (A8-A15), and the upper four address bits (A16-A19). The address/data bits and the highest four address bits are time multiplexed. This technique provides the most efficient use of pins on the processor, permitting the use of a standard 40 lead package. The middle eight address bits are not multiplexed, *i.e.*, they remain valid throughout each bus cycle. In addition, the bus can be demultiplexed at the processor with a single address latch if a standard, non-multiplexed bus is desired for the system.

Each processor bus cycle consists of at least four CLK cycles. These are referred to as T1, T2, T3, and T4 (See Fig. 5.20). The address is emitted from the processor during T1 and data transfer occurs on the bus during T3 and T4. T2 is used primarily for changing the direction of the bus during

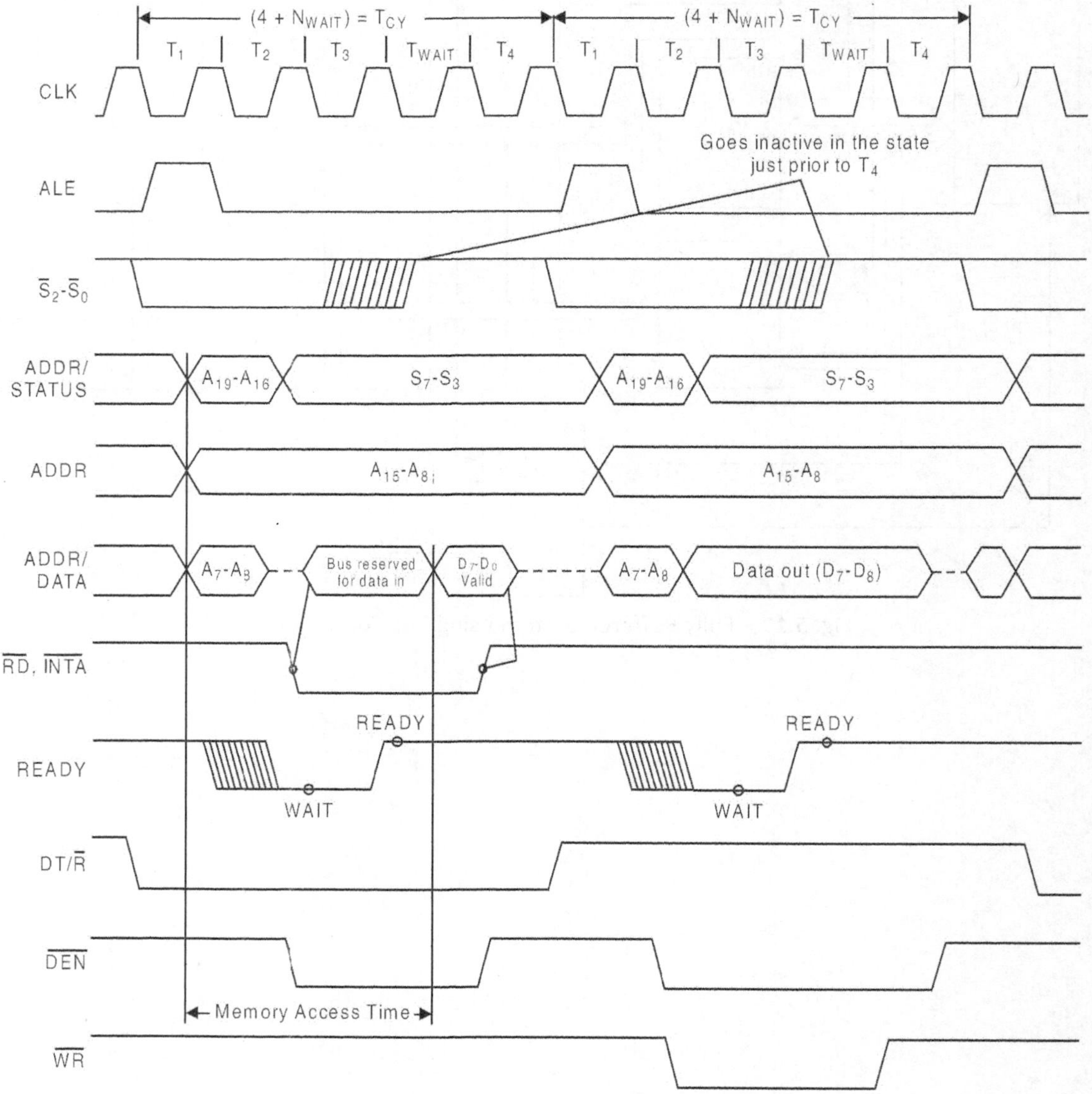

Fig. 5.20 Basic System Timing

read operations. In the event that a "NOT READY" indication is given by the addressed device, "wait" states (Tw) are inserted between T3 and T4. Each inserted "wait" state is of the same duration as a CLK cycle. Periods can occur between 8088 driven bus cycles. These are referred to as "idle" states (Ti), or inactive CLK cycles. The processor uses these cycles for internal housekeeping.

During T1 of any bus cycle, the ALE (address latch enable) signal is emitted (by either the processor or the 8288 bus controller, depending on the MN/$\overline{MX}$ strap). At the trailing edge of this pulse, a valid address and certain status information for the cycle may be latched.

Status bits $\overline{S0}$, $\overline{S1}$, and $\overline{S2}$ are used by the bus controller, in maximum mode, to identify the type of bus transaction according to the following table:

$\overline{S2}$	$\overline{S1}$	$\overline{S0}$	Characteristics
0(LOW)	0	0	Interrupt Acknowledge
0	0	1	Read I/O
0	1	0	Write I/O
0	1	1	Halt
1(HIGH)	0	0	Instruction Fetch
1	0	1	Read Data from Memory
1	1	0	Write Data to Memory
1	1	1	Passive (No Bus Cycle)

Status bits S3 through S6 are multiplexed with high order address bits and are therefore valid during T2 through T4. S3 and S4 indicate which segment register was used for this bus cycle in forming the address according to the following table:

S4	S3	Characteristics
0(LOW)	0	Alternate Data (Extra Segment)
0	1	Stack
1(HIGH)	0	Code or None
1	1	Data

S5 is a reflection of the PSW interrupt enable bit S6 is always equal to 0.

I/O Addressing

In the 8088, I/O operations can address up to a maximum of 64K I/O registers. The I/O address appears in the same format as the memory address on bus lines A15-A0. The address lines A19-A16 are zero in I/O operations. The variable I/O instructions, which use register DX as a pointer, have full address capability, while the direct I/O instructions directly address one or two of the 256 I/O byte locations in page 0 of the I/O address space. I/O ports are addressed in the same manner as memory locations.

Designers familiar with the 8085 or upgrading an 8085 design should note that the 8085 addresses I/O with an 8-bit address on both halves of the 16-bit address bus. The 8088 uses a full 16-bit address on its lower 16 address lines.

5.9.3 External Interface

Processor Reset and Initialization

Processor initialization or start up is accomplished with activation (HIGH) of the RESET pin. The 8088 RESET is required to be HIGH for greater than four clock cycles. The 8088 will terminate operations on the high-going edge of RESET and will remain dormant as long as RESET is HIGH. The low-going transition of RESET triggers an internal reset sequence for approximately 7 clock cycles. After this interval the 8088 operates normally, beginning with the instruction in absolute locations FFFF0H. The RESET input is internally synchronized to the processor clock. At initialization, the HIGH to LOW transition of RESET must occur no sooner than 50 μs after power up, to allow complete initialization of the 8088.

NMI asserted prior to the 2nd clock after the end of RESET will not be honored. If NMI is asserted after that point and during the internal reset sequence, the processor may execute one instruction before responding to the interrupt. A hold request active immediately after RESET will be honored before the first instruction fetch.

All 3-state outputs float to 3-state OFF during RESET. Status is active in the idle state for the first clock after RESET becomes active and then floats to 3-state OFF. ALE and HLDA are driven low.

Interrupt Operations

Interrupt operations fall into two classes: software or hardware initiated. The software initiated interrupts and software aspects of hardware interrupts are specified in the instruction set description. Hardware interrupts can be classified as **nonmaskable** or **maskable**.

Interrupts result in a transfer of control to a new program location. A 256 element table containing address pointers to the interrupt service program locations resides in absolute locations 0 through 3FFH (See Fig. 5.16), which are reserved for this purpose. Each element in the table is 4 bytes in size and corresponds to an interrupt "type." An interrupting device supplies an 8-bit type number, during the interrupt acknowledge sequence, which is used to vector through the appropriate element to the new interrupt service program location.

Non-Maskable Interrupt (NMI)

The processor provides a single non-maskable interrupt (NMI) pin which has higher priority than the maskable interrupt request (INTR) pin. A typical use would be to activate a power failure routine. The NMI is edge-triggered on a LOW to HIGH transition. The activation of this pin causes a type 2 interrupt.

NMI is required to have a duration in the HIGH state of greater than two clock cycles, but is not required to be synchronized to the clock. Any higher going transition of NMI is latched on-chip and will be serviced at the end of the current instruction or between whole moves (2 bytes in the case of word moves) of a block type instruction. Worst case response to NMI would be for multiply, divide, and variable shift instructions. There is no specification on the occurrence of the low-going edge; it may occur before, during, or after the servicing of NMI. Another high-going edge triggers another response if it occurs after the start of the NMI procedure. The signal must be free of logical spikes in general and be free of bounces on the low-going edge to avoid triggering extraneous responses.

Maskable Interrupt (INTR)

The 8088 provides a single interrupt request input (INTR) which can be masked internally by software with the resetting of the interrupt enable (IF) flag bit. The interrupt request signal is level triggered. It is internally synchronized during each clock cycle on the high-going edge of CLK. To be responded to, INTR must be present (HIGH) during the clock period preceding the end of the current instruction or the end of a whole move for a block type instruction. During interrupt response sequence, further interrupts are disabled. The enable bit is reset as part of the response to any interrupt (INTR, NMI, software interrupt, or single step), although the FLAGS register which is automatically pushed onto the stack reflects the state of the processor prior to the interrupt. Until the old FLAGS register is restored, the enable bit will be zero unless specifically set by an instruction.

During the response sequence (See Fig. 5.21), the processor executes two successive (back to back) interrupt acknowledge cycles. The 8088 emits the LOCK signal (maximum mode only) from T2 of the first bus cycle until T2 of the second. A local bus "hold" request will not be honored until the end of the second bus cycle. In the second bus cycle, a byte is fetched from the external interrupt system (e.g., 8259A PIC) which identifies the source (type) of the interrupt. This byte is multiplied by four and used as a pointer into the interrupt vector lookup table. An INTR signal left HIGH will be continually responded to within the limitations of the enable bit and sample period. The interrupt return instruction includes a flags pop which returns the status of the original interrupt enable bit when it restores the flags.

HALT

When a software HALT instruction is executed, the processor indicates that it is entering the HALT state in one of two ways, depending upon which mode is strapped. In minimum mode, the processor issues ALE, delayed by one clock cycle, to allow the system to latch the halt status. Halt status is available on IO/$\overline{\text{M}}$, DT/$\overline{\text{R}}$, and $\overline{\text{SSO}}$. In maximum mode, the processor issues appropriate HALT status on $\overline{\text{S2}}$, $\overline{\text{S1}}$, and $\overline{\text{S0}}$, and the 8288 bus controller issues one ALE. The 8088 will not leave the HALT state when a local bus hold is entered while in HALT. In this case, the processor reissues the HALT indicator at the end of the local bus hold. An interrupt request or RESET will force the 8088 out of the HALT state.

Read/Modify/Write Operations via LOCK

The LOCK status information is provided by the processor when consecutive bus cycles are required during the execution of an instruction. This allows the processor to perform read/modify/write operations on memory (via the "exchange register with memory" instruction), without another system bus master receiving intervening memory cycles. This is useful in multiprocessor system configurations to ac-complish "test and set lock" operations. The $\overline{\text{LOCK}}$ signal is activated (LOW) in the clock cycle following decoding of the LOCK prefix instruction. It is deactivated at the end of the last bus cycle of the instruction following the LOCK prefix While $\overline{\text{LOCK}}$ is active, a request on a $\overline{\text{RQ}}/\overline{\text{GT}}$ pin will be recorded, and then honored at the end of the LOCK.

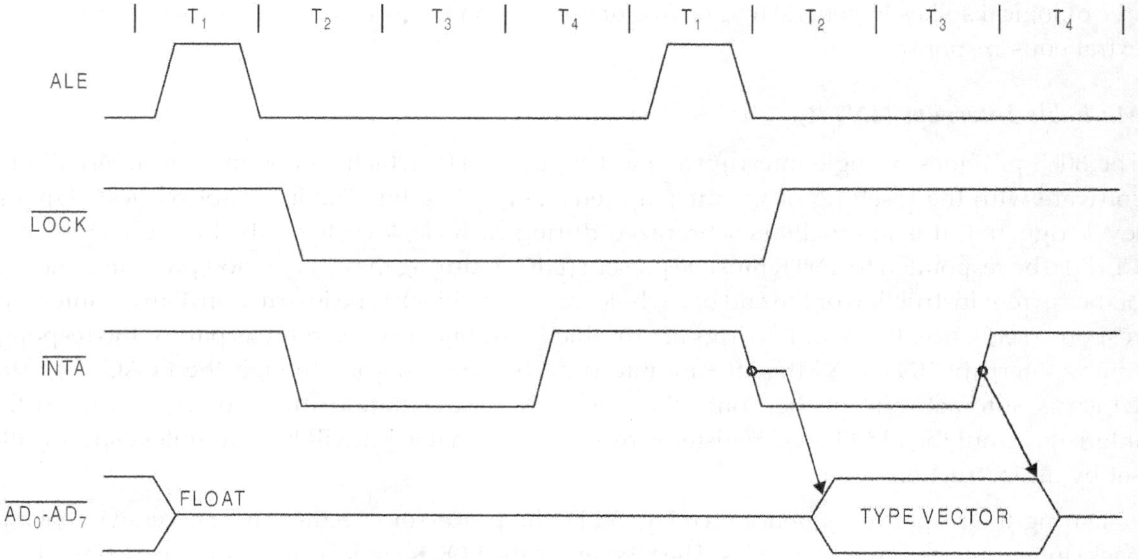

Fig. 5.21 Interrupt Acknowledge Sequence

External Synchronization via TEST

As an alternative to interrupts, the 8088 provides a single software-testable input pin ($\overline{TEST}$). This input is utilized by executing a WAIT instruction. The single WAIT instruction is repeatedly executed until the $\overline{TEST}$ input goes active (LOW). The execution of WAIT does not consume bus cycles once the queue is full.

If a local bus request occurs during WAIT execution, the 8088 3-states all output drivers. If interrupts are enabled, the 8088 will recognize interrupts and process them. The WAIT instruction is then refetched, and reexecuted.

5.9.4 Basic System Timing

In minimum mode, the MN/$\overline{MX}$ pin is strapped to V_{CC} and the processor emits bus control signals compatible with the 8085 bus structure. In maximum mode, the MN/$\overline{MX}$ pin is strapped to GND and the processor emits coded status information which the 8288 bus controller uses to generate MULTIBUS compatible bus control signals.

System Timing—Minimum System

(See Fig. 5.20)

The read cycle begins in T1 with the assertion of the address latch enable (ALE) signal. The trailing (low going) edge of this signal is used to latch the address information, which is valid on the address/data bus (AD0-AD7) at this time, into the 8282/8283 latch. Address lines A8 through A15 do not need to be latched because they remain valid throughout the bus cycle. From T1 to T4 the IO/$\overline{M}$ signal indicates a memory or I/O operation. At T2 the address is removed from the address/data bus and the bus goes to a high impedance state. The read control signal is also asserted at T2. The read ($\overline{RD}$) signal causes the addressed device to enable its data bus drivers to the local bus. Some time later, valid data will be available on the bus and the addressed device will drive the READY line HIGH. When the processor returns the read signal to a HIGH level, the addressed device will again 3-state its bus drivers. If a transceiver is required to buffer the 8088 local bus, signals DT/$\overline{\text{R}}$ and $\overline{\text{DEN}}$ are provided by the 8088.

A write cycle also begins with the assertion of ALE and the emission of the address. The IO/$\overline{\text{M}}$ signal is again asserted to indicate a memory or I/O write operation. In T2, immediately following the address emission, the processor emits the data to be written into the addressed location. This data remains valid until at least the middle of T4. During T2, T3, and Tw, the processor asserts the write control signal. The write ($\overline{\text{WR}}$) signal becomes active at the beginning of T2, as opposed to the read, which is delayed somewhat into T2 to provide time for the bus to float.

The basic difference between the interrupt acknowledge cycle and a read cycle is that the interrupt acknowledge ($\overline{\text{INTA}}$) signal is asserted in place of the read ($\overline{\text{RD}}$) signal and the address bus is floated. (See Fig. 5.21) In the second of two successive $\overline{\text{INTA}}$ cycles, a byte of information is read from the data bus, as supplied by the interrupt system logic (*i.e.*, 8259A priority interrupt controller). This byte identifies the source (type) of the interrupt. It is multiplied by four and used as a pointer into the interrupt vector lookup table.

Bus Timing—Medium Complexity Systems

(See Fig. 5.22)

For medium complexity systems, the MN/$\overline{\text{MX}}$ pin is connected to GND and the 8288 bus controller is added to the system, as well as a latch for latching the system address, and a transceiver to allow for bus loading greater than the 8088 is capable of handling. Signals ALE, $\overline{\text{DEN}}$, and DT/$\overline{\text{R}}$ are generated by the 8288 instead of the processor in this configuration, although their timing remains relatively the same. The 8088 status outputs ($\overline{\text{S2}}$, $\overline{\text{S1}}$, and $\overline{\text{S0}}$) provide type of cycle information and become 8288 inputs. This bus cycle information specifies read (code, data, or I/O), write (data or I/O), interrupt acknowledge, or software halt. The 8288 thus issues control signals specifying memory read or write, I/O read or write, or interrupt acknowledge. The 8288 provides two types of write strobes, normal and advanced, to be applied as required. The normal write strobes have data valid at the leading edge of write. The advanced write strobes have the same timing as read strobes, and hence, data is not valid at the leading edge of write. The transceiver receives the usual T and $\overline{\text{OE}}$ inputs from the 8288's DT/$\overline{\text{R}}$ and $\overline{\text{DEN}}$ outputs.

The pointer into the interrupt vector table, which is passed during the second $\overline{\text{INTA}}$ cycle, can derive from an 8259A located on either the local bus or the system bus. If the master 8289A priority interrupt controller is positioned on the local bus, a TTL gate is required to disable the transceiver when reading from the master 8259A during the interrupt acknowledge sequence and software "poll".

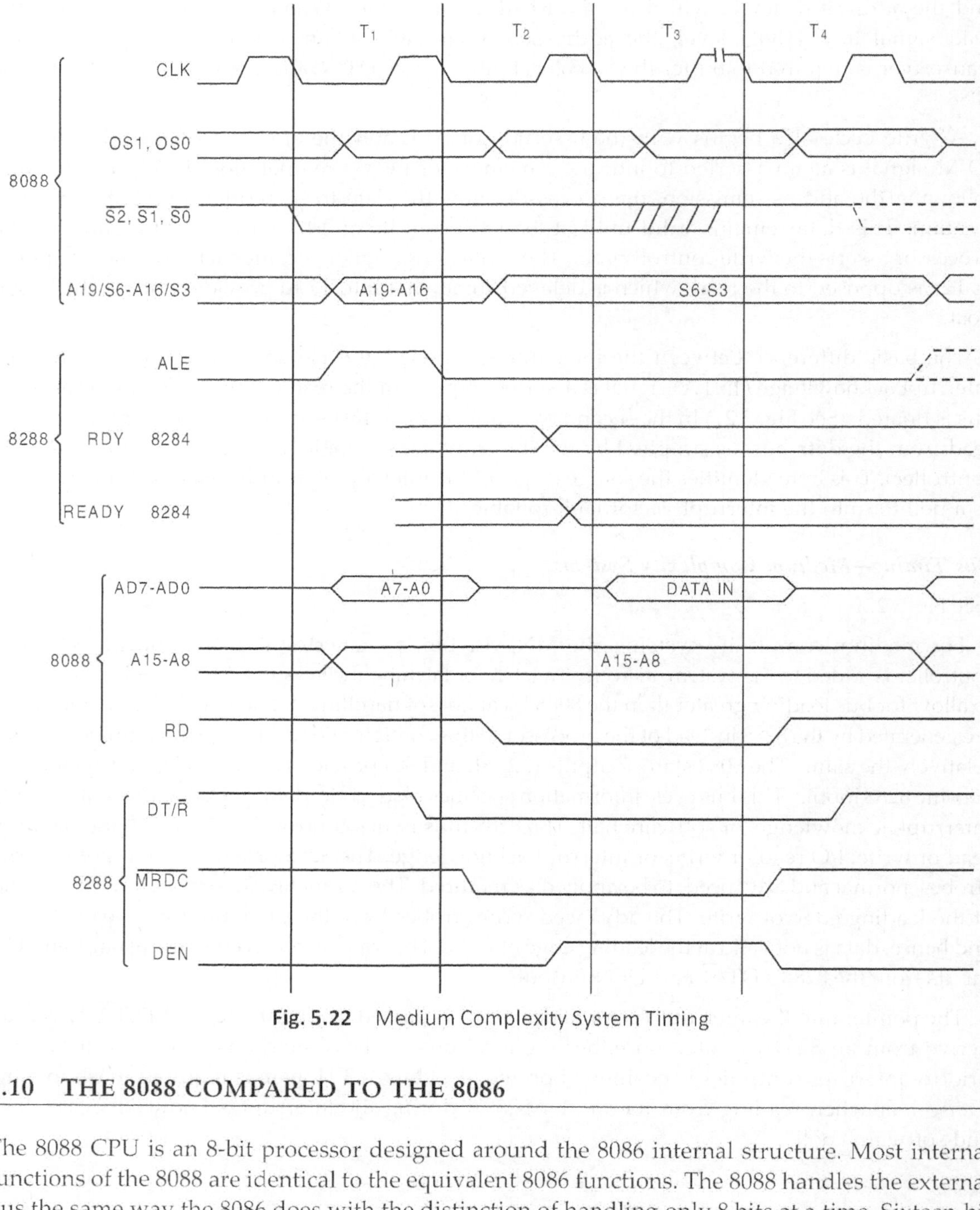

Fig. 5.22 Medium Complexity System Timing

5.10 THE 8088 COMPARED TO THE 8086

The 8088 CPU is an 8-bit processor designed around the 8086 internal structure. Most internal functions of the 8088 are identical to the equivalent 8086 functions. The 8088 handles the external bus the same way the 8086 does with the distinction of handling only 8 bits at a time. Sixteen-bit operands are fetched or written in two consecutive bus cycles. Both processors will appear identical to the software engineer, with the exception of execution time. The internal register structure is identical and all instructions have the same end result The differences between the 8088 and 8086 are outlined below. Internally, there are three differences between the 8088 and the 8086. All changes are related to the 8-bit bus interface.

- The queue length is 4-bytes in the 8088, whereas the 8086 queue contains 6 bytes, or three words. The queue was shortened to prevent overuse of the bus by the BIU when prefetching instructions. This was required because of the additional time necessary to fetch instructions 8 bits at a time.
- To further optimize the queue, the prefetching algorithm was changed. The 8088 BIU will fetch a new instruction to load into the queue each time there is a 1-byte hole (space available) in the queue. The 8086 waits until a 2-byte space is available.
- The internal execution time of the instruction set is affected by the 8-bit interface. All 16-bit fetches and writes from/to memory take an additional four clock cycles. The CPU is also limited by the speed of instruction fetches. This latter problem only occurs when a series of simple operations occur. When the more sophisticated instructions of the 8088 are being used, the queue has time to fill and the execution proceeds as fast as the execution unit will allow.

The 8088 and 8086 are completely software compatible by virtue of their identical execution units. Software that is system dependent may not be completely transferable, but software that is not system dependent will operate equally as well on an 8088 and an 8086.

The hardware interface of the 8088 contains the major differences between the two CPUs. The pin assignments are nearly identical, however, with the following functional changes:

- A8-A15—These pins are only address outputs on the 8088. These address lines are latched internally and remain valid throughout a bus cycle in a manner similar to the 8085 upper address lines.
- BHE has no meaning on the 8088 and has been eliminated.
- $\overline{SSO}$ provides the $\overline{SO}$ status information in the minimum mode. This output occurs on pin 34 in minimum mode only. DT/$\overline{R}$, IO/$\overline{M}$, and $\overline{SSO}$ provide the complete bus status in minimum mode.
- IO/$\overline{M}$ has been inverted to be compatible with the MSC-85 bus structure.
- ALE is delayed by one clock cycle in the minimum mode when entering HALT, to allow the status to be latched with ALE.

5.10.1 8086 and 8088 Microprocessors at a Glance

S.No.	8086	8088
1.	8086 announced in 1978; 8086 is a 16-bit microprocessor with a 16-bit data bus	8088 announced in 1979; 8088 is a 16-bit micro-processor with an 8-bit data bus
2.	Using High-performance Metal Oxide Semiconductor (HMOS) technology	Using High performance Metal Oxide Semi-conductor (HMOS) technology
3.	Contain about 29000 transistors	Contain about 29000 transistors
4.	Are packaged in 40 pin dual-in-line package (DIP)	Are packaged in 40 pin dual-in-line package (DIP)
5.	Address lines A0-A15 and Data lines D0-D15 are multiplexed in 8086. These lines are labelled as AD0-AD15.	Address lines A0-A7 and Data lines D0-D7 are multiplexed in 8088. These lines are labelled as AD0-AD7.

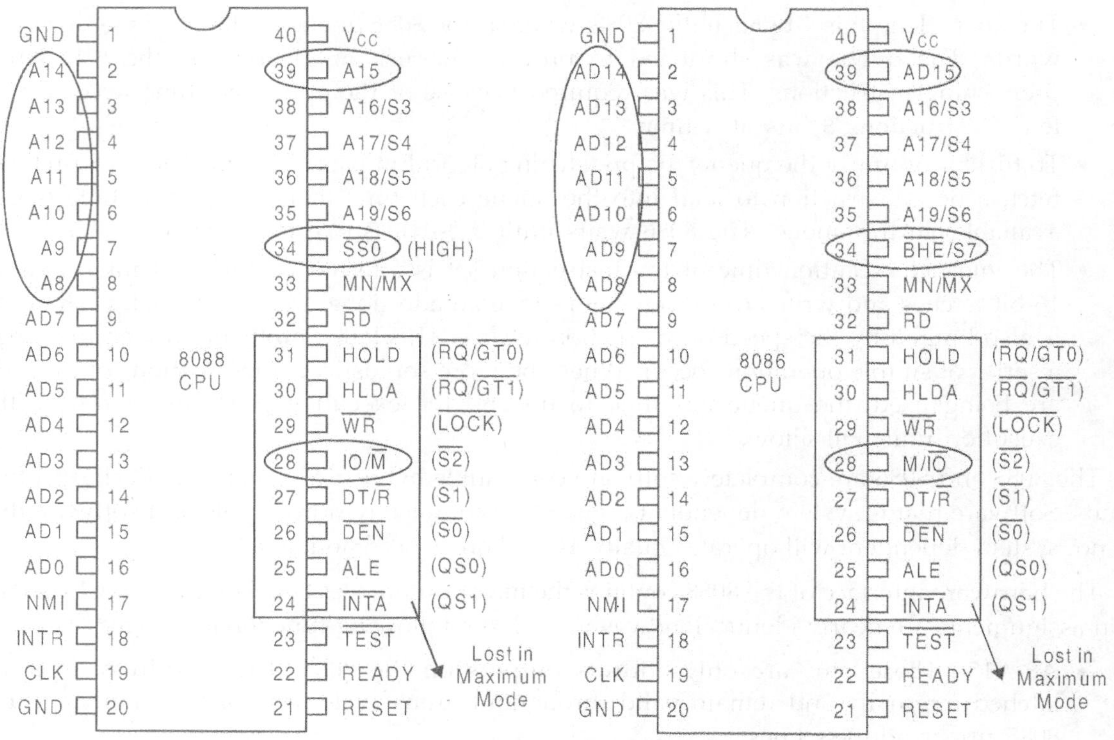

Fig. 5.23 8088 and 8086 Microprocessors

5.11 ENCODING OF ASSEMBLY INSTRUCTION

The encoding of 8086 assembly instruction to machine instruction is necessary because in any kit of 8086, we put machine instruction in stead of instruction itself. Encoding of 8086, instruction is not so simple. Each type of instruction has a fixed template form. To make the machine code, fill up bits according to addressing mode, Bits/words (data type) etc. These we will disscuss some examples that will help you to make machine codes for other instructions.

5.11.1 Template for Fixed Port Instruction

Fig. 5.24 represents the template for 8086 IN instruction. This template is used to build codes for fixed port. As ,represented, two bytes are required for instruction.

1. The upper seven bits of the first byte represents to the processor that is 'IN' instruction. This is fixed and taken from table instruction set of 8086.
2. The bit lebeled W indicates to the processor whether input is byte 'or' word. If input is a byte from an 8-bit port, then take W bit 0. If the input is a word from a 16-bit, port then take W bit 1.
3. The 8-bit fixed port address is to be entered in the second byte of instruction template.
4. As it is totally 16 bit template, so we put first byte in the first memory location, and second byte in the second memory location.

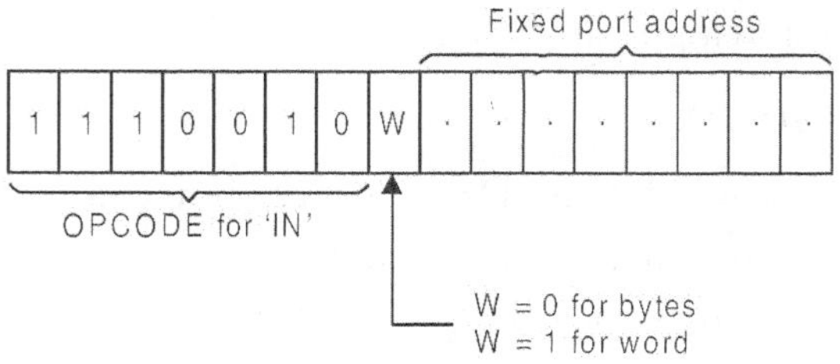

Fig. 5.24 Template for IN

Now let on assembly language instruction IN AL, 07H is to be encoded.

The machine code will be build up as follows:

1. First '7' bits are fixed for 'IN' instruction.
2. The input is type byte from an 8-bit port to AL register, hence W bit is to be mode 0.
3. The 8-bit port address 07H 'or' 0000 0111 binary is to be put in the second byte of instruction.

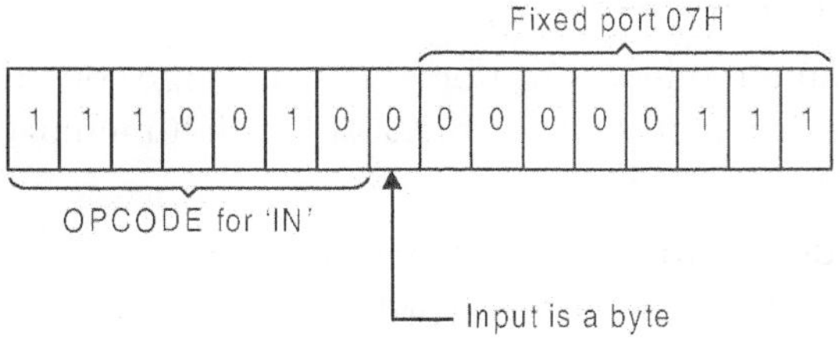

Fig. 5.25 Template for IN AL, 07H

4. So finally we have machine code = E407H.

5.11.2 Template for Data Transfer MOV Instruction

Fig. 5.26 presents the template for 8086 data transfer MOV instruction. This template can be atleast two 'or' to maximum four bytes. This template can be used for data transfer from a register to register, from a register to a memory, 'or' from a memory location to a register.

1. The upper 6-bits of the first byte are reserved for op-code. The 6-bit opcode for the MOV register/memory to/from register instruction is 100010 (that is taken from table of instruction set).
2. The D-bit in the first byte of the template indicates whether the DATA is being moved to the register identified in the REG field of the second byte 'or' from that register. If data is to be moved to the register identified in REG field, then make the D-bit a 1. If the data is to be moved from that register, then make D-bit a '0'.
3. The W-bit in the first byte indicates whether a byte 'or' word is to be moved. If byte is to be moved make 'W' bit 0, if word is to be moved make 'W' bit 1.
4. As in a MOV instruction one operand always should be a register, the 3-bit REG field in the second byte is used to indicate which register is involved. The 3-bit codes of register to be used for REG field are represented in table.
5. In the second byte of the instruction template, the 2-bit field labled MOD and the 3-bit field labled R/M specify the desired addressing mode for the other operand. For this purpose we may use the table 5.10 and 5.11.

For example; consider following instructions.

1. Machine Code for MOV SP, DX

So we have to find machine code for MOV SP, DX that will copy a word from DX register to SP register.

1. The 6-bits op-code for this instruction is 100010.
2. We are moving a word, then W = 1.
3. Since both the operands are registers, data is moving to SP register, then D-bit is 1.
4. SP register is involved, then REG field should be 100.
5. MOD bits are to be put up 11, because both operands are registers.
6. R/M bits are to be put up 010, represents other operand as register DX.

The resultant code for instruction MOV SP, DX will be

100010 1111100010

2. Machine Code for MOV AX, [483AH]

This is a direct addressing mode instruction, and copies the contents of two memory locations in to AX. The effective address 'or' displacement of the first memory location from the start of the data segment is 483AH.

1. The 6-bit op-code for this instruction is 100010.
2. The D-bit will be 1, as moving a word.
3. The W-bit also will be 1, as moving a word.
4. The register involved is AX, the REG field will be 000.
5. The MOD bits will be 00, because the other operand is a direct memory address.
6. R/M bits for mode 00 and 16-bit displacement will be 110.
7. Now at byte 3, we will put low byte of address [483AH] that is 3AH (in binary 00111010). The byte 4, we will have high byte of address [483AH], that is 48H [in binary 01001000].
8. Thus at last we have complete machine code
   = (8B 06 3A 48)H.

That will be written in four successive memory locations.

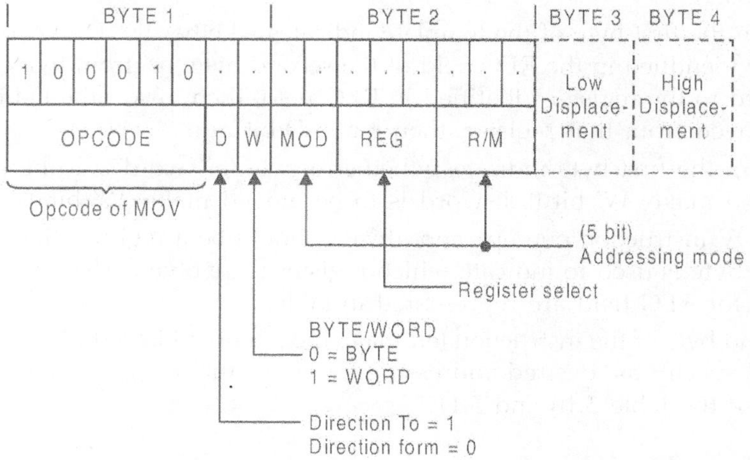

Fig. 5.26 Template for MOV

Table 5.10 Register codes for 8086.

8-Bit register (W = 0)	16-Bit register (W = 1)	Register Code
AL	AX	000
CL	CX	001
DL	DX	010
BL	BX	011
AH	SP	100
CH	BP	101
DH	SI	110
BH	DI	111

Table 5.11 The MOD and R/M bit pattern for 8086.

R/M \ MOD	00	01	10	11	
				W = 0	W = 1
000	(BX) + (SI)	(BX) + (SI) + d8	(BX) + (SI) + d16	AL	AX
001	(BX) + (DI)	(BX) + (DI) + d8	(BX) + (DI) + d16	CL	CX
010	(BP) + (SI)	(BP) + (SI) + d8	(BP) + (SI) + d16	DL	DX
011	(BP) + (DI)	(BP) + (DI) + d8	(BP) + (DI) + d16	BL	BX
100	(SI)	(SI) + d8	(SI) + d16	AH	SP
101	(DI)	(DI) + d8	(DI) + d16	CH	BP
110	d16 (direct address)	(BP) + d8	(BP) + d16	DH	SI
111	(BX)	(BX) + d8	(BX) + d16	BH	DI
	Memory Mode			Register Mode	

d8 = 8-Bit displacement
d16 = 16-Bit displacement

5.12 ASSEMBLER DIRECTIVES

Assembler Directives are defined in form of words and these are directions to the assembler. These are not the instructions for 8086. These directives are used to write and execute programes either turbo-assembler (TASM) 'or' macro-assembler (MASM). In addition to instructions, programmes contains assembler directives which are direction to assembler, called assembler directives for pesudo-instructions. These are various directives used for 8086 asembly language programming.

Segment and End Directives

These directives are used to identify a group of data instructions that to be put together in a particular segment. A group of data items 'or' group of instruction statements contains items between SEGMENT and ENDS directive is called a logical segment. A logical segment is not usually given a physical starting address when it is declared. After the program is assembled and perhaps linked with other assembler modules, it is then assigned the physical address where it will be loaded in memory to run.

EQU Directive

In general, program works with three general categories of data; constants, variables and address. The EQU named equate is used to assign names to constants used in program. The advantages of using EQU is to declare constants at the start of program. Suppose one constant is used at number of places in a program, then it is not required to write constants repeatedly at all the places. If required to change this constant then we need not to make change at number of places but with its uses we can change only at one place when the constant is defined by EQU directive.

DB, DW, DD, DQ and DT Directives

These directives are used to assign names to variables in program as per their length. The DB directive specifies the data is of type byte and stands for define type.

DW—Define Word

The DW directive is used to tell the assembler to define a variable of type word 'or' to reserve storage location of type word in memory location.

DD—Define Double Word

The DD directive is used to declare a variable of type double word 'or' to reverse memory location which can be accessed as type double word 'or' four bytes.

DQ—Define Quad Word

The DQ directive tells the assembler to declare a variable four words in length 'or' to reserve four words of storage in memory.

DT— Define Ten Bytes

The DT directive tells the assembler to define a variable which is ten bytes in length 'or' to reserve ten bytes of storage in memory.

Assume Directive

An 8086, program may have several logical segments that contain codes and several that contains data. At a given time 8086, works directly only with 4 physical segment. The assume directive tells the assembler which logical segment to use for each of these physcial segment at a given time.

Suppose in a program, we define a group of datas by DATA logical segment and codes are defined by CODE logical segment. These logical segment can be defined in terms of physical segment using the ASSUME directive are represented below:

ASSUME DS: DATA, CS: CODE

END—End Program

This directive is put after the last statement of a program to tell the assembler that this is end of program module. After the END directive the assembler will ignore any statement made in the program, the END directive should be used at the end of program modulo only.

PROC—Procedure

The PROC stands for procedure in the directive used to identify the start of a procedure. The PROC directive follows a name given to the procedure. After the PROC directive the term near 'or' the term far is used to specify the type of the procedure. The PROC directive proceeds an ENDP directive.

ENDP-End Procedure

The ENDP directive is used along with the name of procedure to indicate the end of a procedure to the assembler. This directive together with the PROC directive is used to 'bracket' a procedure. A 'PROC' directive is always followed by a 'ENDP' directive.

ENDS-End Segment

The ENDS directive is used with the name of a segment to indicate the end of that logical segment. The ENDS directive is used with the 'SEGMENT' directive to 'bracket' a logical segment containing instructions 'or' data. For example

CODE SEGMENT	:	Start of logical segment containing code.
	:	Instruction statements.
CODE ENDS	:	End of segment named CODE.

PUBLIC-Public

The large programs are usually divided into several small program module. Each module is indivisually assembled, tested and debugged. If all the modules are working perfectly, these object code files are linked together to form the complete program. Inorder for the program module to link together perfectly, any variable name 'or' lable refered to in other modules must be declared public in the module in which it is defined. The 'PUBLIC' directive is used to tell the assembler that a specified named 'or' lable will be accessed from other module.

EXTRN-External

This directive is used to tell the assembler that the names 'or' lables following the directive are in some other assembly module.

GLOBAL

This directive can be used in place of a PUBLIC directive 'or' in place of an EXTRN directive

EVEN-Align on Even Memory Address

This directive tells the assembler to increment the location counter to next even address if it is not already at even address. The 8086 can read a word from memory in one bus cycle if the word is at even address. When the word starts at an odd address, the 8086 must do two bus cycles to get the two bytes of the word. Therefore, a series of words can be read much more quickly if they are at even addresses. If 'EVEN' is used in a data segment, the location counter will simply be incremented to next even address in necessary.

GROUP

This directive is used to tell the assembler to group the logical segment named after directive into one logical group statement.

INCLUDE

The 'INCLUDE' directive is used to tell the assembler to insert a block of source code from the named fill in to the current source module. This shortens the source code.

LABEL

The 'LABEL' directive is used to give a name to the current valve in the location counter. This directive must be followed by a term which specifies the type you want associated with that name. If the label is going to be used as the destination for a jump 'or' a call, then the label must be specified as type near 'or' type far.

LENGTH

'LENGTH' directive is an operator which tells the assembler to determine the number of elements in some named data item, such as a string 'or' an array.

NAME

The directive is used to give a specific name to each assembly module when program consisting of several modules.

OFFSET

The 'OFFSET' is an operator which tells the assembler to determine the offset 'or' displacement of a named data item (variable) 'or' procedure from the start of the segment which contains it. This operator is usually used to load the offset of a variable is to a register so that the variable can be accessed with one of the indexed addressing modes.

PTR-POINTER

The 'PTR' is an operator and used to assign a specific type to a variable 'or' to a label. It is necessary to do this in any instruction where the type of the operand is not clear.

ORG-Originate

This directive allows to set the location counter to a desired valve at any point in the program.

SHORT

'SHORT' is an operator, is used to tell the assembler that only one byte displacement is needed to code a jump instruction. If the jump destination in the program, the assembler will automatically reverse two bytes for the displacement.

TYPE

The 'TYPE' is an operator tells the assembler to determine the type of a specified variable. The assembler actually determines the number of types in the type of the variable. For a byte-type variable the assembler will give a valve of 1. For a word-type variable the assembler will give a value of 2.

5.13 PROGRAMMING OF 8086

Now we will study of 8086 programming. It also has the simple and straight forward steps as 8085 programming. It will include only the declaration of segments and then write a program. Let us start with some simple programes and going to some harder programes.

```
; Write a program to multiply the two 16-bit words in the memory location
; called Multiplicand and Multiplier. The result is stored
; in the memory location, product.
; REGISTERS: USES CS, DS, AX, DX
; PORTS : None used
            DATA-HERE  SEGMENT
                       MULTIPLICAND DW 1048N; First word here.
                       MULTIPLIER DW 382AH; Second word here.
                       PRODUCT DW 2 DUP(0); Result
DATA-HERE ENDS
CODE-HERE SEGMENT
            ASSUME CS  :  CODE-HERE
                   DS  :  DATA-HERE
START:      MOV AX, DATA-HERE       ; Intiliaze DS register
            MOV DS, AX
            MOV AX, Multiplicand    ; Get one word
            MUL MULTIPLIER          ; Multiply with second
            MOV PRODUCT, AX         ; Store LSB of Result
            MOV PRODUCT + 2, DX     ; Store MSB of Result
            INT 3                   ; Wait for command
CODE-HERE ENDS
            END.START
```

```
; Write a program of adding two 16-bit Numbers in the memory location
; called FIRST-NO. and SECOND-NO. The result should be stored in a memory
; location called ADDITION.
; REGISTERS  :  Uses CS, DS, AX, DX
; PORTS      :  None used
DATA_SEG SEGMENT
     FIRST-NO.    DW 2122H          ; First No. storage
     SECOND-NO.   DW 2468H          ; Second No. storage
     ADDITION     DW?               ; Store Addition here
```

```
DATA-SEG    ENDS.
CODE-SEG    SEGMENT
            ASSUME CS : CODE-SEG, DS : DATA_SEG.
START:      MOV AX, DATA-SEG,        ; Initialize DS register
            MOV DS, AX
            MOV AX, FIRST-NO         ; Get-First No.
            ADD AX, SECOND-NO.       ; Add this No. to second No.
            MOV ADDITION, AX         ; Store Addition in Memory
CODE-SEG ENDS
                 END.START
```

; Write 8086 program for multiplication of two 16-bit numbers in the memory

; locations called MULTIPLICAND and MULTIPLIER. The result is 32-bit

; and is to be stored in memory location called RESULT.

```
; REGISTERS  : Uses CS, DS, AX, DX
; PORTS      : None used
DATA-SEG SEGMENT
        MULTIPLICAND    DW 2864H    ; First word storage
        MULTIPLIER      DW 6235H    ; Second word storage
        RESULT          DD?         ; Store Result
DATA-SEG    ENDS.
CODE-SEG    SEGMENT
        ASSUME CS : CODE_SEG, DS : DATA_SEG.
START:      MOV AX, DATA-SEG              ; Initialize DS register
            MOV DS, AX
            MOV AX, MULTIPLICAND          ; Get-First Word
            MOV DX, MULTIPLIER            ; Multiplier in DX
            MUL AX, DX                    ; Multiply the No.
            MOV RESULT, AX                ; Store low word of result
            MOV RESULT +2, DX             ; Store high word of result
CODE_SEG ENDS                             ; End of code segment
            END.START                     ; End of program
```

; Write a program to take average of two Number named ONE-NO. and SECOND-

; NO. and put the result in the memory location.

; AV-NO.

```
;REGISTER    : Uses DS, CS, AX, BL
;PORTS       : None used
DATA SEGMENT
        ONE-NO.        DB 90H       ; First No. storage
        SECOND-NO.     DB 54H       ; Second No. storage
        AV-NO.         DB?          ; Store Average here
DATA     ENDS.
CODE   SEGMENT
        ASSUME CS : CODE, DS : DATA
START:      MOV AX, DATA            ; Intilialize data segment
            MOV DS, AX
            MOV AL, ONE-NO.         ; Get First No.
            ADD AL, SECOND-NO.      ; Add Second No. to it
            MOV AH, 00H             ; Clear all the data of AH
            ADC AH, 00H             ; Put carry in LSB of AH
            MOV BL, 02H             ; Load Divisor to BL.
            DIV BL                  ; Divide AX by BL Quotient in AL and
                                    ; Reminder in AH
            MOV AV-NO. AL           ; Copy result to memory
CODE        ENDS
            END.START
```

; Write an 8086 program to sort in ascending order using bubble sort algorithm
; a given set of byte sized unsigned numbers in memory. The sorted
; elements should replace the original unsorted elements in memory.

```
;NAME BUBBLE SORT
;PAGE 60,80
;TITLE ASCENDING ORDER USING BUBBLE SORT
 MODEL SMALL
 .STACK 64
  .DATA
A DB 34H,78H,56H,47H
SI_ZE DW $-A   ;SI_ZE=NO. OF ELEMENTS
  .CODE
BUBSORT:
```

```
        MOV AX,@DATA
        MOV DS,AX
        MOV BX,SI_ZE
        DEC BX                  ;BX=no of passes needed to complete sorting(n-1) outlup:
        MOV CX,BX               ;CX=no of comparisons to be performed in a pass
        MOV SI,0
INLUP:
        MOV AL,A[SI]
        INC SI
        CMP AL,A[SI]
        JB GO_ON
        XCHG AL,A[SI]
        MOV A[SI-1],AL
GO_ON:
LOOP INLUP; DEC CX, UNTIL CX=0
        DEC BX
        JNZ OUTLUP
        INT 3                   ; BREAKPOINT INTERRUPT
        ALIGN 16
END BUBSORT
```

; Write an 8086 program to sort in descending order, using selection sort
; algorithm a given set of 8-bit unsigned numbers in memory. The sorted
; elements should replace the original unsorted elements in memory. Store
; in a memory location the number of comparisons made.

```
        NAME SELECTIONSORT
        PAGE 60,80
TITLE DESCENDING ORDER USING SELECTION SORT
        .MODEL SMALL
        .STACK 64
        .DATA
A DB 44H,11H,22H,66H
SI_ZE DW $-A                ;SI_ZE=4
NC DW ?                     ;TOTAL. NO. OF. COMPARISIONS. MADE
        .CODE
```

```
SELSORT:
        MOV AX,@DATA
        MOV DS,AX
        MOV DX,SI_ZE            ;in selsort for n elements to sort we need n-1 passes
        DEC DX                  ;DX=3 no of passes required
        MOV NC,0
OUTLUP:
        MOV CX,DX               ;CX=no of comparisons to be performed in a pass
        MOV SI,0
        MOV AH,A[SI]
        MOV BX,SI
INLUP:
        INC SI
        INC NC
        CMP AH,A[SI]
        JB GO_ON
        MOV AH,A[SI]            ;AH=smallest element in the vector
        MOV BX,SI               ;BX=position of the smallest element
GO_ON:
        LOOP INLUP              ;until CX=0
        XCHG AH,A[SI]           ;Exchange the last element pointed by SI,with the AH
        MOV A[BX],AH            ;smallest element pointed by BX
        DEC DX
        JNZ OUTLUP
        INT 3                   ;breakpoint  interrupt
        ALIGN 16
END SELSORT
```

; Write an 8086 program to sort in ascending order using Insertion Sort
; algorithm, a given set of 16-bit unsigned numbers in memory. The sorted
; elements should replace the original unsorted elements in memory.

```
        NAME INSERTIONSORT
        PAGE 60,80
TITLE ASCENDING ORDER USING INSERTION SORT ALGORITHM
        .MODEL SMALL
```

```
        .STACK 64
        .DATA
A DW 78H,34H,12H,56H
SI_ZE DW ($-A)/2                ;SI_ZE=4(NO OF ELEMENTS)
        .CODE
INSORT:
        MOV AX,@DATA
        MOV DS,AX
        MOV CX,2                ;CX=2,insert the second element in the proper position
OUTLUP:
        MOV DX,CX
        DEC DX                  ;DX=CX-1,max no of comparisons needed to insert element
        MOV SI,DX
        ADD SI,SI
        MOV AX,A[SI]
INLUP:
        CMP A[SI-2],AX
        JBE INLUPEXIT
        MOV DI,A[SI-2]
        MOV A[SI],DI
        DEC SI
        DEC SI
        DEC DX
        JNZ INLUP
INLUPEXIT:
        MOV A[SI],AX
        INC CX                  ;INC CX to insert the next element in proper position
        CMP CX,SI_ZE
        JBE OUTLUP
EXIT:
        INT 3                   ;breakpoint interrupt
        ALIGN 16
END INSORT
; Add/Sub of multiword
```

```
        NAME ADDSUB
        PAGE 60,80
TITLE 8086 ALP FOR MULTI WORD ADDITION/SUBTRACTION
        .MODEL SMALL
        .STACK 64
        .DATA
N1 DB 12H,34H,56H,78H,9AH,0BCH,0DEH,0F0H
N2 DB 0BCH,12H,78H,34H,56H,0DEH,0F0H,9AH
S DB 8 DUP(?)
D DB 8 DUP(?)
        .CODE
ADDSUB:
        MOV AX,@DATA
        MOV DS,AX
        MOV CX,8                ;CX is used as loop counter
        MOV BX,7                ;BX contains the offset address of byte in n1 & n2
        CLC
ADDAGN:
        MOV AL,N1[BX]
        ADC AL,N2[BX]
        MOV S[BX],AL            ;STORE SUM IN S[BX]
        DEC BX
        LOOP ADDAGN
        MOV CX, 8
        MOV BX, 7
        CLC
SUBAGN:
        MOV AL, N1 [BX]
        SBB AL, N2 [BX]
        MOV D [BX], AL          ; store difference
        DEC BX
        LOOP SUBAGN
        INT 3
        ALIGN 16
END ADDSUB
```

```
; Write an 8086 alp to get the screen width (no of cols) using BIOS
; interrupt, and calculate the no of rows from the appropriate word
; location in BIOS data area, and clear the screen using BIOS interrupt.
        NAME CLEARSCREEN1
        PAGE 60, 80
TITLE CLEAR SCREEN USING BIOS INTERRUPT
        .MODEL SMALL
        .STACK 64
        .DATA
BYTES DD 0040004CH
ROWS DB?
COLS DB?
MSG1 DB 0DH,0AH,'TOTAL NO OF ROWS(IN HEX)=','$'
MSG2 DB 0DH,0AH,'TOTAL NO OF COLUMNS(IN HEX)=','$'
MSG3 DB 0DH,0AH,'PRESS ANY KEY TO CLEAR SCREEN','$'
HEXCODE DB '0123456789ABCDEF'
        .CODE
DISPLAY PROC
        PUSH AX
        PUSH BX
        PUSH CX
        PUSH DX
        LEA DX,MSG1             ;displays msg1
        MOV AH,09H
        INT 21H
        MOV AL,ROWS             ;AL=no of rows[25]
        MOV CL,10H
        MOV AH,00H
        DIV CL                  ;25/10 | AL=quotient[2] | ah=remainder[5]
        MOV BL,AL               ;AL=2
        MOV DL,HEXCODE[BX]      ;DL=8-bit ASCII code of char to be displayed[2]
        PUSH AX
        MOV AH,02H              ;display char to std o/p dev
        INT 21H
```

```
        POP AX
        MOV BL,AH               ;AH=5
        MOV DL,HEXCODE[BX]      ;DL=8-bit ASCII code of char to be displayed[5]
        MOV AH,02H              ;display char to std o/p dev
        INT 21H
        LEA DX,MSG2             ;displays msg2
        MOV AH,09H
        INT 21H
        MOV AL,COLS             ;AL=no of cols[80]
        MOV CL,10H
        MOV AH,00H
        MOV BH,00H
        DIV CL                  ;80/10 | AL=quotient[8] | ah=remainder[0]
        MOV BL,AL               ;AL=8
        MOV DL,HEXCODE[BX]      ;DL=8-bit ASCII code of char to be displayed[8]
        PUSH AX
        MOV AH,02H              ;display char to std o/p dev
        INT 21H
        POP AX
        MOV BL,AH               ;AH=0
        MOV DL,HEXCODE[BX]      ;DL=8-bit ASCII code of char to be displayed[0]
        MOV AH,02H              ;display char to std o/p dev
        INT 21H
        POP DX
        POP CX
        POP BX
        POP AX
        RET
DISPLAY ENDP                    ;end display procedure
MAIN:
        MOV AX,@DATA
        MOV DS,AX
        MOV AH,0FH
        INT 10H
```

```
        MOV COLS,AH              ;AH=no of char cols on screen
        MOV CL,AH
        MOV CH,0
        PUSH DS
        LDS SI,BYTES
        MOV AX,[SI]              ;AX=total no of bytes on video page(1b ASCII+1b ab)
        POP DS
        SHR AX,1                 ;divides AX by 2 to get total no of chars on page
        DIV CL
        MOV ROWS,AL
        CALL DISPLAY
        LEA DX,MSG3
        MOV AH,09H               ;displays msg3
        INT 21H
        MOV AH,01H               ;I/P char from std I/p dev & echo to std o/p dev
        INT 21H
        MOV DH,0                 ;initialize row coordinate to 0 again:
        MOV BH,0                 ;BH=PAGE 0
        MOV DL,0                 ;initialize column coordinate to 0
        MOV AH,02H               ;set cursor position to (DL,DH)
        INT 10H
        MOV BL,0                 ;colour (set foregnd & bckgnd of char with same colour)
        MOV AL,'X'               ;char to be displayed
        MOV AH,09H               ;write char at cursor position
        INT 10H
        INC DH                   ;INC row coordinate by one position
        CMP DH,ROWS
        JB AGAIN
        MOV AH,4CH ;EXIT
        INT 21H
END MAIN

; GCD of 4 unsigned 16-bit numbers
        NAME GCD
        PAGE 60,80
```

```
TITLE PROGRAM TO FIND GCD OF 4 UNSIGNED 16-BITS NUMBERS
        .MODEL SMALL
        .STACK 64
        .DATA
VALUES DW 0090,0120,4BH,0019H
GCD DW ?
        .CODE
HCF PROC
AGAIN:
        CMP AX,BX
        JE EXIT
        JB BIGBX
DIVAXBX:
        MOV DX,0
        DIV BX
        CMP DX,0
        JE EXIT
        MOV AX,DX
        JMP AGAIN
BIGBX:
        XCHG AX,BX
        JMP DIVAXBX
EXIT:
        MOV GCD,BX
        RET
HCF ENDP
;MAIN PROGRAM
GCD4:
        MOV AX,@DATA
        MOV DS,AX
        MOV AX,VALUES
        MOV BX,VALUES+2
        CALL HCF
        MOV AX,GCD
```

```
        MOV BX,VALUES+4
        CALL HCF
        MOV AX,GCD
        MOV BX,VALUES+6
        CALL HCF
        INT 3
        ALIGN 16
END GCD4
; LCM of 2 16-bit unsigned numbers
        NAME LCM
        PAGE 60,80
TITLE PROGRAM TO FIND LCM OF 2 16-BIT UNSIGNED NUMBERS
        .MODEL SMALL
        .STACK 64
        .DATA
VALUES DW 0025,0015
LCM DW 2 DUP(?)
        .CODE
L_C_M:
        MOV AX,@DATA
        MOV DS,AX
        MOV DX,0
        MOV AX,VALUES ;DX_AX=25
        MOV BX,VALUES+2 ;BX=15
AGAIN:
        PUSH AX
        PUSH DX
        DIV BX
        CMP DX,0                ;remainder of the division is stored in DX
        JE EXIT
        POP DX
        POP AX
        ADD AX,VALUES
        JNC NOINCDX
```

```
        INC DX
NOINCDX:
        JMP AGAIN
EXIT:
        POP LCM+2
        POP LCM
        INT 3
        ALIGN 16
END L_C_M
```

; Write an 8086 program to search for a given 8-bit value using linear search
; in an array of 8 bit numbers. Message should be displayed on CRT
; indicating whether the search was a failure or a success. If it is a
; success case, the position of the element in the array is to be displayed

```
        NAME LINEARSEARCH
        PAGE 60,80
TITLE LINEAR SEARCH PROGRAM
        .MODEL SMALL
        .STACK 64
        .DATA
ARRAY DB 55H,33H,44H,66H,22H
LEN DW $-ARRAY ;LENGTH=5
SCRKEY EQU 33H
ASC1 EQU (SCRKEY/10H)+'0' ;ASC1='3'
ASC2 EQU (SCRKEY MOD 10H)+'0' ;ASC2='5'
SUCMSG DB 'ELEMENT ',ASC1,ASC2,' FOUND AT POSITION:'
RESULT DB ?,0CH,0AH,'$'
FAILMSG DB 'ELEMENT ',ASC1,ASC2,' NOT FOUND',0CH,0AH,'$'
        .CODE
LIN:
        MOV AX,@DATA
        MOV DS,AX
        MOV ES,AX
        CLD                    ;direction flag D=0
        MOV DI,0               ;DI=0,autoincrement DI
```

```
        MOV AL,SCRKEY              ;AL=35
        MOV CX,LEN
        REPNE SCASB                ;SCASB==>[AL]-[[DI]]
        JZ SUCCESS
        LEA DX,FAILMSG
        JMP DISPLAY
SUCCESS:
        MOV BX,DI                  ;BX=position of the scrkey found in array
        ADD BL,'0'                 ;convert this position into ASCII for display purpose
        MOV RESULT,BL
        LEA DX,SUCMSG
DISPLAY:
        MOV AH,09H
        INT 21H
        MOV AH,4CH
        INT 21H
        ALIGN 16
END LIN
```

; Write an 8086 alp to search for a given 16-bit value using binary search
; in an array of 16-bit numbers, which are in ascending order. Message
; should be displayed on CRT indicating whether the search was a failure
; or a success. If it is a success case, the position of the element in the
; array is to be displayed.

```
        NAME BINARYSEARCH
        PAGE 60,80
TITLE BINARY SEARCH PROGRAM TO SEARCH A 16-BIT VALUE
        .MODEL SMALL
        .STACK 64
        .DATA
CR EQU 13
LF EQU 10
ARRAY DW 1122H,2345H,3344H,4455H,5566H
LEN DW ($-ARRAY)/2 ;LENGTH=5
SCRKEY EQU 2345H
```

```
ASC1 EQU (SCRKEY/1000H)+'0'      ;ASC1='2'
ASC2 EQU (SCRKEY/100H) MOD 10H + '0'          ;ASC2='3'
ASC3 EQU (SCRKEY/10H) MOD 10H " '0'           ;ASC3='4'
ASC4 EQU (SCRKEY MOD 10H) + '0'               ;ASC4='5'
SUCMSG DB 'GIVEN ELEMENT'
       DB 'FOUND AT POSITION:'
RESULT DB ?,CR,LF,'$'
FAILMSG DB 'GIVEN ELEMENT'
       DB 'NOT FOUND',CR,LF,'$'
       .CODE
BINSCR:
       MOV AX,@DATA
       MOV DS,AX
       MOV BX,1
       MOV DX,LEN              ;DX=5
       MOV CX,SCRKEY           ;CX=2345
AGAIN:
       CMP BX,DX
       JA FAILURE
       MOV AX,BX
       ADD AX,DX
       SHR AX,1
       MOV SI,AX
       DEC SI
       ADD SI,SI
       CMP CX,ARRAY[SI]
       JAE BIGER
       DEC AX
       MOV DX,AX
       JMP AGAIN
BIGER:
       JE SUCCESS
       INC AX
       MOV BX,AX
```

```
        JMP AGAIN
SUCCESS:
        ADD AL,'0'
        MOV RESULT,AL
        LEA DX,SUCMSG
        JMP DISPLAY
FAILURE:
        LEA DX,FAILMSG
DISPLAY:
        MOV AH,09H
        INT 21H
QUIT:
        MOV AH,4CH
        INT 21H
        ALIGN 16
END BINSCR
```

; Using BIOS routine, write an 8086 alp to find memory size of the PC you
; are using. Using appropriate message, the display should indicate memory
; size in Kilo bytes using 4 hex digits. Also check the result with the
; appropriate word in BIOS data area using debug/code view.

```
        NAME MEMORYSIZE
        PAGE 60, 80
TITLE PROGRAM TO FIND MEMORY SIZE USING INT 12H
        .MODEL SMALL
        .STACK 64
        .DATA
MSG DB 'MEMORY SIZE IN KILO BYTES='
ASCRES DB 4 DUP (?),'HEX', 0CH, 0AH,'$'
RES DW?
HEXCODE DB '0123456789ABCDEF'
        .CODE
HEX_ASC PROC
        MOV DL,10H
        MOV AH,0
```

```
        MOV BX,0
        DIV DL
        MOV BL,AL
        MOV DH,HEXCODE[BX]
        MOV BL,AH
        MOV DL,HEXCODE[BX]
        RET
HEX_ASC ENDP
MAIN:
        MOV AX,@DATA
        MOV DS,AX
        INT 12H
        MOV RES,AX
        MOV AL,BYTE PTR RES
        CALL HEX_ASC
        MOV ASCRES+2,DH
        MOV ASCRES+3,DL
        MOV AL,BYTE PTR RES+1
        CALL HEX_ASC
        MOV ASCRES,DH
        MOV ASCRES+1,DL
        MOV DX,OFFSET MSG
        MOV AH,09H
        INT 21H
        MOV AH,4CH
        INT 21H
        ALIGN 16
        END MAIN
```

; Write an 8086 alp to multiply two 3x3 matrices of signed 8-bit
; integers. Display result using DEBUG or CODEVIEW.Assume that each
; of the elements of the product matrix can be stored in 8-bits.

```
        NAME MATRIXMUL
        PAGE 60,80
TITLE 8086 ALP FOR MATRIX MULTIPLICATION OF 3X3 MATRICES
```

```
        .MODEL SMALL
        .STACK 64
        .DATA
AR1 DB 2,2,2                    ;ROW1 OF ARRAY A
AR2 DB 2,2,2                    ;ROW2 ,,
AR3 DB 2,2,2                    ;ROW3 ,,
BC1 DB 1,1,1                    ;COLUMN1 OF ARRAY B
BC2 DB 1,1,1                    ;COLUMN2 ,,
BC3 DB 1,1,1                    ;COLUMN3 ,,
C DB 9 DUP(?)                   ;RESULT MATRIX
L2 DB?
L1 DB?
        .CODE
MAIN:
        MOV AX,@DATA
        MOV DS,AX
        MOV ES,AX
        MOV BP,0
        MOV L2,3
        LEA SI,AR1
REP2:
        LEA DI,BC1
        MOV L1,3
REP1:
        CALL MATMUL
        MOV DS:C[BP],DL
        INC BP
        ADD DI,3
        DEC L1
        JNZ REP1
        ADD SI,3
        DEC L2
        JNZ REP2
        INT 3
```

```
MATMUL PROC
        MOV CX, 3
        MOV BX, 0
        MOV DL, 0
AGAIN:
        MOV AL, [SI] [BX]
        IMUL BYTE PTR [DI] [BX]
        ADD DL, AL
        INC BX
        LOOP AGAIN
        RET
MATMUL ENDP
        ALIGN 16
END MAIN
```

; Write an 8086 alp to compute ^nCr, given n and r, using recursion.

; Display result using DEBUG.

```
        NAME nCr
        PAGE 60, 80
TITLE COMPUTATION OF NCR USING RECURSION
        .MODEL SMALL
        .STACK 64
        .DATA
N DB 4
R DB 2
RES DB?
        .CODE
MAIN:
        MOV AX,@DATA
        MOV DS, AX
        MOV AL, N
        MOV BL, R
        CALL NCR
        INT 3
```

```
NCR PROC
        CMP AL,BL          ;IF N=R THEN NCR=1
        JE P8
        CMP BL,0           ;IF R=0 THEN NCR=1
        JE P8
        CMP BL,1           ;IF R=1 THEN NCR=N
        JE P10
        DEC AL             ;N-1
        CMP BL,AL          ;IF R=N-1 THEN NCR=N
        JE P9
        PUSH AX            ;(N-1)CR
        PUSH BX            ;
        CALL NCR           ;
        POP BX
        POP AX
        DEC BL             ;
        PUSH AX            ;
        PUSH BX            ;
        CALL NCR           ;(N-1)C(R-1)
        POP BX
        POP AX
        RET
P8:     INC RES
        RET
P9:     INC RES
P10:    ADD RES,AL
        RET
        ALIGN 16
NCR ENDP
END MAIN
```

5.14 PROCEDURES

A series of instructions written separately to carry out a specific subtask in a main program in called subprogramme 'or' procedure. This is also called subroutine.

The main purpose of procedure in when we have to execute a small programme ('or' at part of main programe) again and again. So to write that part of programmes again and again, in main programme will waste the memory space and also the execution time of programmes are increased. Sometimes, procedures are written to break a large programes in to small modules. Doing this make easy to write, test and debugg a program. A 'CALL' instruction in the main program calls the procedure and then procedure is executed. At the end of procedure a RET, return instruction, sends execution back to the next instruction after the CALL in the main program.

The details of CALL and RET instruction are given below:

5.14.1 CALL Instruction

The 'CALL' instruction cells the procedures in the main program. When 8086 CALL instruction executes, it performs two operations.

1. It stores the address of the instruction after the CALL instruction on the stack. This is return address to which execution return after the procedure executes. If the procedure is a near procedure, then only the instruction pointer content will be saved on the stack. If the procedure is far, then both the instruction pointer and the code segment register contents will be saved on the stack.
2. Another operation of the call instruction is to change the contents of instruction pointer (IP) and, sometimes, the code-segment (CS) register to contain the starting address of the procedure.

5.14.2 RET Instruction

RET instruction return the execution back to the main program.

1. In case of near procedure, CALL instruction saves the instruction pointer valve for the instruction after the CALL on the stack. The RET at the end of the procedure copies this valve from the stack back to the instruction pointer. This way execution returns from procedure to the main program.
2. In the case of far procedure, the RET instruction at the end of procedure copies the contents of the IP and CS register from stack back to the IP and CS register respectively to return execution to the main program.

5.14.3 Far Procedure

A 'far procedure' is one that is located in a segment which has a different name from the segment containing the CALL Instruction. It should be clear during the description of CALL and RET instruction. It should be clear that to get the starting address of a far procedure, the 8086 must change the content of both the code segment register and instruction pointer.

```
CODE        SEGMENT
            ASSUME CS : DS:DATA
                   SS : STACK-SEG
                      :
                   CALL MULTIPLY-32
```

```
                CODE ENDS
PROCEDURES SEGMENT
     MULTIPLY-32 PROC FAR
                ASSUME CS : PROCEDURES
                                :
                 MULTIPLY : 32 ENDP
              PROCEDURES : ENDS
```

So it should also be cleared that during the returning from procedure to main program, the content of code segment register and instruction pointer must be poped off the stack to return to the calling program (that was also disscussed in RET instruction).

If we are using assembler to assemble a program containing a far procedure, these are a few additional directives you have to give the assembler.

Now we will discuss some cases of far procedure.

Accessing a Procedure in Another Segment

Suppose that in program, we want to put all of the main line program in one logical segment and we want to put several procedure in other logical segment to keep them separate to the main line program.Program discussed above is example of this. Here our main line instructions are written in a segment named CODE. A procedure called MULTIPLY-32 is a segment named PROCEDURES.

As we see that when assembler will get a line.

MULTIPLY-32 PROCFAR it will automoatically finds that procedure called is a FAR procedure and it will apply intersegment CALL and intersegment RET instructions.

Thus we use 'FAR' directive for this purpose.

Now remaining thing is that the program gets assembled correctly. To make sure that the assembler is using right code segment for each part of program, we use ASSUME Directive. At the starting of main line program, we use

ASSUME CS : CODE,

total assembler to complete the offsets of the following instructions from the segment base, named CODE. At the start of procedure, we use

ASSUME CS : PROCEDURES.

For same purpose.

5.14.4 Nested Procedure

When a procedure calls another procedure with in itself, this is called nesting of procedures. In NESTED Procedure, first of all one procedure is called by CALL Instruction then again another procedure is called by CALL Instruction. Now, a RET instruction at the end of the lower level procedure returns execution to the higher level procedure. Another RET instruction at the end of the higher level procedure returns execution to the main program.

5.15 THE STACK

Like 8085, in 8086 also, stack is a part of memory set aside for storage of return addresses and also to save the content of register for the main program during the execution of a procedure. The stack is also used to hold the data 'or' addresses that will be acted upon by a procedure.

The stack segment (SS) register and stack pointer (SP) are used in associated with the stack. The SS register contains the base location of the stack, and the stack pointer (SP) contains the address of the top of the stack (where the valves are pushed by PUSH Instruction). Any time we are using stack, we have to initialize stack pointer (SP) and stack segment (SS) registers, that will be shown by following program.

Program Fragment for Stack

```
; Program fragment for stack and initialization of SS.
; Program fragment is for 8086 processor.
        STACK-SEG   SEGMENTSTACK     ; Stack segment start
               DB   30   DUP (0)     ; Set 30 byte locations
        STACK-SEG   ENDS             ; Stack segment Ends.
         CODE-SEG   SEGMENT          ; Code segment start
ASSUME CS : CODE-SEG, SS : STACK-SEG
          MOV AX,   STACK-SEG        ; Initialize stack segment register
        MOV SS, AX                   ; Start of program
          :     :                      :
          :     :                      :
         CODE-SEG   ENDS             ; Code segment Ends.
                    END              ; End of program
```

The logical segment defined by STACK-SEG SEGMENT STACK and STACK-SEG ENDS is used as stack. The STACK directive is used to indicate the assembler that this segment will be used as last in, First-out stack. As we know that we have stack segment of 64K bytes in 8086; But we donot use 64K byte in all the programes; But use according to our requirement. Here the DB 30 DUP(0) statement set aside 30 decimal bytes. In the ASSUME statement, the term SS : STACK-SEG tells the assembler that any reference in the program to the stack means the segment STACK-SEG. This terms tells the assembler that SS will contain the start address of STACK-SEG, but does not intilialize SS with this valve. The instructions MOV AX, STACK-SEG, and MOV SS, AX intialize the SS register with upper 16-bit of starting address of stack segment.

5.16 PROC DIRECTIVE

The PROC directive declares the begining of a procedure. The PROC directive is proceeded by name of the procedure and followed by type of procedure is NEAR 'or' FAR. In a program as the PROC is used to start a procedure, the ENDP directive is used to end the procedure. The format of a program segment using PROC and ENDP directive will be as follows.

```
CODE-SEG        SEGMENT
                ASSUME CS : CODE-SEG, DS : DATA-SEG
                       SS : STACK-SEG
                MOV     AX,DATA-SEG  ; INTIALIZATION
                 :        :          :
                 :        :          :
                DELAY   PROC NEAR    ; Start of a near procedure
                 :        :            named delay
                 :        :          :
                DELAY   ENDP         ; End of procedure
                CODE-SEG ENDS.
```

Similarly, if the procedure is FAR the statement DELAY PROC FAR is written in place of statement DELAY PROC NEAR.

5.17 PARAMETER PASSING

When we call a procedure, we need to make some data valves 'or' addresses available to the procedure. Similarly we need a procedure to make some processed data valves 'or' address available to main program. These addresses and data valves passed back and forth between the main program and the procedure are named parameters. Parameter passing is an important concept in assembly language programming. Basically, these are four main ways of passing parameters to and from a procedure.

1. In registers.
2. In dedicated memory locations.
3. Passing parameters using pointers.
4. In stack locations.

To describe these methods of parameters passing we can write small programs. For this we can write a program to convert a BCD number to a hexadecimal. The algorithm for this conversion involves the following steps.

1. Separate nibbles.
2. Save lower nibble.
3. Multiply upper limit by OAH.
4. Add lower nibble to the result of multiplication.

We will write the conversion program as procedure, so that it may be called any- where in the main programme.

Parameters Passing in Registers

This method of parameters passing is stated in the program to convert a BCD into its hexadecimal equivalent.

The 2-digit BCD (8-Bit Binary) number is passed to the procedure in the AL register and the hexadecimal equivalent is passed back to the calling program in the AL register.

The program starts by pushing the flag register and other register used in procedure. But never push 'or' POP AX register because we are using it to pass a valve to procedure and expecting the procedure to pass a different valve back to the calling program in it.

Program

```
; This is an 8086 program
; Procedure to convert a 2-ditgit BCD in to hexadcimal
; Program shows how to use AL register to pass parameter to the procedure
; Register used: CS, DS, SS, AL, AH, BL, BH, CL, CH.
; Procedure used: BCD-BIN
DATA-SEG SEGMENT
            BCD-IN   DB ?   ; Storage for BCD valve
           HEX-NUM   DB ?   ; Storage for hexadecimal valve
DATA-SEG ENDS
         STACK-SEG   SEGMENT STACK
                DB   30 ?
         STACK-SEG   ENDS
          CODE-SEG   SEGMENT
       ASSUME CS :   CODE-SEG, DS : DATA-SEG,
              SS :   STACK-SEG
START    MOV AX, DATA-SEG
         MOV DS, AX
         MOV AX, STACK-SEG
         MOV SS, AX
         MOV AL, BCD-IN          ; Get BCD in to AL
         CALL BCD-BIN            ; Call conversion procedure BCD-BIN
         MOV HEX-NUM, AL         ; Store the result
; Procedure BCD-BIN for conversion
BCD-BIN  PROC NEAR               ; Near Procedure
         PUSH F                  ; Push Flag
         PUSH BX
         PUSH CX                 ; Push Register.
```

```
; Start of conversion
        MOV AH, AL          ; Save copy of BCD in AH.
        AND AH, 0FH         ; Separate and save lower BCD digit.
        MOV BL, AH
        AND AL, 0FOH        ; Separate upper nibble.
        MOV CL, 04          ; Move upper BCD; digit to low nibble.
        ROR AL, CL          ; Position for multiply.
        MOV BH, 0AH         ; Load conversion factor in BH.
        MUL BH              ; Upper BCD digit in; AL multiplied by
                            ; 0AH in BH, Result in AX
        ADD AL, BL          ; Add lower BCD to; result of MUL
                            ; Final result in AL.
; End of conversion
        POP   CX
        POP   BX
        POP   F
        RET
BCD-BIN ENDP
CODE-SEG ENDS
        END START
```

Remaining three methods are for your own practise, thus the parameter passing takes place.

5.18 RECURSIVE PROCEDURE

When a procedure calls itself, it is called recursive procedure. These procedures are used to work with complex data structures like tree.

Recursive procedure can be understood better by taking an example of computing factorial of an number N. Here, we will write a program to compute the factorial of a number N ranging from 1 to 9. Mathematically, the factorial of a number is the product of number and all of the positive integers less than number.

For example: Factorial of 7

$$= 7 \times 6 \times 5 \times 4 \times 3 \times 2 \times 1.$$

$$= 7 \times \text{Factorial of } 6.$$

The basic algorithm can be defined as if N = 1, then factorial of 1 = 1 and factorial of N = N × (Factorial of N-1).

Program

```
; This is an 8086 program to compute
; The factorial of a number N from 1 to 9.
          NUMBER EQV03 ?        ;          Number N of which
                                ;          factorial is to be
                                ;          computed
CODE-SEG  SEGMENT
          ASSUME CS    : CODE-SEG, DS : DATA-SEG,
          SS           : STACK-SEG.
START     MOV AX, STACK-SEG; Initialize SS.
          MOV SS, AX
          MOV SP, OFFSET STACK TOP ; Initialize SP
          SUB SP, 0004H         ; Space for factorial on stack
          MOV AX, Number        ; Put No. to be passed on stack
          CALL FACTORIAL        ; Call procedure to compute factorial.
          POP   AX              ; Get Result
          :     :               :
          :     :               :
; FACTORIAL-Procedure to compute factorial N.
FACTORIAL                 PROC NEAR
          PUSH F
          PUSH AX
          PUSH DX
          PUSH BP
          MOV BP, SP            ; Pointing BP to top of stack
          MOV AX, [BP+10]       ; Number from stack to AX.
          CMP AX, 000IH         ; Check No.
          JNE GO-ON             ; If Not 1, go-on.
          MOV WORD PTR [BP+12], 0001 H ; If 1 load 1
          MOV WORD PTR [BP+14], 0000 H ; go to main program.
GO-ON     SUB SP, 0004H         ; Space in stack for factorial
          DEC AX                ; Decrement No. in AX by 1.
          PUSH AX               ; Save this No. in stack.
          CALL FACTORIAL        ; Again go to take factorial of this No.
```

```
            MOV BP, SP               ; Again point BP to top of stack.
            MOV, AX, [BP+2]          ; Last (N-1) from stack to AX.
            MUL WORD PTR [BP+16] ; Multiply previous N.
            MOV [BP+18], AX          ; Copy New factorial to stack.
            MOV [BP+20], DX
            ADD SP, 0006H            ; Point to pushed Register.
EXIT:       POP   BP                 ; Restore Registers and Flags
            POP   DX
            POP   AX
            POP   F
            RET
FACTORIAL                ENDP              ;        End of procedure.
CODE-SEG                 ENDS
STACK-SEG             SEGMENT STACK
        DW               300 ?
STACK-TOP             LABEL WORD
STACK-SEG                ENDS
            END START
```

5.19 ASSEMBLER MACRO

There are two major disadvantages of using procedure, first need of stack, another overhead time required to call and return from procedure.

If the group of instructions (to be repeatedly written in a program) is very short 'or' it is not appropriate to write a procedure, we can use a Macro. A Macro is a group of instructions can be put in a bracket and given a name at the start of program, whenever we CALL a macro in our program, the assembler will insert the defined group of instruction in place of the CALL.

Now here we will see how a Macro can be written to save all the register.

```
PUSH-ALL   MACRO
           PUSH F
           PUSH AX
           PUSH BX
           PUSH CX
           PUSH DX
           PUSH SI
           PUSH DI
```

PUSH DS

PUSH ES

PUSH SS

ENDM.

The PUSH-ALL MACRO statement shows the starting of macro and ENDM shows the ending of macro.

Now during our main program; where ever we will write

PUSH-ALL MACRO

Automatically all the instructions under this macro will be placed at that place.

REVIEW QUESTIONS

1. Draw the pin diagram of 8086. Explain the function of each pin in detail.
2. Draw the functional diagram of 8086. Explain each block in detail.
3. Explain the maximum and minimum mode operation of 8086.
4. What is the need of 8282/8283 Octal latch and 8286/8287 Bus tranceiver in the Minimum Mode operation of 8086?
5. In maximum mode operation of 8086; how does the control signals are generate?
6. Differentiate 8086 and 8088.
7. Write a 8086 program for the addition of two sixteen bit number and answer should be in 16 bits.
8. What do you understand for directives? Explain some 8086 directives in detail.
9. Explain Assembler Macro.
10. Explain following instructions in reference of 8086.
    (*a*) $\overline{RQ}/\overline{GT0}$ and $\overline{RQ}/\overline{GT1}$
    (*b*) $\overline{LOCK}$
    (*c*) QS1, QS0
    (*d*) $\overline{DEN}$
11. Explain the addressing modes of 8086.
12. Explain the interrupts of 8086 in detail.
13. Explain the 8088 block diagram in detail.
14. Explain parameter passing in 8086. What is the importance of parameter passing?
15. What do you understand by Near and Far procedure?
16. Explain CALL and RET for Near and Far procedure.

6

Basic I/O Interfacing

6.1 INTRODUCTION

For various applications, a microprocessor needs to be interfaced to the

(*a*) input devices such as keyboard

(*b*) output devices such as printer.

But as we know that speed of processor is very high but the speed of input and output devices is very low. So if we want to interface a external input/output device, we need some intelligent technique. In this chapter, we will discuss such interfacings devices such as:

1. DMA controller-8237
2. Programmable peripheral interface 8255A
3. USART-8251A
4. Programmable Interval Timer-8254
5. Programmable Interrupt Controller-8259A
6. Programmable Keyboard/Display Interface-8279.

6.2 PROGRAMMED INPUT/OUTPUT (I/O)

As the name suggest that in this type of input/output interfacing, software programme (instructions) are used to get data into 'or' out of CPU.

In programmed I/O for transferring data, a memory resident routine requests the device for data transfer to 'or' from one of the microprocessor registers.

Always remember that programmed I/O devices are generally used when a relatively small amount of data is transferred using relatively slow I/O devices. For example:

(*a*) Analog to digital (A/D) converters.

(*b*) Digital to Analog (D/A) converters.

(*c*) Peripheral floating point arithimetic unit.

In programmed I/O, usually one byte 'or' word of data is transferred at a time. To correctly time the data transfer, programmed I/O depends upon clock timing 'or' handshaking.

Now we will discuss programmed input and programmed output. We will also discuss Interrupt Driven I/O.

6.2.1 Programmed Input

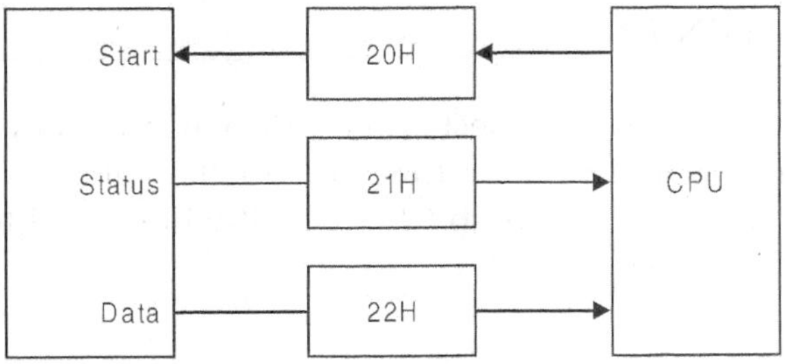

Fig. 6.1 Programmed Input

It is very simple to understand

(*a*) Here CPU want to get data from a peripheral device.

(*b*) CPU sends start bit to the input device; (suppose we are using a device of past address (20H)).

(*c*) Now this input device send a status signal (we are using CPU address 21H). If status bit is high, CPU waits; as status bit goes low the data is transferred from device to CPU. This work is totally performed using instructions.

6.2.2 Programmed Output

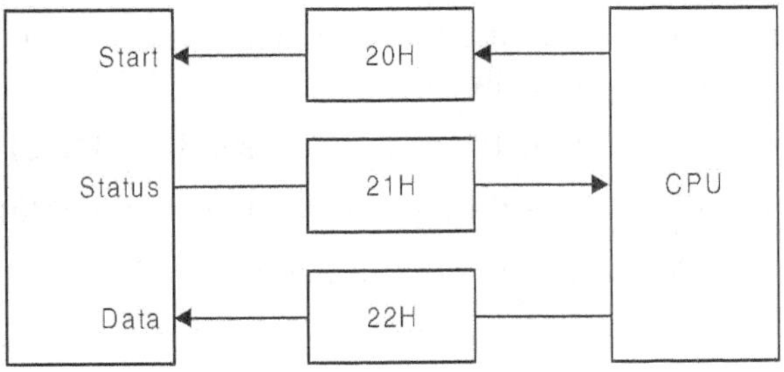

Fig. 6.2 Programmed Output

(*a*) Here CPU is waiting for data from a peripheral device.

(*b*) Now by sending a start bit; CPU sends a signal that it wants to transfer data to the output device.

(*c*) Now by sending a status signal peripheral device tells to the CPU whether it is ready to receive the data 'or' not. If status signal is low, it means device is ready to receive data; otherwise CPU has to wait until it get low.

Programmed Input/Output method is also called '*Polled Method*'. This method has some disadvantages as following:

(*i*) CPU has to check the status of peripheral device again and again to provide the service to a user.

(*ii*) It is wastage of CPU processing power.

(*iii*) Due to this process; we have unnecessary delay.

To avoid these disadvantages; we use the Interrupt Driven I/O devices.

6.3 INTERRUPT DRIVEN I/O

Interrupt driven I/O devices send its request to the CPU when I/O devices require any service. CPU will not waste its processing power 'or' time to check the status. Now it is also possible that a number of interrupt can ask for service to CPU; that it will take all the requests but provide service only to that has the highest priority.

6.4 DIRECT MEMORY ACCESS

If we want to access the memory of system directly without any help of CPU; It is called direct memory access technique. In programmed I/O for any communication, we need fetching and executing the instruction by CPU; but some communication (such as if we want to save a data in floppy disk) that does not require any CPU processing so

(*a*) We can save time of CPU by directly coping the data from system to floppy.

(*b*) The speed of communication is also high because of very less number of instructions (generally no instruction during communication).

For this communication, simply CPU has to release the control of buses; and the communcation process should be correct. For that purpose we need a DMA Controller Device also.

Now how a CPU will come to know that DMA is on demand and there are some signals in each processor for that purpose. For example, in 8085:

HOLD—Now DMA controller initiate the process by sending '**HOLD**' signal to CPU.

As the HOLD goes high, CPU releases the control of buses (if possible) to the DMA controller.

HLDA—CPU tells to the DMA controller that it has released the control of buses by sending hold acknowledgement (HLDA) signal.

6.5 8257/8257-5 PROGRAMMABLE DMA CONTROLLER

- 4-Channel DMA Controller
- Priority DMA Request Logic
- Channel Inhibit Logic
- Terminal Count and Modulo 128 Outputs
- Single TTL Clock
- Single + 5V Supply
- Auto Load Mode

The Intel 8257 is a 4 channel direct memory access (DMA) controller. It is specifically designed to simplify the transfer of data at high speeds for the Intel microcomputer systems. Its primary function is to generate, upon a peripheral request, a sequential memory address which will allow the peripheral to read or write data directly to or from memory. Acquisition of the system bus in accomplished via the CPU's hold function. The 8257 has priority logic that resolves the peripherals requests and issues a composite hold request to the CPU. It maintains the DMA cycle count for each channel and outputs a control signal to notify the peripheral that the programmed number of DMA cycles is complete. Other output control signals simplify sectored data transfers. The 8257 represents a significant savings in component count for DMA-based microcomputer systems and greatly simplifies the transfer of data at high speed between peripherals and memories.

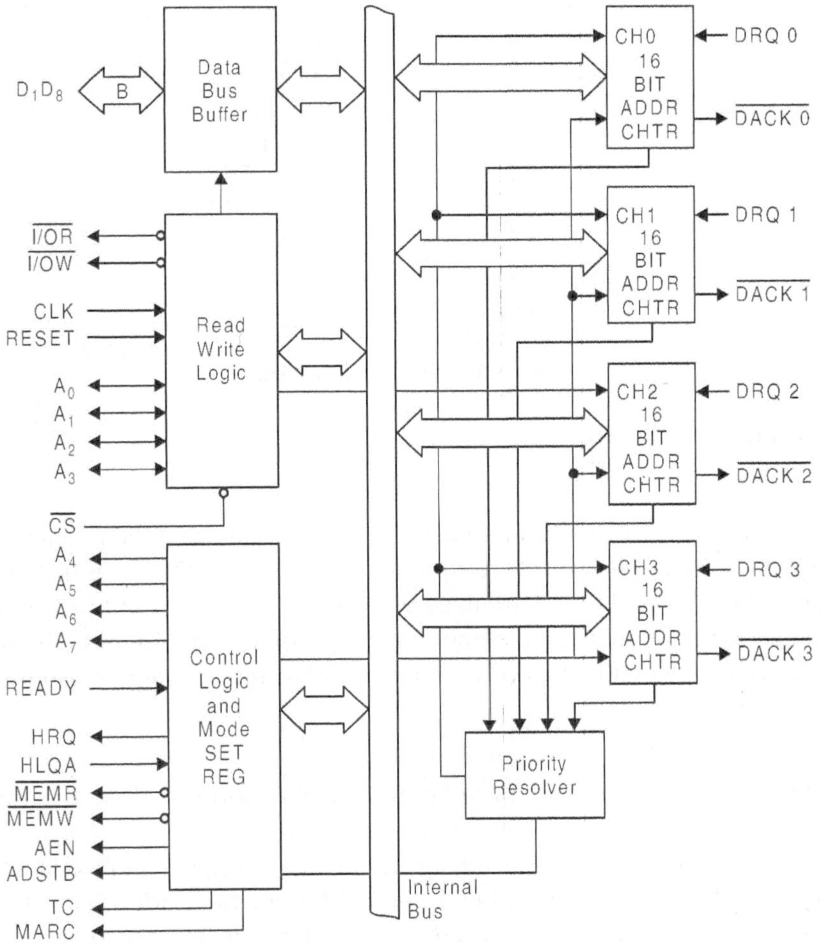

Fig. 6.3 Block Diagram

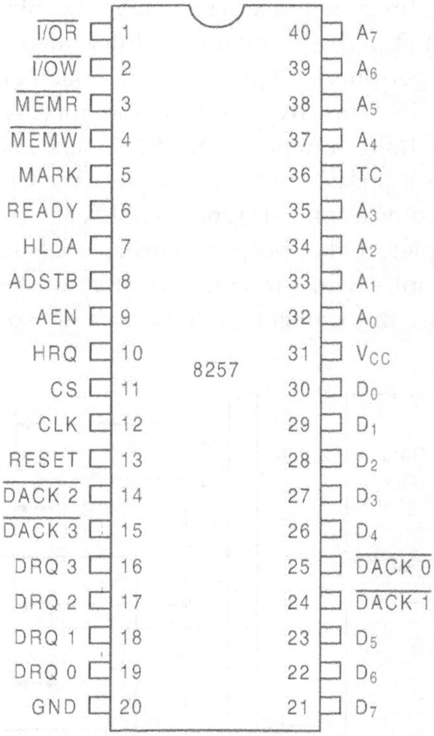

Fig. 6.4 Pin Configuration

6.5.1 Functional Description

General

The 8257 is a programmable, Direct Memory Access (DMA) device which, when coupled with a single Intel 8212 I/O port device, provides a complete four-channel DMA controller for use in Intel microcomputer systems. After being initialized by software, the 8257 can transfer a block of data, containing up to 16,384 bytes, between memory and a peripheral device directly, without further intervention required of the CPU. Upon receiving a DMA transfer request from an enabled peripheral, the 8257:

1. Acquires control of the system bus.
2. Acknowledges that requesting peripheral which is connected to the highest priority channel.
3. Outputs the least significant eight bits of the memory address onto system address lines A_0-A_7, outputs the most significant eight bits of the memory address to the 8212 I/O port via the data bus (the 8212 places these address bits on lines A_8-A_{15}), and
4. Generates the appropriate memory and I/O read/write control signals that cause the peripheral to receive or deposit a data byte directly from or to the addressed location in memory.

The 8257 will retain control of the system bus and repeat the transfer sequence, as long as a peripheral maintains its DMA request. Thus, the 8257 can transfer a block of data to/from a high speed peripheral (*e.g.*, a sector of data on a floppy disk) in a single "burst". When the specified number of data bytes have been transferred, the 8257 activates its Terminal Count (TC) output, informing the CPU that the operation is complete.

The 8257 offers three different modes of operation: (1) DMA read, which causes data to be transferred from memory to a peripheral: (2) DMA write, which causes data to be transferred from a peripheral to memory; and (3) DMA verify, which does not actually involve the transfer of data.

When an 8257 channel is in the DMA verify mode, it will respond the same as described for transfer operations, except that no memory or I/O read/write control signals will be generated, thus preventing the transfer of data. The 8257, however, will gain control of the system bus and will acknowledge the peripheral's DMA request for each DMA cycle. The peripheral can use these acknowledge signals to enable an internal access of each byte of a data block in order to execute some verification procedure, such as the accumulation of a CRC (Cyclic Redundancy Code) check word. For example, a block of DMA verify cycles might follow a block of DMA read cycles (memory to peripheral) to allow the peripheral to verify its newly acquired data.

6.5.2 Block Diagram Description

1. DMA Channels

The 8257 provides four separate DMA channels (labeled CH-0 to CH-3). Each channel includes two sixteen-bit registers: (1) a DMA address register, and (2) a terminal count register.

Both registers must be initialized before a channel is enabled. The DMA address register is loaded with the address of the first memory location to be accessed. The value loaded into the low-order 14-bits of the terminal count register specifies the number of DMA cycles minus one before the Terminal Count (TC) output is activated. For instance, a terminal count of 0 would cause the TC output to be active in the first DMA cycle for that channel. In general, if N = the number of desired DMA cycles, load the value N-1 into the low-order 14-bits of the terminal count register. The most significant two bits of the terminal count register specify the type of DMA operation for that channel.

These two bits are not modified during a DMA cycle, but can be changed between DMA blocks.

Each channel accepts a DMA Request (DRQn) input and provides a DMA Acknowledge (DACKn) output

(DRQ 0-DRQ 3). DMA Request: These are individual asynchronous channel request Inputs used by the peripherals to obtain a DMA cycle. If not in the rotating priority mode then DRQ 0 has the highest priority and DRQ 3 has the lowest. A request can be generated by raising the request line and holding it high until DMA acknowledge. For multiple DMA cycles (Burst Mode) the request line is held high until the DMA acknowledge of the last cycle arrives.

(DACK 0-DACK 3). DMA Acknowledge: An active low level on the acknowledge output informs the peripheral connected to that channel that it has been selected for a DMA cycle. The DACK output acts as a "chip select" for the peripheral device requesting service. This line goes active (low) and inactive (high) once for each byte transferred even if a burst of data is being transferred.

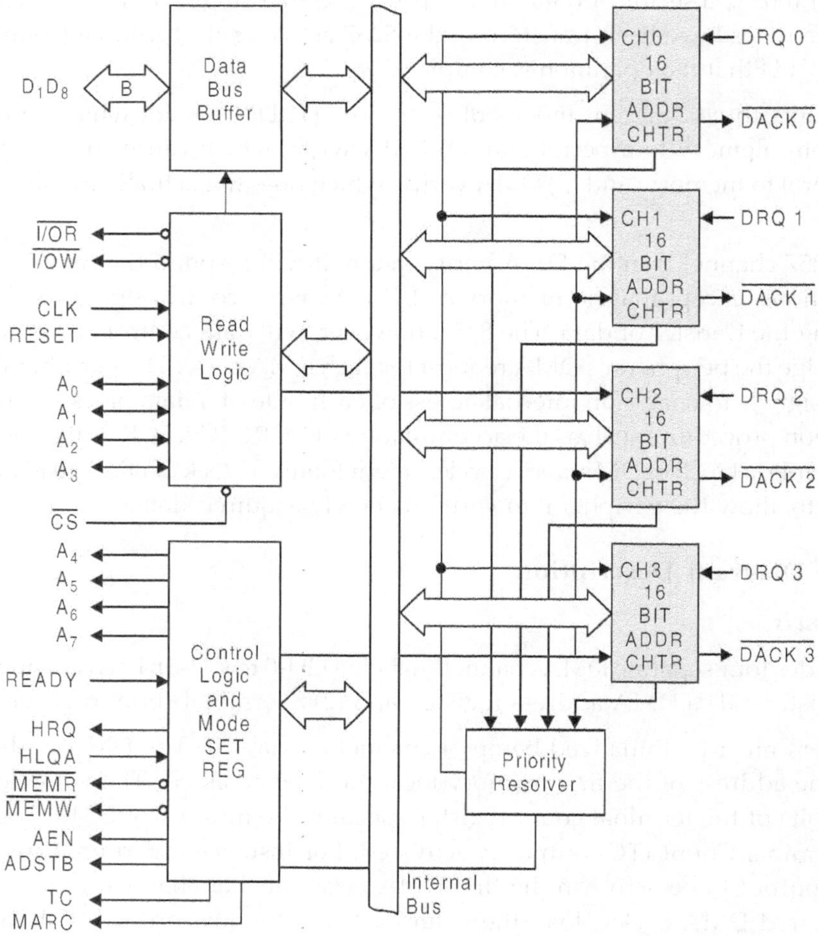

Fig. 6.5 8257 Block Diagram Showing DMA Channels

2. Data Bus Buffer

This three-state, bi-directional, eight bit buffer interfaces the 8257 to the system data bus.

(D_0-D_7). *Data Bus Lines:* These are bi-directional three-state lines. When the 8257 is being programmed by the CPU eight-bits of data for a DMA address register, a terminal count register or the Mode Set register are received on the data bus. When the CPU reads a DMA address register, a terminal count register or the Status register, the data is sent to the CPU over the data bus. During DMA cycles (when the 8257 is the bus master), the 8257 will output the most significant eight-bits of the memory address (from one of the DMA address registers) to the 8212 latch via the data bus. These address bits will be transferred at the beginning of the DMA cycle; the bus will then be released to handle the memory data transfer during the balance of the DMA cycle.

Bit 15	Bit 14	Type of DMA Operation
0	0	Verify DMA Cycle
0	1	Write DMA Cycle
1	0	Read DMA Cycle
1	1	(Illegal)

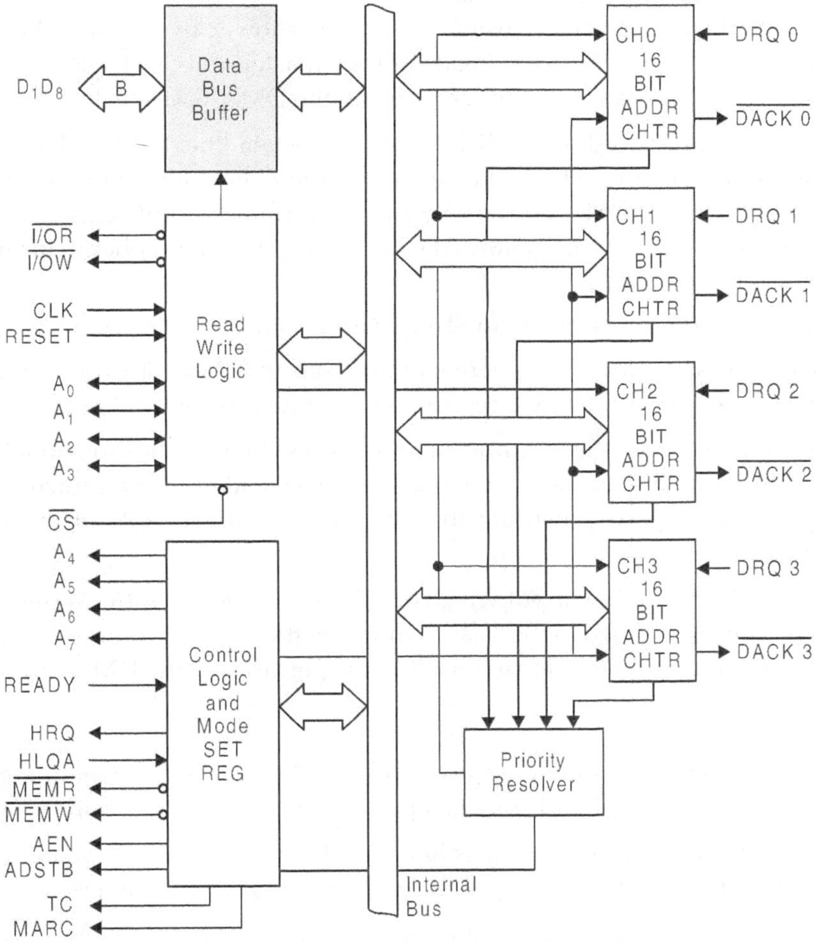

Fig. 6.6 8257 Block Diagram Showing Data Bus Buffer

3. Read/Write Logic

When the CPU is programming or reading one of the 8257's registers (*i.e.*, when the 8257 is a "slave" device on the system bus), the Read/Write Logic accepts the I/O Read ($\overline{I/OR}$) or I/O Write ($\overline{I/OW}$) signal, decodes the least significant four address bits, (A_0–A_3), and either writes the contents of the data bus into the addressed register (if/ $\overline{I/OW}$ is true) or places the contents of the addressed register onto the data bus.

During DMA cycles (*i.e.*, when the 8257 is the bus "master"), the Read/write Logic generates the I/O read and memory write (DMA write cycle) or I/O Write and memory read (DMA read cycle) signals which control the data link with the peripheral that has been granted the DMA cycle.

Note that during DMA transfers Non-DMA I/O devices should be de-selected (disabled) using "AEN" signal to inhibit I/O device decoding of the memory address as an erroneous device address.

($\overline{I/OR}$). *I/O Read:* An active-low, bi-directional three-state line. In the "slave" mode, it is an input which allows the 8-bit status register or the upper/lower byte of a 16-bit DMA address register or terminal count register to be read. In the "master" mode, $\overline{I/OR}$ is a control output which is used to access data from a peripheral during the DMA write cycle.

($\overline{I/OW}$). *I/O Write:* An active-low, bi-directional three-state line. In the "slave" mode, it is an input which allows the contents of the data bus to be loaded into the 8-bit mode set register or the upper/lower byte of a 16-bit DMA address register or terminal count register. In the "master" mode $\overline{I/OW}$ is a control output which allows data to be output to a peripheral during a DMA read cycle.

(CLK). *Clock Input:* Generally from an Intel 8224 Clock Generator Device.

(RESET). *Reset:* An asynchronous input (generally from an 8224 or 8085 device) which disables all DMA channels by clearing the mode register and 3-states all control lines.

(A_0-A_3). *Address Lines:* These least significant four address lines are bi-directional. In the "slave" mode they are inputs which select one of the registers to be read or programmed. In the "master" mode, they are outputs which constitute the least significant four bits of the 16-bit memory address generated by the 8257.

($\overline{CS}$). *Chip Select:* An active-low input which enables the I/O Read or I/O Write input when the 8257 is being read or programmed in the "slave" mode. In the "master" mode, $\overline{CS}$ is automatically disabled to prevent the chip from selecting itself while performing the DMA function.

4. Control Logic

This block controls the sequence of operations during all DMA cycles by generating the appropriate control signals and the 16-bit address that specifies the memory location to be accessed.

(A_4-A_7). *Address Lines:* These four address lines are three-state outputs which constitute bits 4 through 7 of the 16-bit memory address generated by the 8257 during all DMA cycles.

(READY). *Ready:* This asynchronous input is used to elongate the memory read and write cycles in the 8257 with wait states if the selected memory requires longer cycles. READY must conform to specified setup and hold times.

(HRQ). *Hold Request:* This output requests control of the system bus. In systems with only one 8257, HRQ will normally be applied to the HOLD input on the CPU. HRQ must conform to specified setup and hold times.

(HLDA). *Hold Acknowledge:* This input from the CPU indicates that the 8257 has acquired control of the system bus.

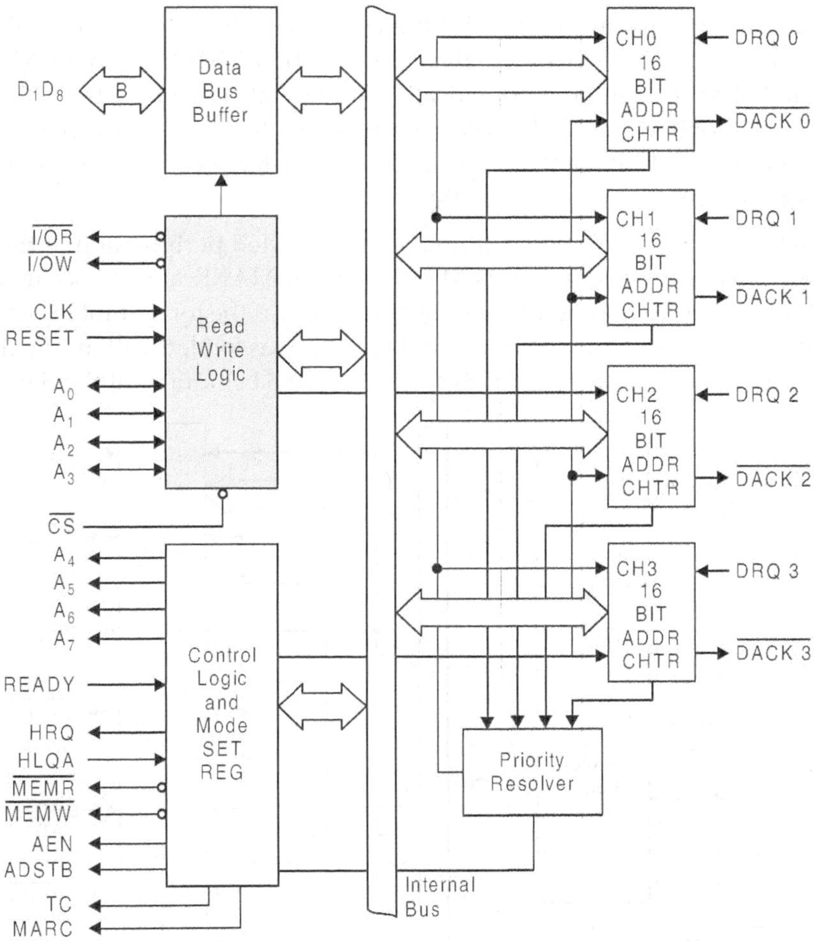

Fig. 6.7 8257 Block Diagram Showing Read/Write Logic Function

($\overline{MEMR}$). *Memory Read:* This active-low three-state output is used to read data from the addressed memory location during DMA Read cycles.

($\overline{MEMW}$). *Memory Write:* This active-low three-state output is used to write data into the addressed memory location during DMA Write cycles.

(ADSTB). *Address Strobe:* This output strobes the most significant byte of the memory address into the 8212 device from the data bus.

(AEN). *Address Enable:* This output is used to disable (float) the System Data Bus and the System Control Bus. It may also be used to disable (float) the System Address Bus by use of an enable on the Address Bus drivers in systems to inhibit non-DMA devices from responding during DMA cycles. It may be further used to isolate the 8257 data bus from the System Data Bus to facilitate the transfer of the 8 most significant DMA address bits over the 8257 data I/O pins without subjecting the System Data Bus to any timing constraints for the transfer. When the 8257 is used in an I/O device structure (as opposed to memory mapped), this AEN output should be used to disable the selection of an I/O device when the DMA address is on the address bus. The I/O device selection should be determined by the DMA acknowledge outputs for the 4 channels.

(TC). *Terminal Count:* This output notifies the currently selected peripheral that the present DMA cycle should be the last cycle for this data block. If the TC STOP bit in the Mode Set register is set, the selected channel will be automatically disabled at the end of that DMA cycle. TC is activated when the 14-bit value in the selected channel's terminal count register equals zero. Recall that the low-order 14-bits of the terminal count register should be loaded with the values (n-1), where n = the desired number of the DMA cycles.

(MARK). *Modulo 128 Mark:* This output notifies the selected peripheral that the current DMA cycle is the 128th cycle since the previous MARK output. MARK always occurs at 128 (and all multiples of 128) cycles from the end of the data block. Only if the total number of DMA cycles (n) is evenly divisible by 128 (and the terminal count register was loaded with n-1) will MARK occur at 128 (and each succeeding multiple of 128) cycles from the beginning of the data block.

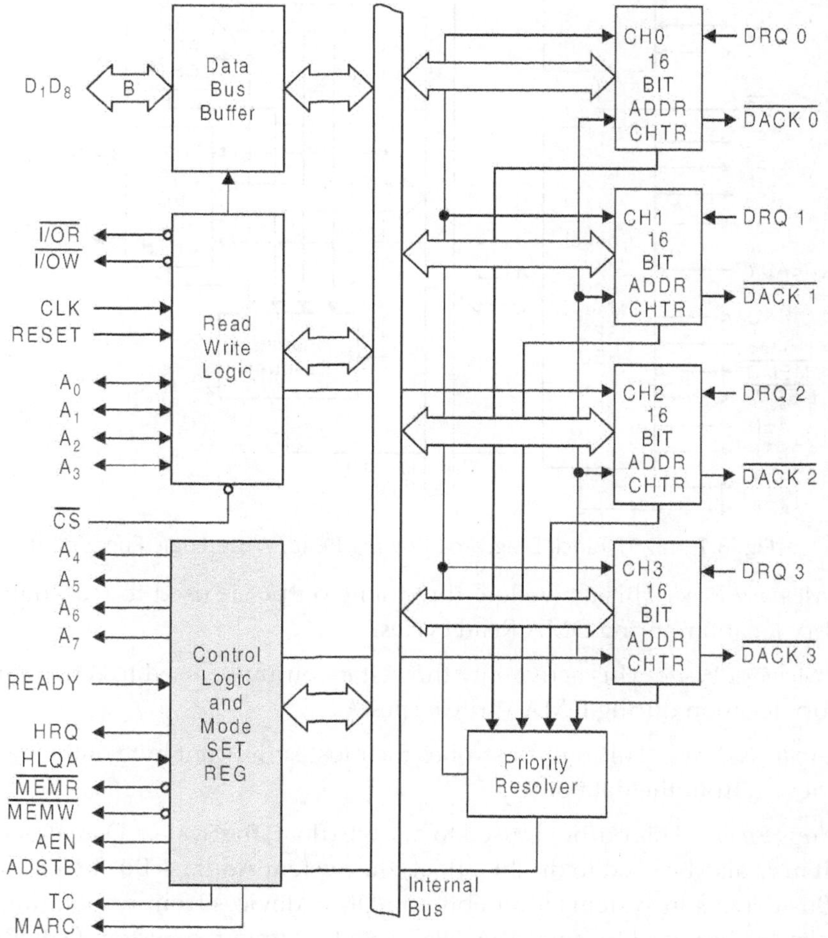

Fig. 6.8 8257 Block Diagram Showing Control Logic and Mode Set Register

5. Mode Set Register

When set, the various bits in the Mode Set register enable each of the four DMA channels, and allow four different options for the 8257:

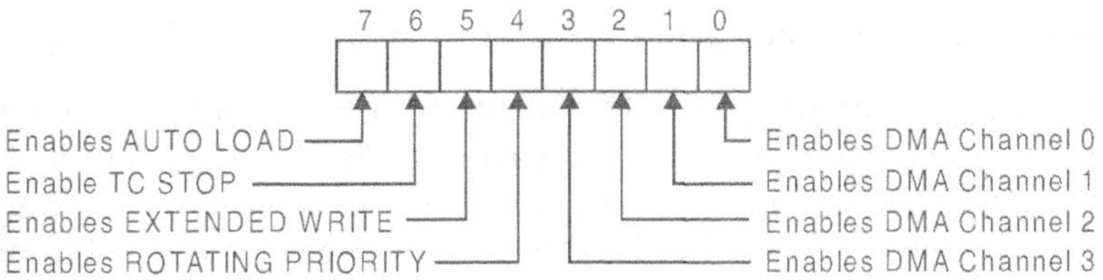

The Mode Set register is normally programmed by the CPU after the DMA address register(s) and terminal count register(s) are initialized. The Mode Set Register is cleared by the RESET input, thus disabling all options, inhibiting all channels, and preventing bus conflicts on power-up. A channel should not be left enabled unless its DMA address and terminal count registers contain valid values; otherwise, an inadvertent DMA request (DRQn) from a peripheral could initiate a DMA cycle that would destroy memory data.

The various options which can be enabled by bits in the Mode Set register are explained below:

Rotating Priority Bit 4. In the Rotating Priority Mode, the priority of the channels has a circular sequence. After each DMA cycle, the priority of each channel changes. The channel which had just been serviced will have the lowest priority.

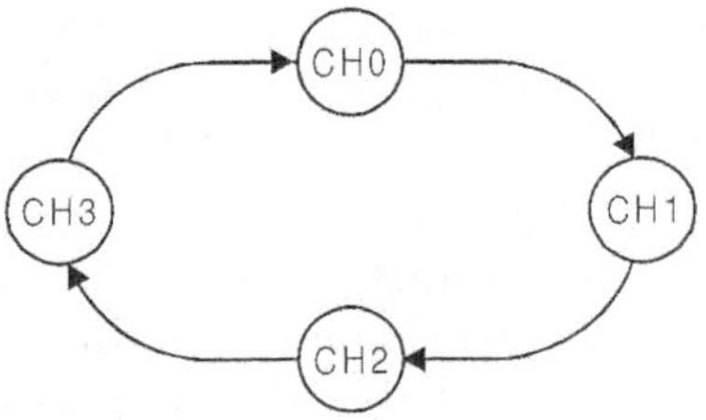

If the ROTATING PRIORITY bit is not set (set to a zero), each DMA channel has a fixed priority. In the fixed priority mode, Channel 0 has the highest priority and Channel 3 has the lowest priority. If the ROTATING PRIORITY bit is set to a one. The priority of each channel changes after each DMA cycle (not each DMA request). Each channel moves up to the next highest priority assignment, while the channel which has just been serviced moves to the lowest priority assignment:

	Channel > just Serviced	**CH-0**	**CH-1**	**CH-2**	**CH-3**
Priority >	Highest	CH-1	CH-2	CH-3	CH-0
Assignments		CH-2	CH-3	CH-0	CH-1
		CH-3	CH-0	CH-1	CH-2
	Lowest	CH-0	CH-1	CH-2	CH-3

Note that rotating priority will prevent any one channel from monopolizing the DMA mode; consecutive DMA cycles will service different channels if more than one channel is enabled and requesting service. There is no overhead penalty associated with this mode of operation. All DMA operations began with Channel 0 initially assigned to the highest priority for the first DMA cycle.

Extended Write Bit 5. If the EXTENDED WRITE bit is set, the duration of both the $\overline{\text{MEMW}}$ and $\overline{\text{I/OW}}$ signals is extended by activating them earlier in the DMA cycle. Data transfers within microcomputer systems proceed asynchronously to allow use of various types of memory and I/O devices with different access times. If a device cannot be accessed within a specific amount of time it returns a "not ready" indication to the 8257 that causes the 8257 to insert one or more wait states in its internal sequencing. Some devices are fast enough to be accessed without the use of wait states, but if they generate their READY response with the leading edge of the $\overline{I/OW}$ or / $\overline{MEMW}$ signal (which generally occurs late in the transfer sequence), they would normally cause the 8257 to enter a wait state because it does not receive READY in time. For systems with these types of devices, the Extended Write option provides alternative timing for the I/O and memory write signals which allows the devices to return an early READY and prevents the unnecessary occurrence of wait states in the 8257, thus increasing system throughput.

TC Stop Bit 6. If the TC STOP bit is set, a channel is disabled (*i.e.*, its enable bit is reset) after the Terminal Count (TC) output goes true, thus automatically preventing further DMA operation on that channel. The enable bit for that channel must be re-programmed to continue or begin another DMA operation. If the TC STOP bit is not set, the occurrence of the TC output has no effect on the channel enable bits. In this case, it is generally the responsibility of the peripheral to cease DMA requests in order to terminate a DMA operation.

Auto Load Bit 7. The Auto Load mode permits Channel 2 to be used for repeat block or block chaining operations, without immediate software intervention between blocks. Channel 2 registers are initialized as usual for the first data block; Channel 3 registers, however, are used to store the block re-initialization parameters (DMA starting address, terminal count and DMA transfer mode). After the first block of DMA cycles is executed by Channel 2 (*i.e.*, after the TC output goes true), the parameters stored in the Channel 3 registers are transferred to Channel 2 during an "update" cycle. Note that the TC STOP feature, described above, has no effect on Channel 2 when the Auto Load bit is set.

If the Auto Load bit is set, the initial parameters for Channel 2 are automatically duplicated in the Channel 3 registers when Channel 2 is programmed. This permits repeat block operations to be set up with the programming of a single channel. Repeat block operations can be used in applications such as CRT refreshing. Channels 2 and 3 can still be loaded with separate values if Channel 2 is loaded before loading Channel 3. Note that in the Auto Load mode, Channel 3 is still available to the user if the Channel 3 enable bit is set, but use of this channel will change the values to be auto loaded into Channel 2 at update time. All that is necessary to use the Auto Load feature for chaining operations is to reload Channel 3 registers at the conclusion of each update cycle with the new parameters for the next data block transfer.

Each time that the 8257 enters an update cycle, the update flag in the status register is set and parameters in Channel 3 are transferred to Channel 2, non-destructively for Channel 3. The actual re-initialization of Channel 2 occurs at the beginning of the next channel 2 DMA cycle after the TC cycle. This will be the first DMA cycle of the new data block for Channel 2. The update flag is cleared at the conclusion of this DMA cycle. For chaining operations, the update flag in the status register can be monitored by the CPU to determine when the re-initialization process has been completed so that the next block parameters can be safely loaded into Channel 3.

6. Status Register

The eight-bit status register indicates which channels have reached a terminal count condition and includes the update flag.

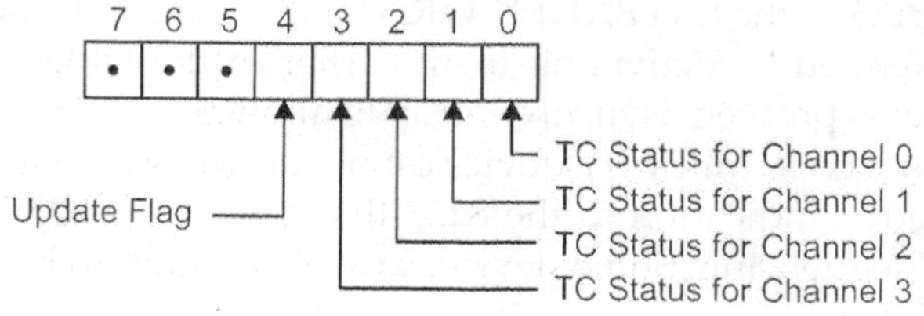

The TC status bits are set when the Terminal Count (TC) output is activated for that channel. These bits remain set until the status register is read or the 8257 is reset. The UPDATE FLAG, however, is not affected by a status register read operation. The UPDATE FLAG can be cleared by resetting the 8257, by changing to the non-auto load mode (*i.e.*, by resetting the AUTO LOAD bit in the Mode Set register) or it can be left to clear itself at the completion of the update cycle. The purpose of the UPDATE FLAG is to prevent the CPU from inadvertently skipping a data block by overwriting a starting address or terminal count in the Channel 3 registers before those parameters are properly auto-loaded into Channel 2.

The user is cautioned against reading the TC status register and using this information to reenable channels that have not completed operation. Unless the DMA channels are inhibited a channel could reach Terminal Count (TC) between the status read and the mode write. DMA can be inhibited by a hardware gate on the HRQ line or by disabling channels with a mode word before reading the TC status.

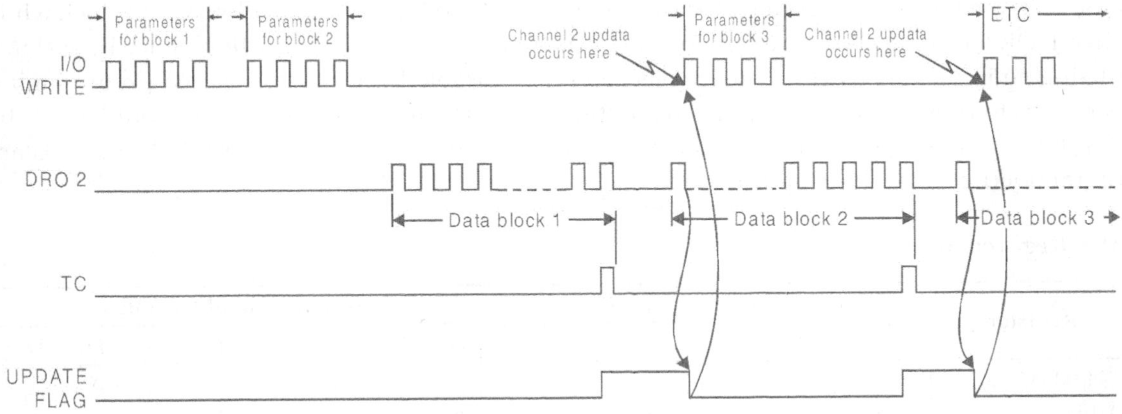

Fig. 6.9 Autoload Timing

6.5.3 Programming and Reading the 8257 Registers

There are four pairs of "channel registers": each pair consisting of a 16-bit DMA address register and a 16-bit terminal count register (one pair for each channel). The 8257 also includes two "general registers": one 8-bit Mode Set register and one 8-bit Status register. The registers are loaded or read when the CPU executes a write or read instruction that addresses the 8257 device and the appropriate register within the 8257. The 8228 generates the appropriate read or write control signal (generally $\overline{I/OR}$ or $\overline{I/OW}$ while the CPU places a 16-bit address on the system address bus, and either outputs the data to be written onto the system data bus or accepts the data being read from the data bus. All or some of the most significant 12 address bits A_4-A_{15} (depending on the systems memory, I/O configuration) are usually decoded to produce the chip select ($\overline{CS}$) input to the 8257. An I/O Write Input (or Memory Write in memory mapped I/O configurations) specifies that the addressed register is to be programmed, while an I/O Read Input specifies that the addressed register is to be read. Address bit 3 specifies whether a "channel register" ($A_3 = 0$) or the Mode Set (program only)/Status (read only) register ($A_3 = 1$) is to be accessed.

The least significant three address bits, A_0-A_2, indicate the specific register to be accessed. When accessing the Mode Set or Status register, A_0-A_2 are all zero. When accessing a channel register bit A_0 differentiates between the DMA address register ($A_0 = 0$) and the terminal count register ($A_0 = 1$). while bits A_1 and A_2 specify one of the four channels.

Control Input	$\overline{CS}$	$\overline{I/OW}$	$\overline{I/OR}$	A_3
Program Half of a Channel Register	0	0	1	0
Read Half of a Channel Register	0	1	0	0
Program Mode Set Register	0	0	1	1
Read Status Register	0	1	0	1

Because the "channel registers" are 16-bits, two program instruction cycles are required to load or read an entire register. The 8257 contains a first/last (F/L) flip flop which toggles at the completion of each channel program or read operation. The F/L flip flop determines whether the upper or lower byte of the register is to be accessed. The F/L flip flop is reset by the RESET input and whenever the Mode Set register is loaded. To maintain proper synchronization when accessing the "channel registers" all channel command instruction operations should occur in pairs, with the lower byte of a register always being accessed first. Do not allow $\overline{CS}$ to clock while either $\overline{I/OR}$ or $\overline{I/OW}$ is active, as this will cause an erroneous F/L flip flop state. In systems utilizing an interrupt structure, interrupts should be disabled prior to any paired programming operations to prevent an Interrupt from splitting them. The result of such a split would leave the F/L F/F in the wrong state. This problem is particularly obvious when other DMA channels are programmed by an interrupt structure.

8257 Register Selection

Register	BYTE	Address Inputs				F/L	*Bi-Directional Data Bus							
		A_3	A_2	A_1	A_0		D_7	D_6	D_5	D_4	D_3	D_2	D_1	D_0
CH-0 DMA	LSB	0	0	0	0	0	A_7	A_6	A_5	A_4	A_3	A_2	A_1	A_0
Address	MSB	0	0	0	0	1	A_{15}	A_{14}	A_{13}	A_{12}	A_{11}	A_{10}	A_9	A_8
CH-0 Terminal	LSB	0	0	0	1	0	C_7	C_6	C_5	C_4	C_3	C_2	C_1	C_0
Count	MSB	0	0	0	1	1	Rd	Wr	C_{13}	C_{12}	C_{11}	C_{10}	C_9	C_8
CH-1 DMA	LSB	0	0	1	0	0								
Address	MSB	0	0	1	0	1	Same as channel 0							
CH-1 Terminal	LSB	0	0	1	1	0								
Count	MSB	0	0	1	1	1								
CH-2 DMA	LSB	0	1	0	0	1								
Address	MSB	0	1	0	0	1	Same as channel 0							
CH-2 Terminal	LSB	0	1	0	1	0								
Count	MSB	0	1	0	1	1								
CH-3 DMA	LSB	0	1	1	0	0								
Address	MSB	0	1	0	1	1	Same as channel 0							
CH-3 Terminal	LSB	0	1	1	1	0								
Count	MSB	0	1	1	1	1								
MODE SET (Program only)	–	1	0	0	0	0	AL	TCS	EW	RP	EN3	EN2	EN1	EN0
STATUS (Read only)	–	1	0	0	0	0	0	0	0	UP	TC3	TC2	TC1	TC0

*A_0-A_{15}: DMA Starting Address, C_0-C_{13}: Terminal Count value (N-1), Rd and Wr: DMA Verify (00), Write (01) or Read (10) cycle selection, AL: Auto Load, TCS: TC STOP, EW: EXTENDED WRITE, RP: ROTATING PRIORITY, EN3-EN0: CHANNEL ENABLE MASK, UP: UPDATE FLAG, TC3-TC0: TERMINAL COUNT STATUS BITS.

6.5.4 DMA Operation

Single Byte Transfers

A single byte transfer is initiated by the I/O device raising the DRQ line of one channel of the 8257. If the channel is enabled, the 8257 will output a HRQ to the CPU. The 8257 now waits until a HLDA is received insuring that the system bus is free for its use. Once HLDA is received the $\overline{\text{DACK}}$ line for the requesting channel is activated (LOW). The $\overline{\text{DACK}}$ line acts as a chip select for the requesting I/O device. The 8257 then generates the read and write commands and byte transfer occurs between the selected I/O device and memory. After the transfer is complete, the $\overline{\text{DACK}}$ line is set HIGH and the HRQ line is set LOW to indicate to the CPU that the bus is now free for use. DRQ must remain HIGH until $\overline{\text{DACK}}$ is issued to be recognized and must go LOW before S4 of the transfer sequence to prevent another transfer from occurring.

Consecutive Transfers

If more than one channel requests service simultaneously, the transfer will occur in the same way a burst does. No overhead is incurred by switching from one channel to another. In each cycle, the DRQ lines are sampled and the highest priority request is recognized during the next transfer. A burst mode transfer in a lower priority channel will be overridden by a higher priority request. Once the high priority transfer has completed, control will return to the lower priority channel if its DRQ is still active. No extra cycles are needed to execute this sequence. The HRQ line remains active until all DRQ lines go LOW.

Control Override

The continuous DMA transfer mode described above can be interrupted by an external device by lowering the HLDA line. After each DMA transfer the 8257 samples the HLDA line to insure that it is still active. If it is not active, the 8257 completes the current transfer, releases the HRQ line (LOW) and returns to the idle state. If DRQ lines are still active the 8257 will raise the HRQ line in the third cycle and proceed normally.

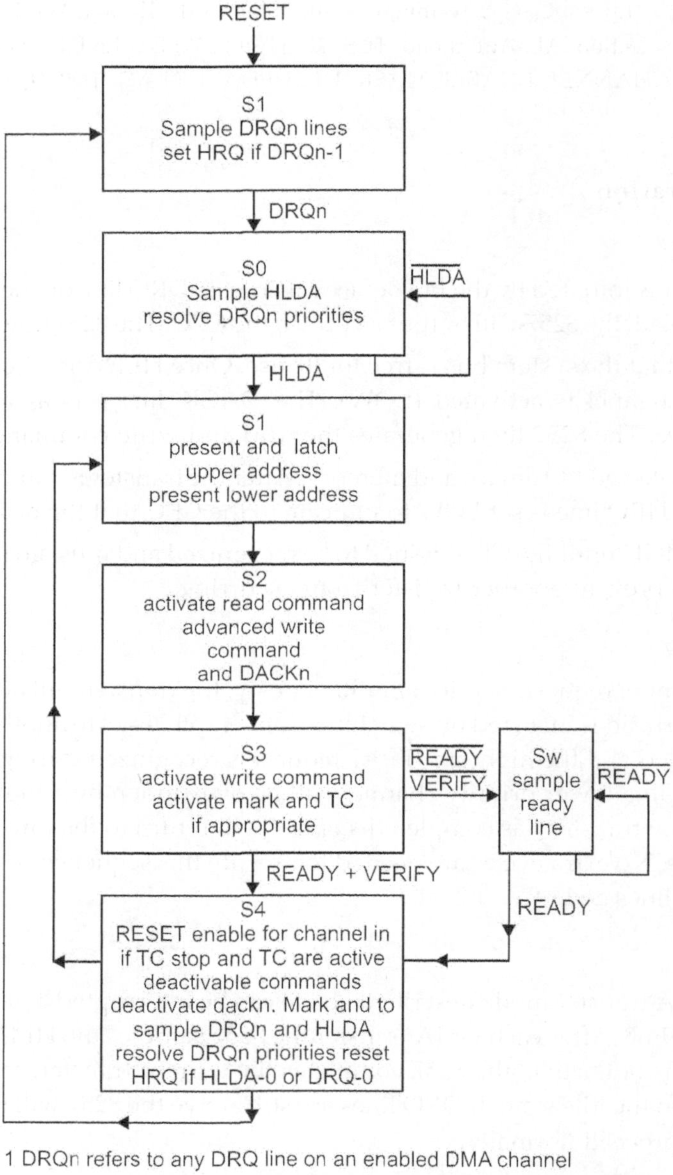

Fig. 6.10 DMA Operation State Diagram

Not Ready

The 8257 has a Ready input similar to the 8080A and the 8085A. The Ready line is sampled in State 3. If Ready is LOW, the 8257 enters a wait state. Ready is sampled during every wait state. When Ready returns HIGH, the 8257 proceeds to State 4 to complete the transfer. Ready is used to interface memory or I/O devices that cannot meet the bus set up times required by the 8257.

Speed

The 8257 uses four clock cycles to transfer a byte of data. No cycles are lost in the master to master transfer maximizing bus efficiency. A 2MHz clock input will allow the 8257 to transfer at a rate of 500K bytes/second.

Memory Mapped I/O Configurations

The 8257 can be connected to the system bus as a memory device instead of as an I/O device for memory mapped I/O configurations by connecting the system memory control lines to the 8257's I/O control lines and the system I/O control lines to the 8257's memory control lines.

Note that with this connection, the programming of the Read (bit 15) and Write (bit 14) bits in the terminal count register will have a different meaning.

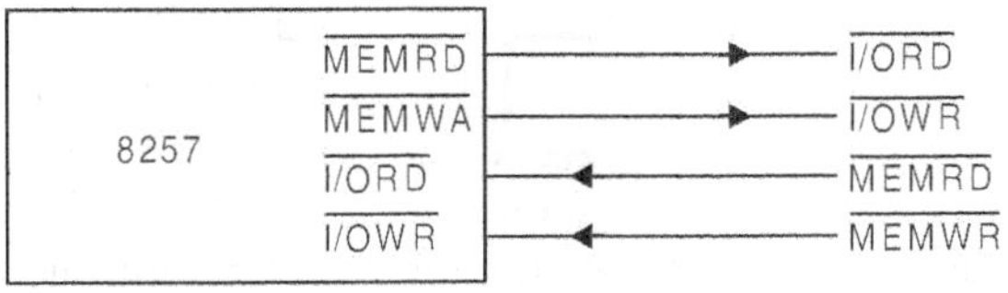

Fig. 6.11 System interface for Memory Mapped I/O

Bit 15 Read	Bit 14 Write	
0	0	DMA Verify Cycle
0	1	DMA Read Cycle
1	0	DMA Write Cycle
1	1	Illegal

Fig. 6.12 TC Register for Memory Mapped I/O Only

System Application Examples

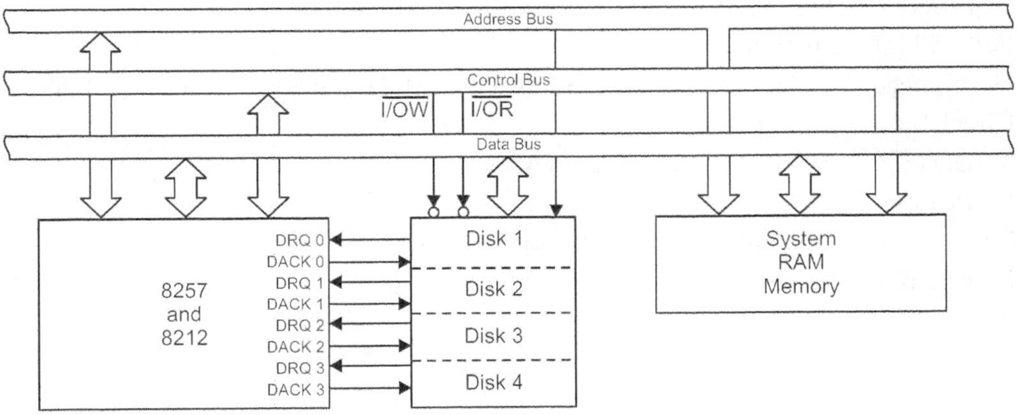

Fig. 6.13 Floppy Disk Controller (4 Drives)

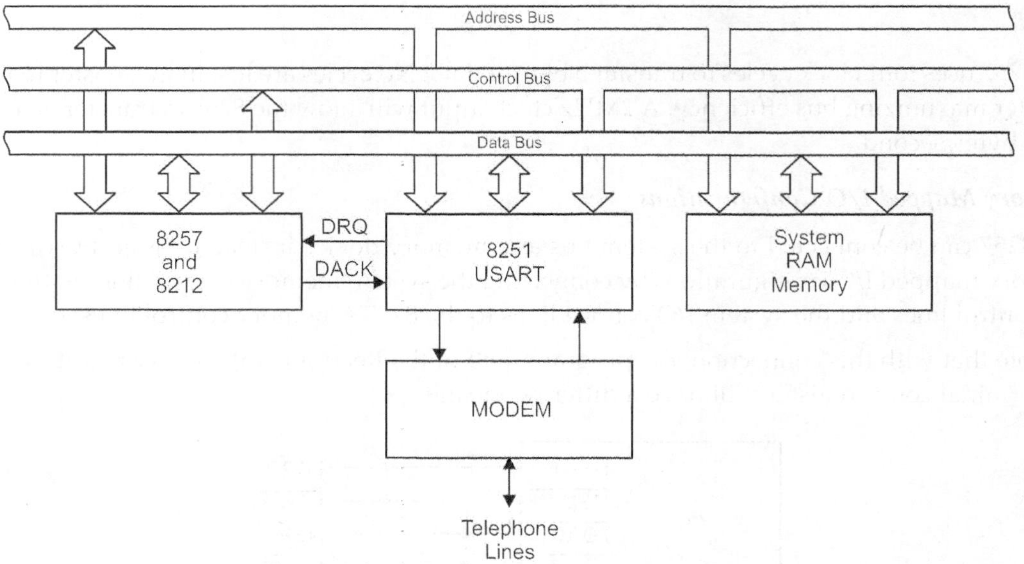

Fig. 6.14 High-Speed Communication Controller

6.6 82C55A

6.6.1 CMOS Programmable Peripheral Interface

The 82C55A is a high performance CMOS is manufactured using a self-aligned silicon gate CMOS process. It is a general purpose programmable I/O device which may be used with many different microprocessors. There are 24 I/O pins which may be individually programmed in 2 groups of 12 and used in 3 major modes of operation. The high performance and industry standard configuration of the 82C55A make it compatible with the 80C86, 8QC88 and other microprocessors.

Features

- 24 Programmable I/O Pins
- Fully TTL Compatible
- High Speed, No "Wait State" Operation with 5MHz and 8MHz 80C86 and 80C88
- Direct Bit Set/Reset Capability
- Enhanced Control Word Read Capability
- L7 Process
- 2.5mA Drive Capability on All I/O Ports
- Low Standby Power.................. 10μA

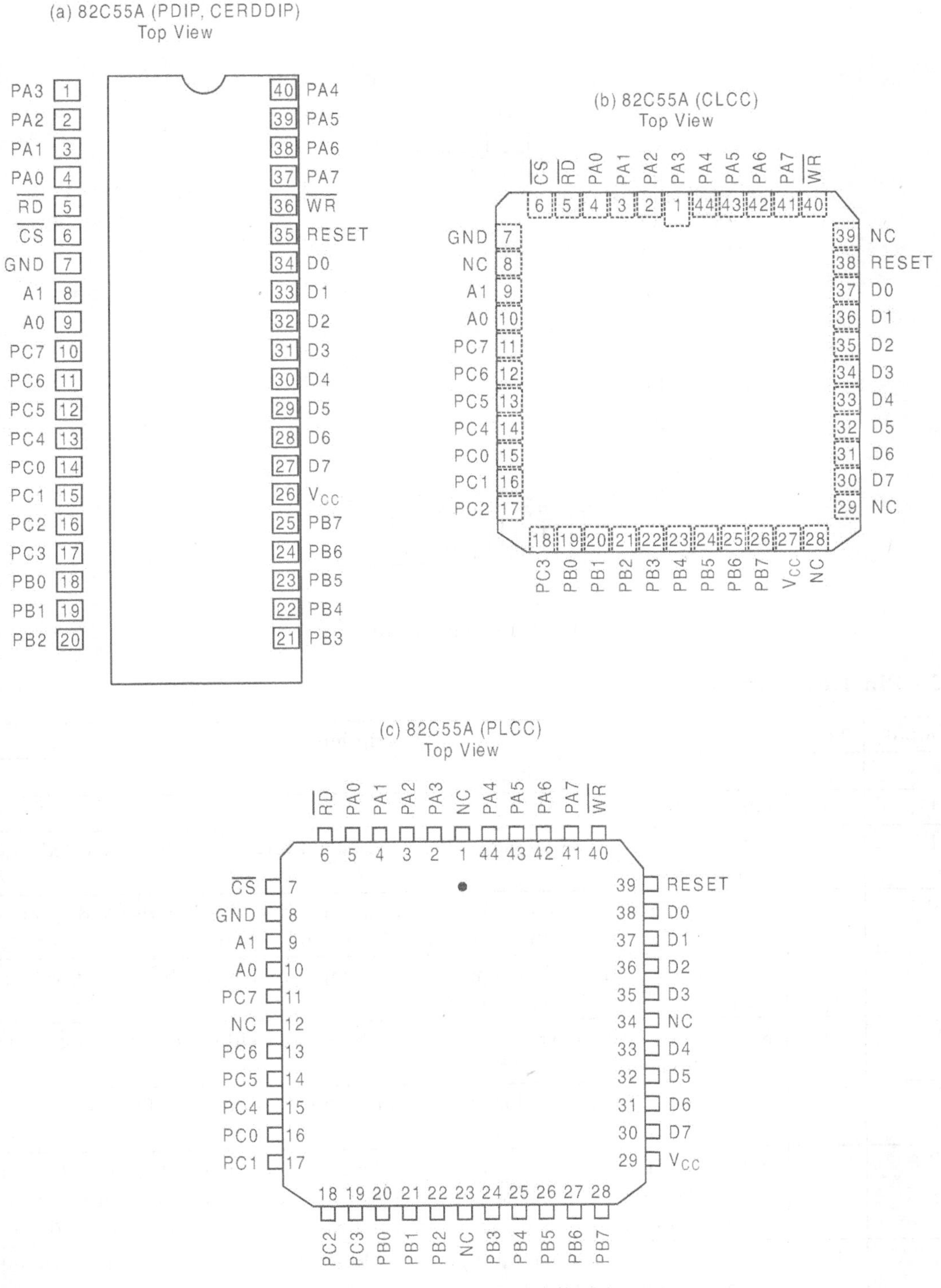

Fig. 6.15 Pinouts (a) (b) (c)

(d) 82C55A (MQFP)
Top View

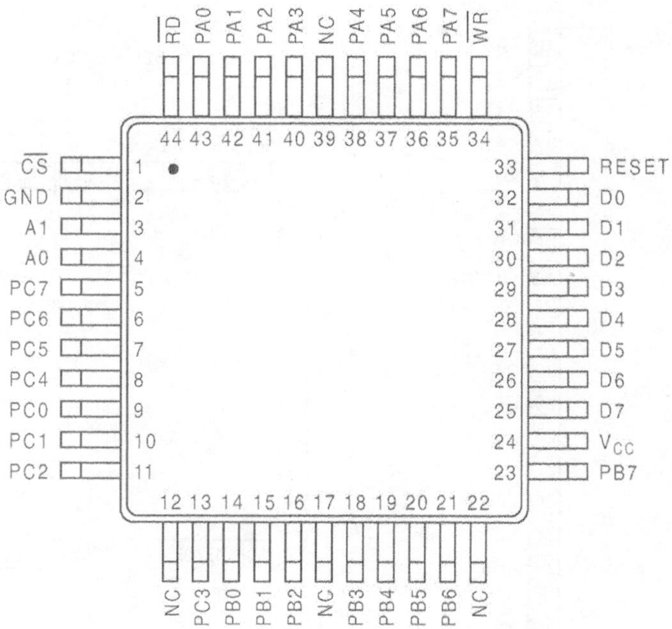

Fig. 6.15 Pinouts (d)

6.6.2 Pin Description

Symbol	Type	Description
V_{CC}		V_{CC}: The +5V power supply pin.
GND		GROUND
D0-D7	I/O	*DATA BUS:* The Data Bus lines are bidirectional three-state pins connected to the system data bus.
RESET	I	*RESET:* A high on this input clears the control register and all ports (A, B, C) are set to the input mode with the "Bus Hold" circuitry turned on.
$\overline{CS}$	I	*CHIP SELECT:* Chip select is an active low input used to enable the 82C55A onto the Data Bus for CPU communications.
$\overline{RD}$	I	*READ:* Read is an active low input control signal used by the CPU to read status information or data via the data bus.
$\overline{WR}$	I	*WRITE:* Write is an active low input control signal used by the CPU to load control words and data into the 82C55A.
A0-A1	I	*ADDRESS:* These input signals, in conjunction with the $\overline{RD}$ and $\overline{WR}$ inputs, control the selection of one of the three ports or the control word register. A0 and A1 are normally connected to the least significant bits of the Address Bus A0, A1.
PA0-PA7	I/O	*PORT A:* 8-bit input and output port. Both bus hold high and bus hold low circuitry are present on this port.
PB0-PB7	I/O	*PORT B:* 8-bit input and output port. Bus hold high circuitry is present on this port.
PC0-PC7	I/O	*PORT C:* 8-bit input and output port. Bus hold circuitry is present on this port.

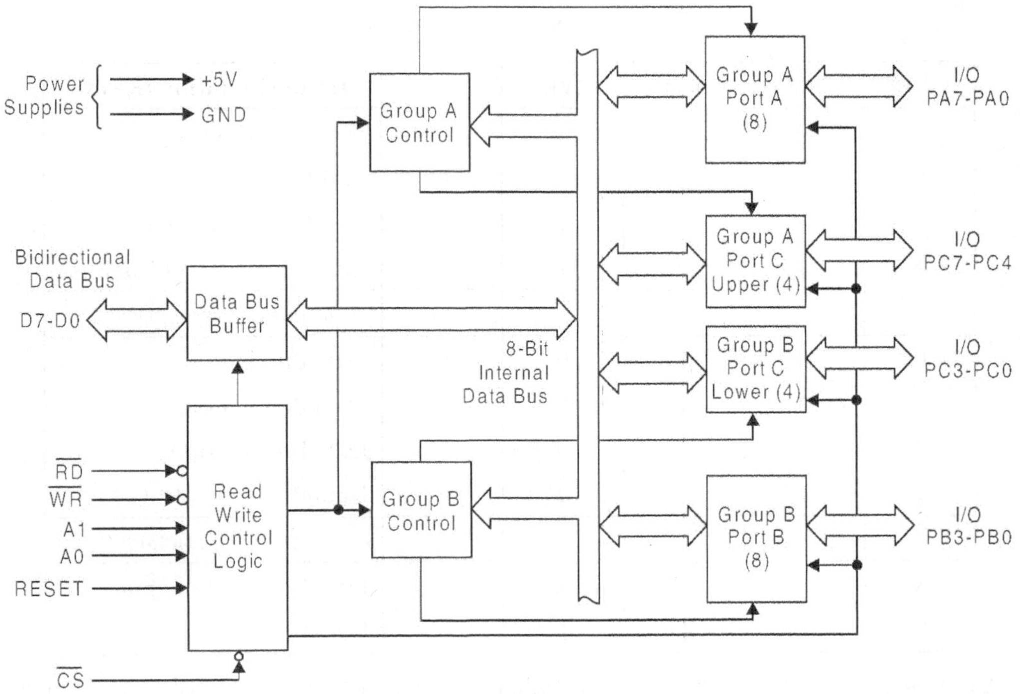

Fig. 6.16 Functional Diagram

6.6.3 Functional Description

Data Bus Buffer

This three-state bidirectional 8-bit buffer is used to interface the 82C55A to the system data bus. Data is transmitted or received by the buffer upon execution of input or output instructions by the CPU. Control words and status information are also transferred through the data bus buffer.

Read/Write and Control Logic

The function of this block is to manage all of the internal and external transfers of both Data and Control or Status words. It accepts inputs from the CPU Address and Control busses and in turn, issues commands to both of the Control Groups.

($\overline{CS}$) *Chip Select.* A "low" on this input pin enables the communication between the 82C55A and the CPU.

($\overline{RD}$) Read. A "low" on this input pin enables 82C55A to send the data or status information to the CPU on the data bus. In essence, it allows the CPU to "read from" the 82C55A.

($\overline{WR}$) *Write.* A "low" on this input pin enables the CPU to write data or control words into the 82C55A.

(A0 and A1) Port Select 0 and Port Select 1. These input signals, in conjunction with the RD and WR inputs, control the selection of one of the three ports or the control word register. They are normally connected to the least significant bits of the address bus (A0 and A1).

82C55A Basic Operation

A1	A0	$\overline{RD}$	$\overline{WR}$	$\overline{CS}$	**Input Operation (Read)**
0	0	0	1	0	Port A → Data Bus
0	1	0	1	0	Port B → Data Bus
1	0	0	1	0	Port C → Data Bus
1	1	0	1	0	Control Word → Data Bus
					Output Operation (Write)
0	0	1	0	0	Data Bus → Port A
0	1	1	0	0	Data Bus → Port B
1	0	1	0	0	Data Bus → Port C
1	1	1	0	0	Data Bus → Control
					Disable Function
X	X	X	X	1	Data Bus → Three-State
X	X	1	1	0	Data Bus → Three-State

(RESET) Reset. A "high" on this input initializes the control register to 9Bh and all ports (A, B, C) are set to the input mode. "Bus hold" devices internal to the 82C55A will hold the I/O port inputs to a logic "1" state with a maximum hold current of 400μA.

Group A and Group B Controls

The functional configuration of each port is programmed by the systems software. In essence, the CPU "outputs" a control word to the 82C55A. The control word contains information such as "mode", "bit set", "bit reset", etc., that initializes the functional configuration of the 82C55A.

Each of the Control blocks (Group A and Group B) accepts "commands" from the Read/Write Control logic, receives "control words" from the internal data bus and issues the proper commands to its associated ports.

Control Group A - Port A and Port C upper (C7 - C4)

Control Group B - Port B and Port C lower (C3 - C0)

The control word register can be both written and read as shown in the "Basic Operation" table. Figure 6.20 shows the control word format for both Read and Write operations. When the control word is read, bit D7 will always be a logic "1", as this implies control word mode information.

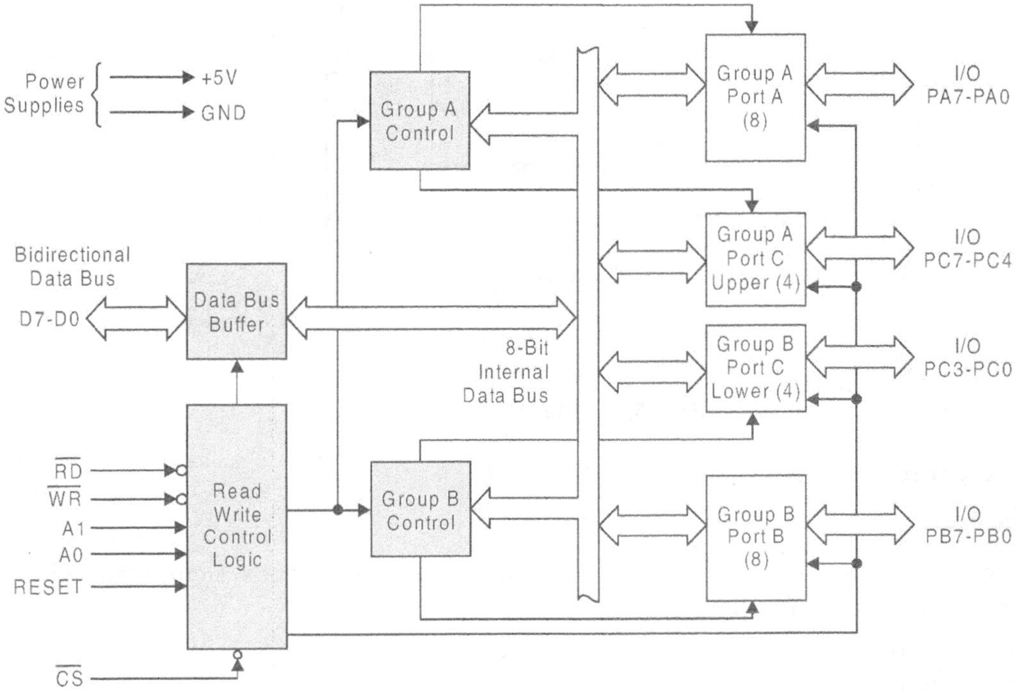

Fig. 6.17 82C55A Block Diagram. Data Bus Buffer, READ/WRITE, Group A and B Control Logic Functions

Ports A, B, and C

The 82C55A contains three 8-bit ports (A, B, and C). All can be configured to a wide variety of functional characteristics by the system software but each has its own special features or "personality" to further enhance the power and flexibility of the 82C55A.

Port A. One 8-bit data output latch/buffer and one 8-bit data input latch. Both "pull-up" and "pull-down" bus-hold devices are present on Port A. See Fig. 6.18(*a*).

Port B. One 8-bit data input/output latch/buffer and one 8-bit data input buffer. See Fig. 6.18(*b*).

Port C. One 8-bit data output latch/buffer and one 8-bit data input buffer (no latch for input). This port can be divided into two 4-bit ports under the mode control. Each 4-bit port contains a 4-bit latch and it can be used for the control signal output and status signal inputs in conjunction with ports A and B. See Fig. 6.18(*b*).

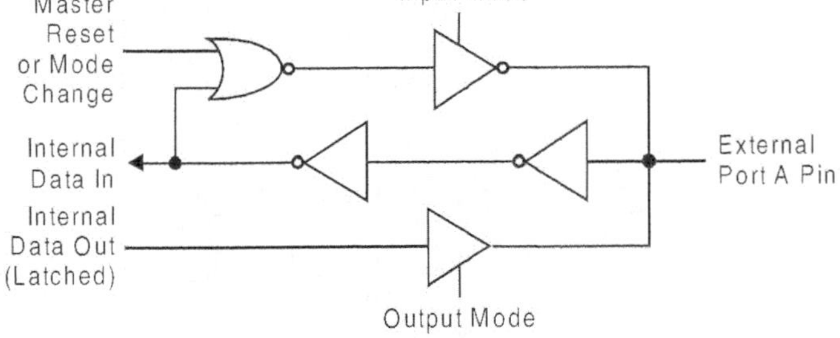

(*a*) Port A Bus-Hold Configuration.

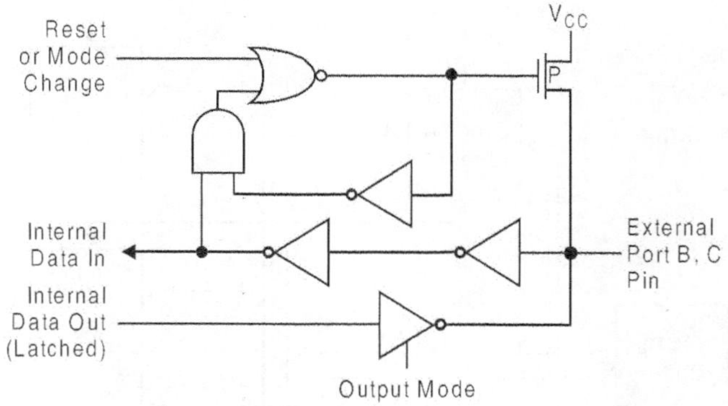

(b) Port B and C Bus-Hold Configuration.

Fig. 6.18 Bus-Hold Configuration

6.6.4 Operational Description

Mode Selection

There are three basic modes of operation than can be selected by the system software:

Mode 0—Basic Input/Output

Mode 1—Strobed Input/Output

Mode 2—Bidirectional Bus

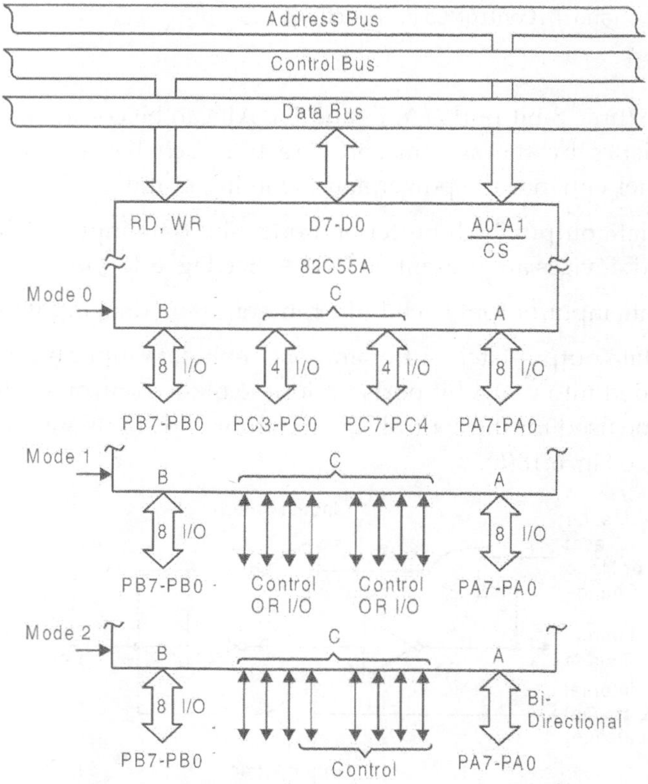

Fig. 6.19 Basic Mode Definitions and Bus Interface

When the reset input goes "high", all ports will be set to the input mode with all 24 port lines held at a logic "one" level by internal bus hold devices. After the reset is removed, the 82C55A can remain in the input mode with no additional initialization required. This eliminates the need to pull-up or pull-down resistors in all-CMOS designs. The control word register will contain 9Bh. During the execution of the system program, any of the other modes may be selected using a single output instruction. This allows a single 82C55A to service a variety of peripheral devices with a simple software maintenance routine. Any port programmed as an output port is initialized to all zeros when the control word is written.

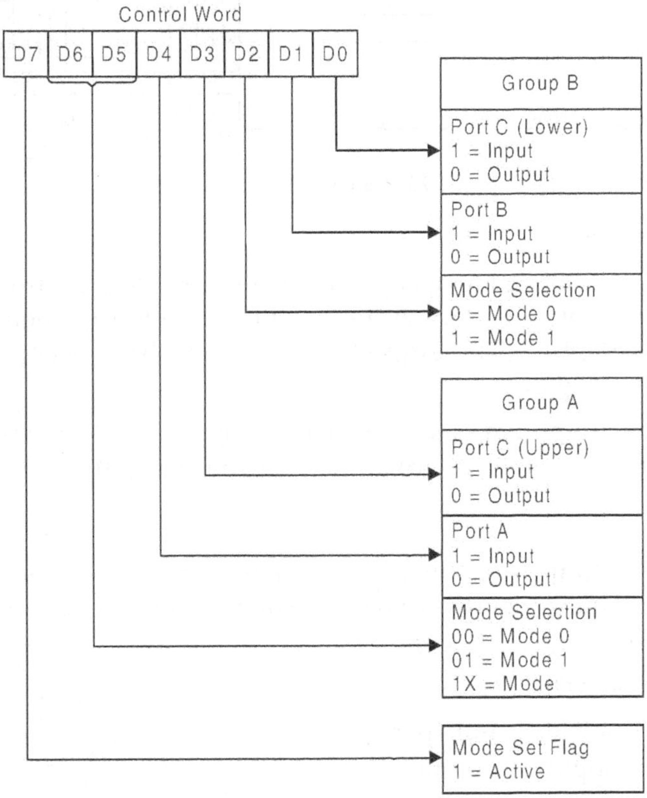

Fig. 6.20 Mode Definition Format

The modes for Port A and Port B can be separately defined, while Port C is divided into two portions as required by the Port A and Port B definitions. All of the output registers, including the status flip-flops, will be reset whenever the mode is changed. Modes may be combined so that their functional definition can be "tailored" to almost any I/O structure. For instance: Group B can be programmed in Mode 0 to monitor simple switch closings or display computational results, Group A could be programmed in Mode 1 to monitor a keyboard or tape reader on an interrupt-driven basis.

Single Bit Set/Reset Feature

Any of the eight bits of Port C can be Set or Reset using a single Output instruction. This feature reduces software requirements in control-based applications.

When Port C is being used as status/control for Port A or B, these bits can be set or reset by using the Bit Set/Reset operation just as if they were output ports.

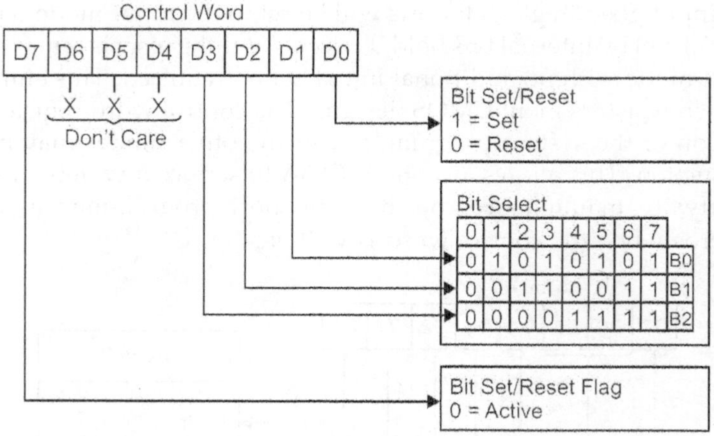

Fig. 6.21 Bit Set/Reset Format

Interrupt Control Functions

When the 82C55A is programmed to operate in mode 1 or mode 2, control signals are provided that can be used as interrupt request inputs to the CPU. The interrupt request signals, generated from port C, can be inhibited or enabled by setting or resetting the associated INTE flip-flop, using the bit set/reset function of port C.

This function allows the programmer to enable or disable a CPU interrupt by a specific I/O device without affecting any other device in the interrupt structure.

Operating Modes

Mode 0 (Basic Input/Output). This functional configuration provides simple input and output operations for each of the three ports. No handshaking is required, data is simply written to or read from a specific port.

Mode 0 Basic Functional Definitions:

- Two 8-bit ports and two 4-bit ports
- Any port can be input or output
- Outputs are latched
- Inputs are not latched
- 16 different Input/Output configurations possible

MODE 0 Port Definition

A		B		Group A		#	Group B	
D4	D3	D1	D0	PORT A	PORT C (Upper)		PORT B	PORT C (Lower)
0	0	0	0	Output	Output	0	Output	Output
0	0	0	1	Output	Output	1	Output	Input
0	0	1	0	Output	Output	2	Input	Output
0	0	1	1	Output	Output	3	Input	Input
0	1	0	0	Output	Input	4	Output	Output

0	1	0	1	Output	Input	5	Output	Input
0	1	1	0	Output	Input	6	Input	Output
0	1	1	1	Output	Input	7	Input	Input
1	0	0	0	Input	Output	8	Output	Output
1	0	0	1	Input	Output	9	Output	Input
1	0	1	0	Input	Output	10	Input	Output
1	0	1	1	Input	Output	11	Input	Input
1	1	0	0	Input	Input	12	Output	Output
1	1	0	1	Input	Input	13	Output	Input
1	1	1	0	Input	Input	14	Input	Output
1	1	1	1	Input	Input	15	Input	Input

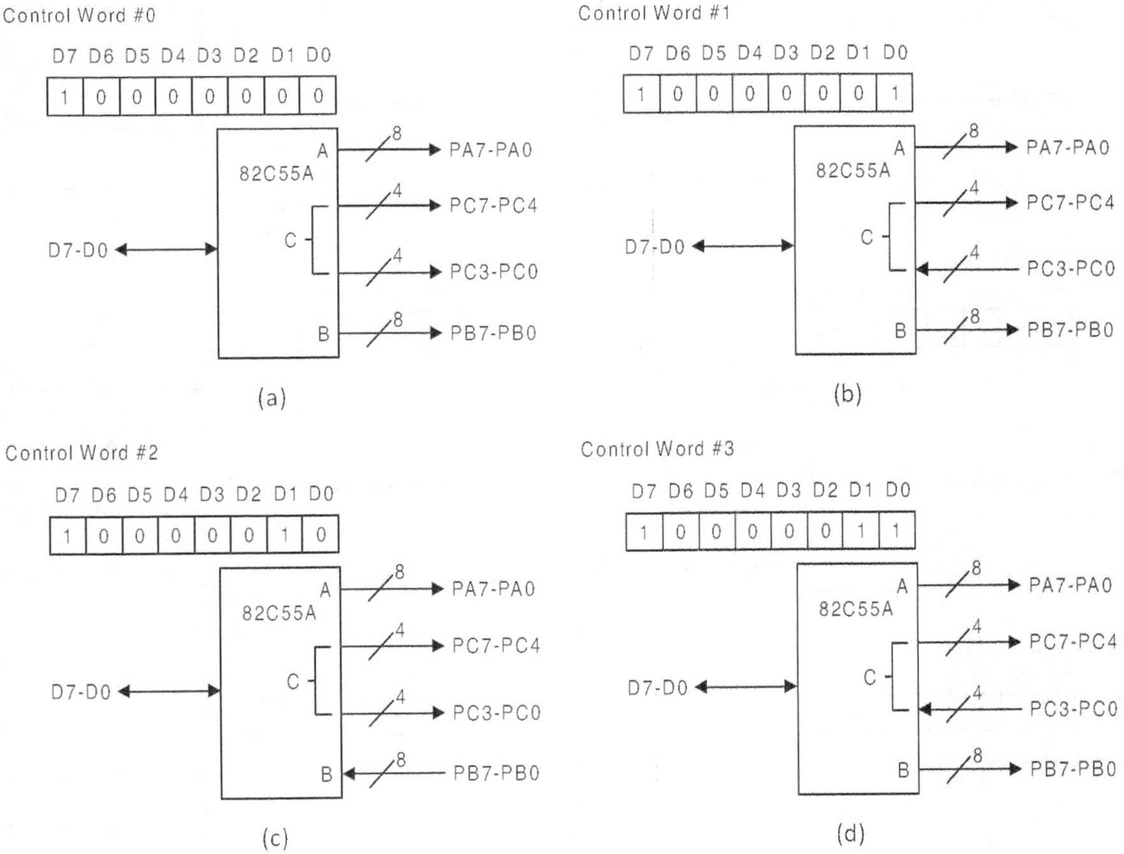

Fig. 6.22 Mode 0 Configurations (Contd...)

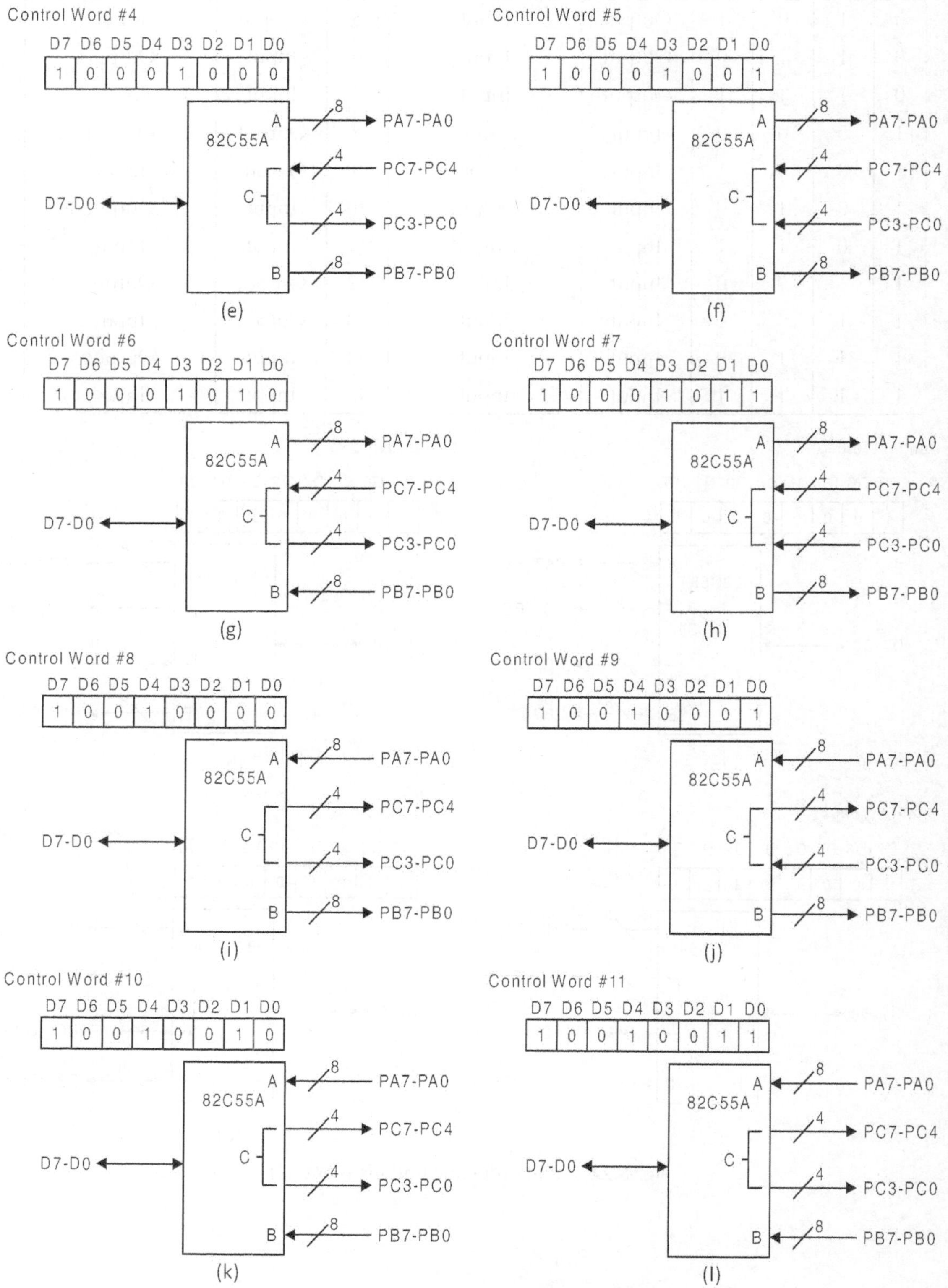

Fig. 6.22 Mode 0 Configurations (Contd...)

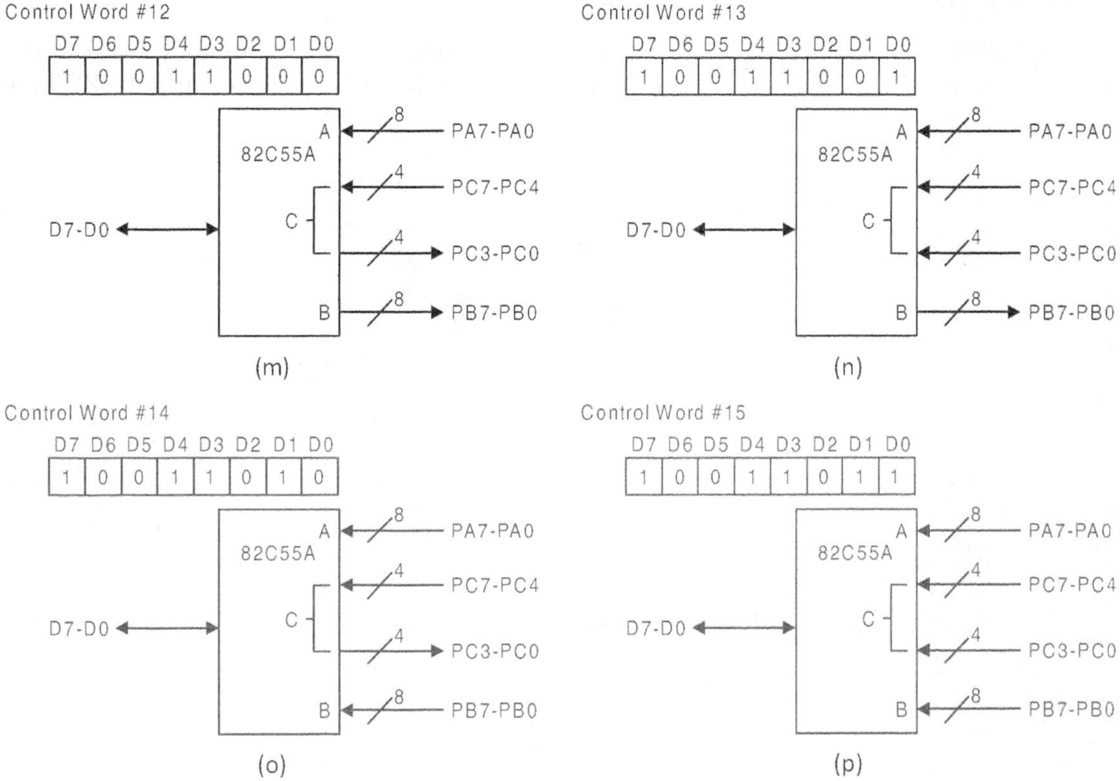

Fig. 6.22 Mode 0 Configurations

Operating Modes

Mode 1 (Strobed Input/Output). This functional configuration provides a means for transferring I/O data to or from a specified port in conjunction with strobes or "hand shaking" signals. In mode 1, port A and port B use the lines on port C to generate or accept these "hand shaking" signals.

Mode 1 Basic Function Definitions:

- Two Groups (Group A and Group B).
- Each group contains one 8-bit port and one 4-bit control/data port.
- The 8-bit data port can be either input or output. Both inputs and outputs are latched.
- The 4-bit port is used for control and status of the 8-bit port.

Input Control Signal Definition

(Figures 6.23 and 6.24)

$\overline{STB}$ (Strobe Input)

A "low" on this input loads data into the input latch.

IBF (Input Buffer Full F/F)

A "high" on this output indicates that the data has been loaded into the input latch: in essence, an acknowledgement. IBF is set by $\overline{STB}$ input being low and is reset by the rising edge of the RD input.

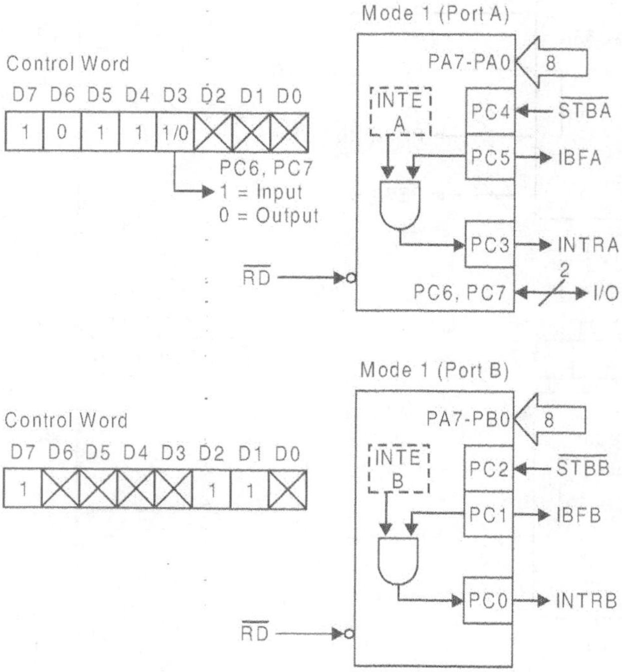

Fig. 6.23 Mode 1 Input

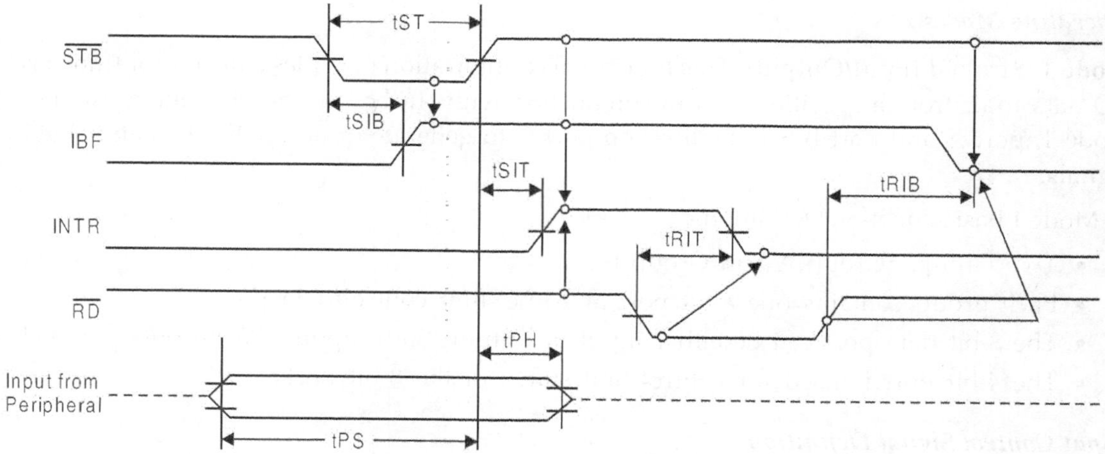

Fig. 6.24 Mode 1 (Strobed Input)

INTR (Interrupt Request)

A "high" on this output can be used to interrupt the CPU when an input device is requesting service. INTR is set by the condition: $\overline{STB}$ is a "one", IBF is a "one" and INTE is a "one". It is reset by the falling edge of $\overline{RD}$. This procedure allows an input device to request service from the CPU by simply strobing its data into the port.

INTE A

Controlled by bit set/reset of PC4.

INTE B

Controlled by bit set/reset of PC2.

Output Control Signal Definition

(Figure 6.25 and 6.26)

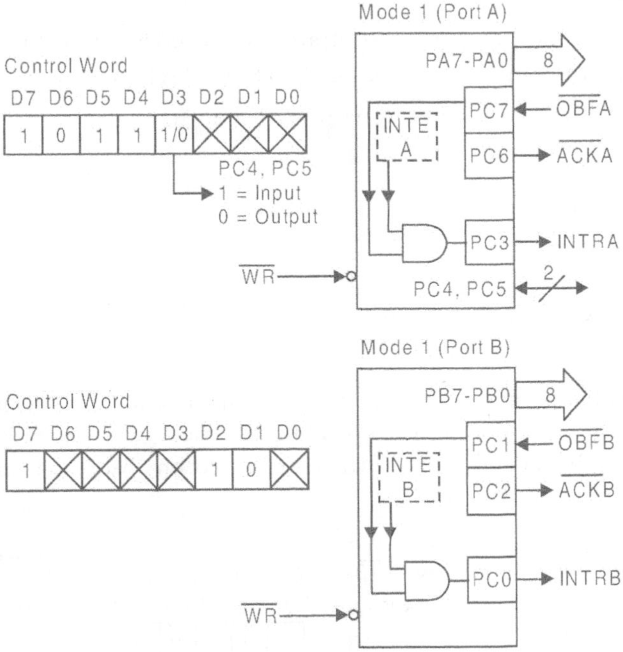

Fig. 6.25 Mode 1 Output

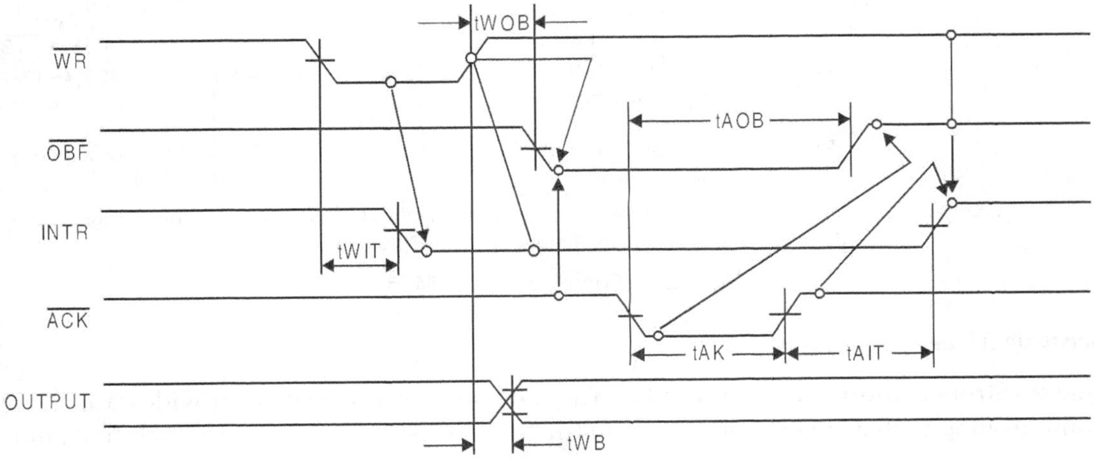

Fig. 6.26 Mode 1 (Strobed Output)

$\overline{OBF}$ -(Output Buffer Full F/F). The $\overline{OBF}$ output will go "low" to indicate that the CPU has written data out to the specified port. This does not mean valid data is sent out of the port at this time since $\overline{OBF}$ can go true before data is available. Data is guaranteed valid at the rising edge of $\overline{OBF}$, (See Note 1). The $\overline{OBF}$ F/F will be set by the rising edge of the WR input and reset by $\overline{ACK}$ input being low.

$\overline{ACK}$ -(Acknowledge Input). A "low" on this input informs the 82C55A that the data from Port A or Port B is ready to be accepted. In essence, a response from the peripheral device indicating that it is ready to accept data, (See Note 1).

INTR-(Interrupt Request). A "high" on this output can be used to interrupt the CPU when an output device has accepted data transmitted by the CPU. INTR is set when $\overline{ACK}$ is a "one", $\overline{OBF}$ is a "one" and INTE is a "one". It is reset by the falling edge of $\overline{WR}$.

INTE A

Controlled by Bit Set/Reset of PC6.

INTE B

Controlled by Bit Set/Reset of PC2.

Note:

1 To strobe data into the peripheral device, the user must operate the strobe line in a hand shaking mode. The user needs to send $\overline{OBF}$ to the peripheral device, generates an $\overline{ACK}$ from the peripheral device and then latch data into the peripheral device on the rising edge of $\overline{OBF}$.

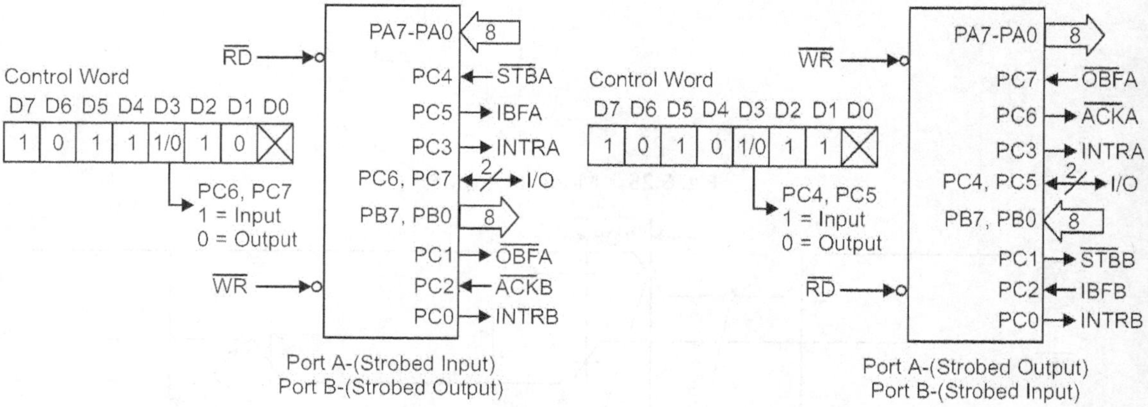

Combinations of Mode 1: Port A and Port B can be individually defined as input or output in Mode 1 to support a wide variety of strobed I/O applications.

Fig. 6.27 Combinations of Mode 1

Operating Modes

Mode 2 (Strobed Bidirectional Bus I/O): This functional configuration provides a means for communicating with a peripheral device or structure on a single 8-bit bus for both transmitting and receiving data (bidirectional bus I/O). "Hand shaking" signals are provided to maintain proper bus flow discipline similar to Mode 1. Interrupt generation and enable/disable functions are also available.

Mode 2 Basic Functional Definitions:

- Used in Group A only
- One 8-bit, bidirectional bus Port (Port A) and a 5-bit control Port (Port C)
- Both inputs and outputs are latched
- The 5-bit control port (Port C) is used for control and status for the 8-bit, bidirectional bus port (Port A)

Bidirectional Bus I/O Control Signal Definition

(Figures 6.28, 6.29, 6.30, 6.31)

INTR-(Interrupt Request). A high on this output can be used to interrupt the CPU for both input or output operations.

Output Operations

$\overline{\text{OBF}}$ -*(Output Buffer Full).* The $\overline{\text{OBF}}$ output will go "low" to indicate that the CPU has written data out to port A.

$\overline{\text{ACK}}$ -*(Acknowledge).* A "low" on this input enables the three-state output buffer of port A to send out the data. Otherwise, the output buffer will be in the high impedance state.

INTE 1-(The INTE flip-flop associated with $\overline{OBF}$). Controlled by bit set/reset of PC4.

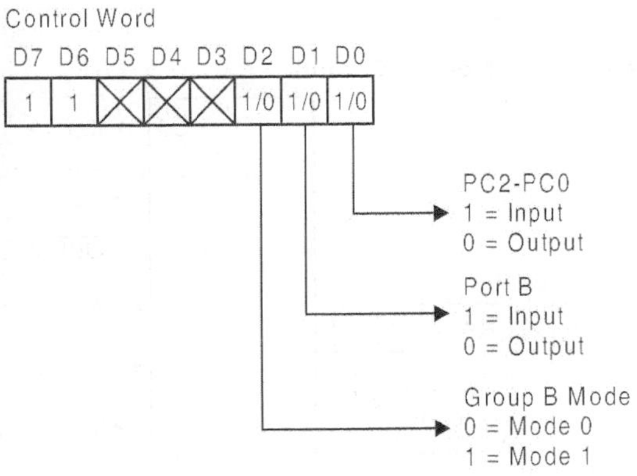

Fig. 6.28 Mode Control Word

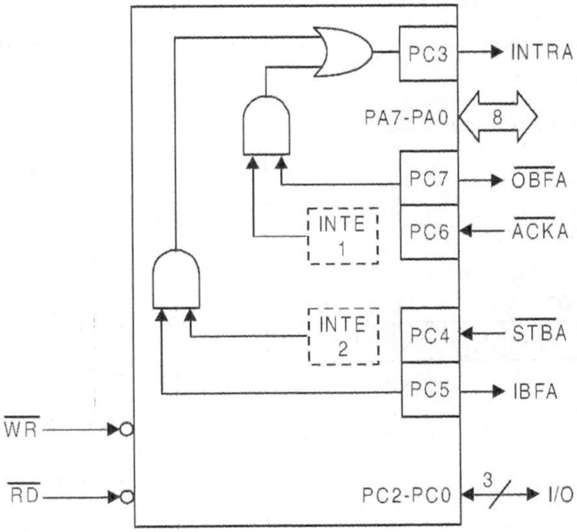

Fig. 6.29 Mode 2

Input Operations

$\overline{\text{STB}}$ -(Strobe Input). A "low" on this input loads data into the input latch.

IBF-(Input Buffer Full F/F). A "high" on this output indicates that data has been loaded into the input latch.

INTE 2-(The INTE flip-flop associated with IBF). Controlled by bit set/reset of PC4.

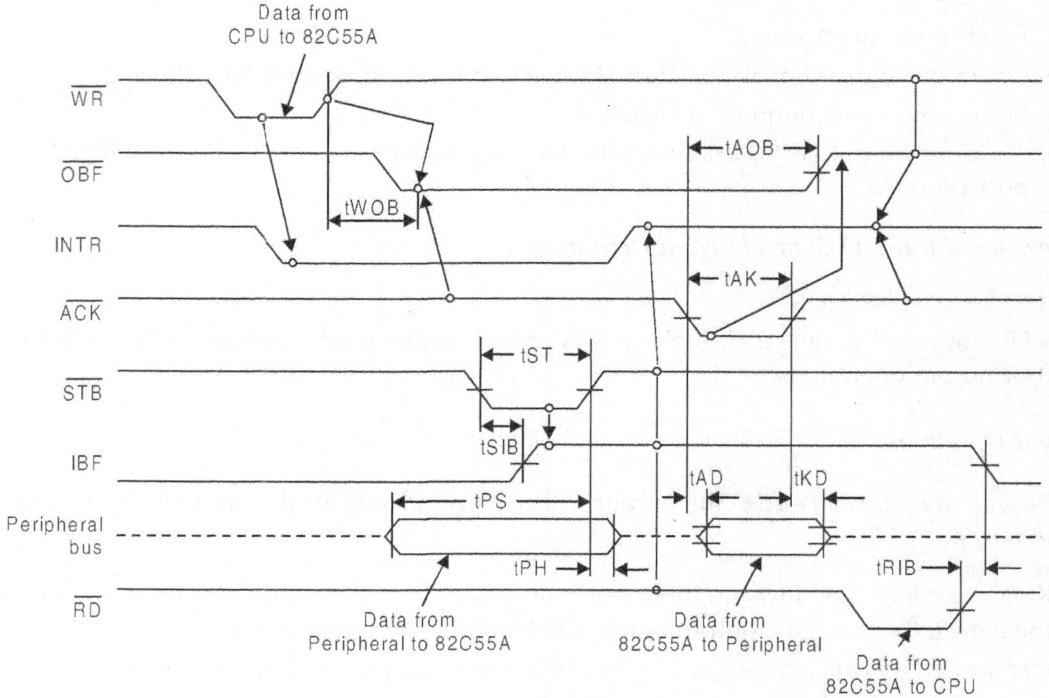

Note: Any sequence $\overline{WR}$ occurs before $\overline{ACK}$ and $\overline{STB}$ occurs before RD is permissible.
$(INTR = IBF \bullet MASK \bullet \overline{STB} \bullet \overline{RD} + \overline{OBF} \bullet MASK \bullet \overline{ACK} \bullet \overline{WR})$

Fig. 6.30 Mode 2 (Bidirectional)

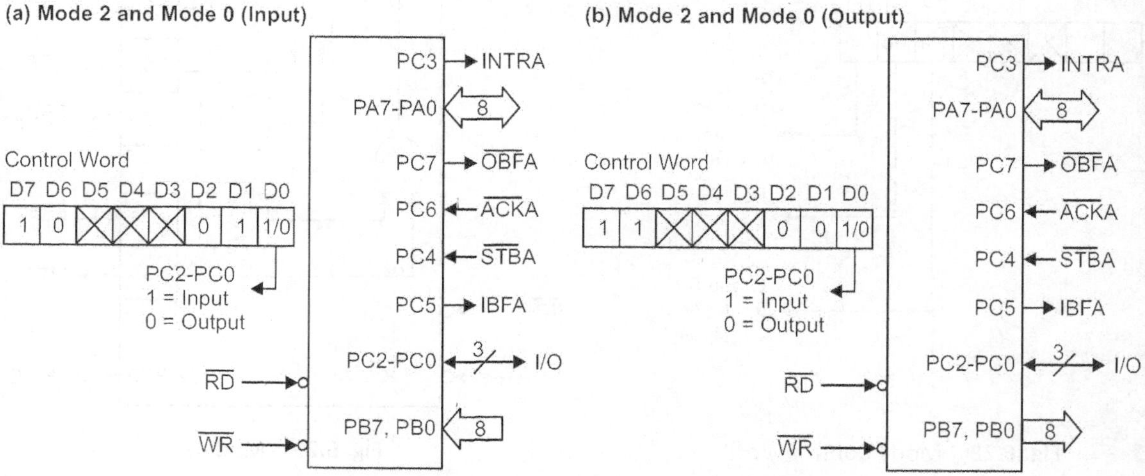

Fig. 6.31 Mode 2 Combinations (Contd...)

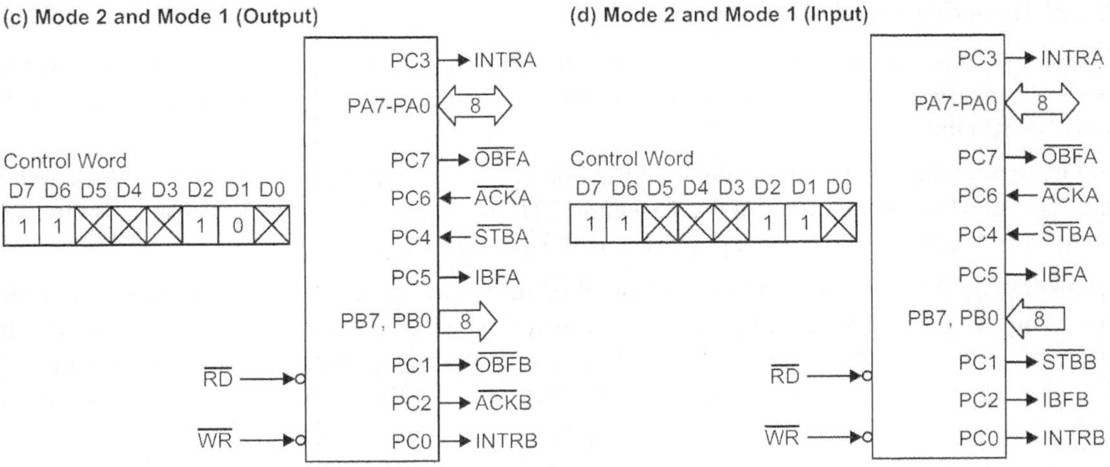

Fig. 6.31 Mode 2 Combinations

Mode Definition Summary

	Mode 0		Mode 1		Mode 2
	In	Out	In	Out	Group a only
PA0	In	Out	In	Out	↔
PA1	In	Out	In	Out	↔
PA2	In	Out	In	Out	↔
PA3	In	Out	In	Out	↔
PA4	In	Out	In	Out	↔
PA5	In	Out	In	Out	↔
PA6	In	Out	In	Out	↔
PA7	In	Out	In	Out	↔
PB0	In	Out	In	Out	
PB1	In	Out	In	Out	
PB2	In	Out	In	Out	
PB3	In	Out	In	Out	
PB4	In	Out	In	Out	
PB5	In	Out	In	Out	
PB6	In	Out	In	Out	
PB7	In	Out	In	Out	
PC0	In	Out	INTRB	INTRB	I/O
PC1	In	Out	IBFB	$\overline{\text{OBFB}}$	I/O
PC2	In	Out	$\overline{\text{STBB}}$	$\overline{\text{ACKB}}$	I/O
PC3	In	Out	INTRA	INTRA	INTRA
PC4	In	Out	$\overline{\text{STBA}}$	I/O	$\overline{\text{STBA}}$
PC5	In	Out	IBFA	I/O	IBFA
PC6	In	Out	I/O	$\overline{\text{ACKA}}$	$\overline{\text{ACKA}}$
PC7	In	Out	I/O	$\overline{\text{OBFA}}$	$\overline{\text{OBFA}}$

(PB0–PB7: Mode 0 or Mode 1 Only)

Special Mode Combination Considerations

There are several combinations of modes possible. For any combination, some or all of Port C lines are used for control or status. The remaining bits are either inputs or outputs as defined by a "Set Mode" command.

During a read of Port C. the state of all the Port C lines, except the $\overline{ACK}$ and $\overline{STB}$ lines, will be placed on the data bus. In place of the $\overline{ACK}$ and $\overline{STB}$ line states, flag status will appear on the data bus in the PC2, PC4, and PC6 bit positions as illustrated by Fig 6.31.

Through a "Write Port C" command, only the Port C pins programmed as outputs in a Mode 0 group can be written. No other pins can be affected by a "Write Port C" command, nor can the interrupt enable flags be accessed. To write to any Port C output programmed as an output in Mode 1 group or to change an interrupt enable flag, the "Set/Reset Port C Bit" command must be used.

With a "Set/Reset Port C Bit" command, any Port C line programmed as an output (including IBF and $\overline{OBF}$) can be written, or an interrupt enable flag can be either set or reset. Port C lines programmed as inputs, including $\overline{ACK}$ and $\overline{STB}$ lines, associated with Port C are not affected by a "Set/Reset Port C Bit" command. Writing to the corresponding Port C bit positions of the $\overline{ACK}$ and $\overline{STB}$ lines with the "Set Reset Port C Bit" command will affect the Group A and Group B interrupt enable flags.

Current Drive Capability

Any output on Port A, B or C can sink or source 2.5mA. This feature allows the 82C55A to directly drive Darlington type drivers and high-voltage displays that require such sink or source current.

Input Configuration

D7	D6	D5	D4	D3	D2	D1	D0
I/O	I/O	IBFA	INTEA	INTRA	INTEB	IBFB	INTRB
Group A					Group B		

Output Configuration

D7	D6	D5	D4	D3	D2	D1	D0
OBFA	INTEA	I/O	I/O	INTRA	INTEB	ODFB	INTRB
Group A					Group B		

Fig. 6.32 Mode 1 Status Word Format

D7	D6	D5	D4	D3	D2	D1	D0
OBFA	INTE1	IBFA	INTE2	INTRA	X	X	X
Group A					Group B (Defined by Mode 2 or Mode 1 selection		

Fig. 6.33 Mode 2 Status Word Format

Reading Port C Status (Figures 6.32 and 6.33).

In Mode 0, Port C transfers data to or from the peripheral device. When the 82C55A is programmed to function in Modes 1 or 2, Port C generates or accepts "hand shaking" signals with the peripheral device. Reading the contents of Port C allows the programmer to test or verify the "status" of each peripheral device and change the program flow accordingly.

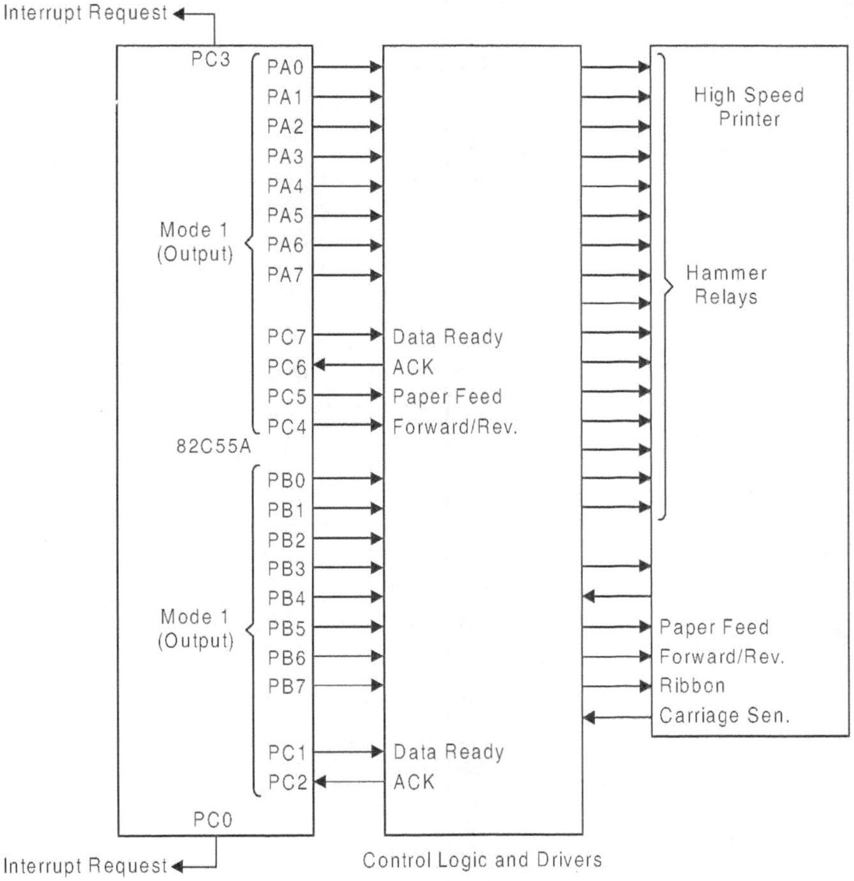

Fig. 6.34 Printer Interface

There is not a special instruction to read the status information from Port C. A normal read operation of Port C is executed to perform this function.

6.6.5 Applications of the 82C55A

The 82C55A is a very powerful tool for interfacing peripheral equipment to the microcomputer system. It represents the optimum use of available pins and is flexible enough to interface almost any I/O device without the need for additional external logic.

Interrupt Enable Flag	Position	Alternate Port C Pin signal (Mode)
INTE B	PC2	$\overline{\text{ACKB}}$ (Output Mode 1) or $\overline{\text{STBB}}$ (Input Mode 1)
INTE A2	PC4	$\overline{\text{STBA}}$ (Input Mode 1 or Mode 2)
INTE A1	PC6	$\overline{\text{ACKA}}$ (Output Mode 1 or Mode 2)

Fig. 6.35 Interrupt Enable Flags in Modes 1 and 2

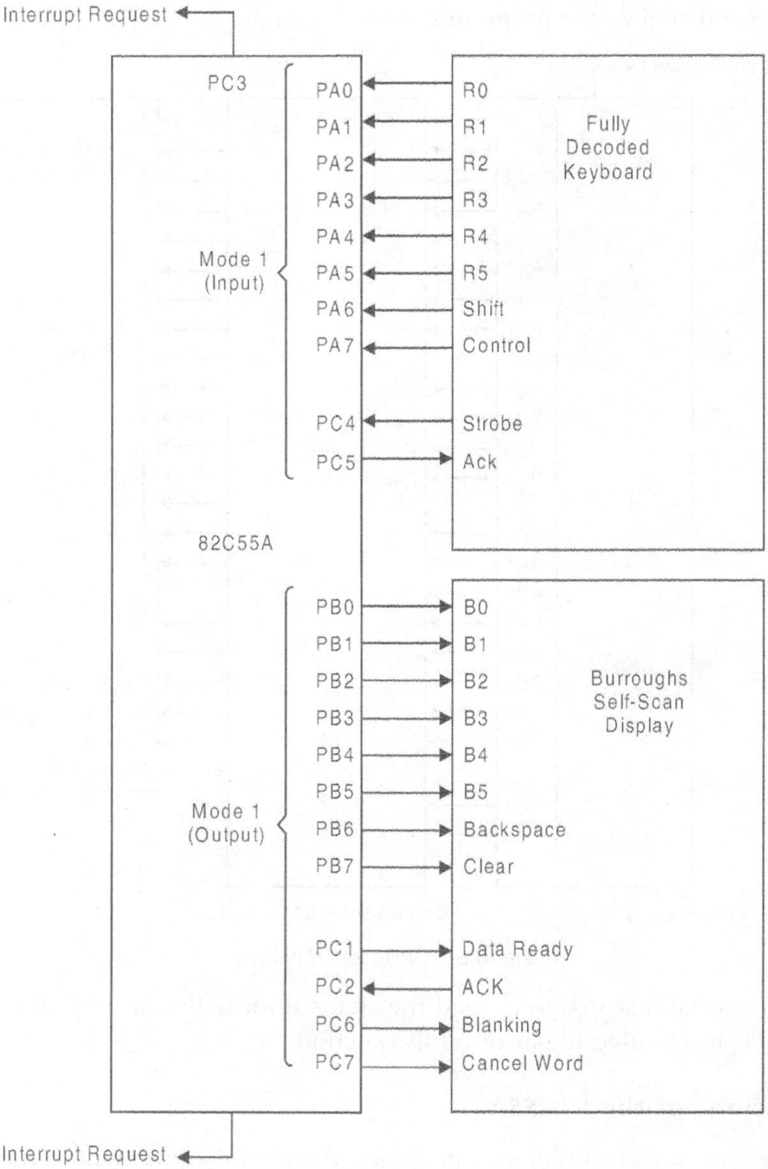

Fig. 6.36 Keyboard and Display Interface

Each peripheral device in a microcomputer system usually has a "service routine" associated with it. The routine manages the software interface between the device and the CPU. The functional definition of the 82C55A is programmed by the I/O service routine and becomes an extension of the system software. By examining the I/O devices interface characteristics for both data transfer and timing, and matching this information to the examples and tables in the detailed operational description, a control word can easily be developed to initialize the 82C55A to exactly "fit" the application. Figure 6.35 through 24 present a few examples of typical applications of the 82C55A.

6.7 8251A USART

- Synchronous and Asynchronous Operation
- Synchronous 5-8 Bit Characters; Internal or External Character Synchronization; Automatic Sync Insertion
- Asynchronous 5-8 Bit Characters; Clock Rate—1, 16 or 64 Times Baud Rate; Break Character Generation; 1, 1½, or 2 stop Bits; False Start Bit Detection; Automatic Break Detect and Handling
- Synchronous Baud Rate—DC to 64K Baud
- Asynchronous Baud Rate—DC to 19.2K Baud
- Full-Duplex, Double-Buffered Transmitter and Receiver
- Error Detection—Partly, Overrun and Framing
- Compatible with an Extended Range of Intel Microprocessors
- 28—Pin DIP Package
- All Inputs and Outputs are TTL Compatible

The 8251A is the industry standard Universal Synchronous/Asynchronous Receiver/Transmitter (USART), designed for data communications with microprocessor families such as MCS-48, 80, 85, and IAPX-86, 88. The 8251A is used as a peripheral device and is programmed by the CPU to operate using virtually any serial data transmission technique presently in use (including IBM "bi-sync"). The USART accepts data characters from the CPU in parallel format and the converts them into a continuous serial data stream for transmission, Simultaneously, it can receive serial data streams and convert them into parallel data characters for the CPU. The USART will signal the CPU whenever it can accept a new character for transmission or whenever it has received a character for the CPU. The CPU can read the complete status of the USART at any time. These include data transmission errors and control signals such as SYNDET, TxEMPTY. The chip is fabricated using Inlet's high performance HMOS technology.

6.7.1 Features and Enhancements

The 8251A is an advanced design of the industry standard USART. The Intel 8251. The 8251A operates with an extended range of Intel microprocessors and maintains compatibility with the 8251, Familiarization time is minimal because of compatibility and involves only knowing the additional features and enhancements, and reviewing the AC and DC specifications of the 8251A.

The 8251A incorporates all the key features of the 8251 and has the following additional features and enhancements:

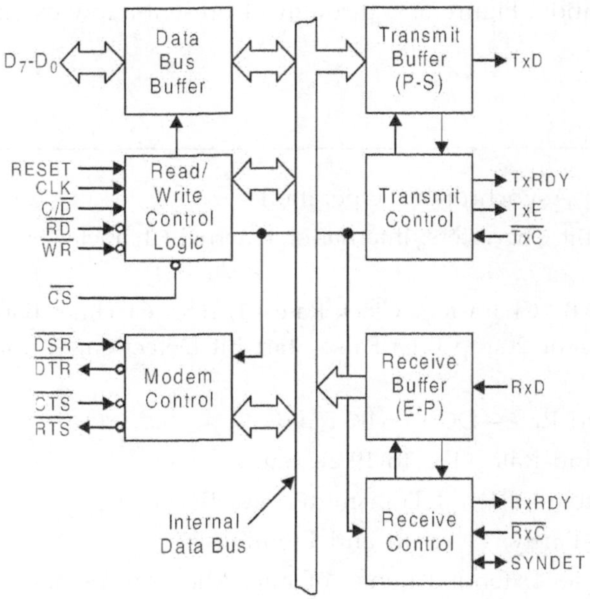

Fig. 6.37 Block Diagram

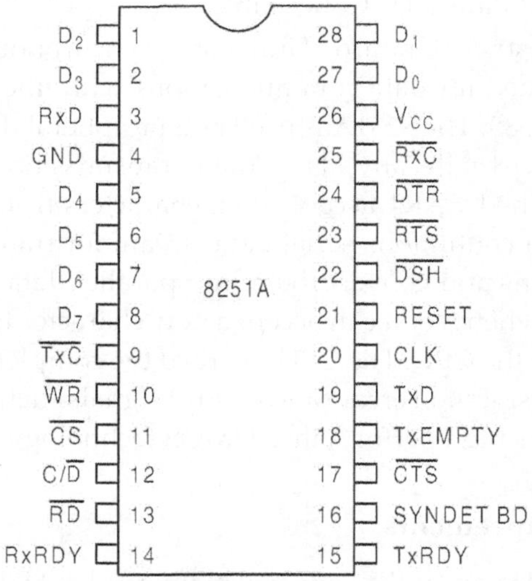

Fig. 6.38 Pin Configuration

- 8251A has double-buffered data paths with separate I/O registers for control, status, Data in, and Data Out, which considerably simplifies control programming and minimizes CPU overhead.
- In asynchronous operations, the receiver detects and handles "break" automatically, relieving the CPU of this task.
- A refined Rx initialization prevents the receiver from starting when in "break" state, preventing unwanted Interrupts from a disconnected USART.
- At the conclusion of a transmission, TxD line will always return to the marking state unless SBRK is programmed.

- Tx Enable logic enhancement prevents a Tx Disable command from halting transmission until all data previously written has been transmitted The logic also prevents the transmitter from turning off in the middle of a word.
- When External Sync Detect is programmed, internal Sync Detect is disabled, and an External Sync Detect status is provided via a flip-flop which clears itself upon a status read.
- Possibility of false sync detect is minimized by ensuring that if double character sync is programmed, the characters be contiguously detected and also by clearing the Rx register to all ones whenever Enter Hunt command to issued in Sync mode.
- As long as the 8251A to not selected, the $\overline{RD}$ and $\overline{WR}$ do not affect the internal operation of the device.
- The 8251A status can be read at any time but the status update will be inhibited during status read.
- The 8251A it free from extraneous glitches and has enhanced AC and DC characteristics, providing higher speed and better operating margins.
- Synchronous Baud rate from DC to 64K.

6.7.2 Functional Description

General

The 8251A is a Universal Synchronous/Asynchronous Receiver/Transmitter designed for a wide range of Intel microcomputers such as 8048, 8080, 8085, 8086 and 8088. Like other I/O devices in a microcomputer system, its functional configuration is programmed by the system's software for maximum flexibility. The 8251A can support most serial data techniques in use, including IBM "bi-sync".

In a communication environment an interface device must convert parallel format system data into serial format for transmission and convert incoming serial format data into parallel system data for reception. The interface device must also delete or insert bits or characters that are functionally unique to the communication technique. In essence, the interface should appear "transparent" to the CPU, a simple input or output of byte-oriented system data.

Data Bus Buffer

This 3-state bidirectional, 8-bit buffer is used to interface the 8251A to the system Data Bus. Data to transmitted or received by the buffer upon execution of INput or OUTput instructions of the CPU. Control words, Command words and Status information are also transferred through the Data Bus Buffer. The Command Status, Data-in and Data-Out registers are separate, 8-bit registers communicating with the system bus through the Data Bus Buffer.

This functional block accepts inputs from the system Control bus and generates control signals for overall device operation. It contains the Control Word Register and Command Word Register that store the various control formats for the device functional definition.

RESET (Reset)

A "high" on this input forces the 8251A into an "Idle" mode. The device will remain at "Idle" until a new set of control words is written into the 8251A to program its functional definition. Minimum RESET pulse width is 6 t_{CY} (clock must be running).

A command reset operation also puts the device into the "Idle" stale.

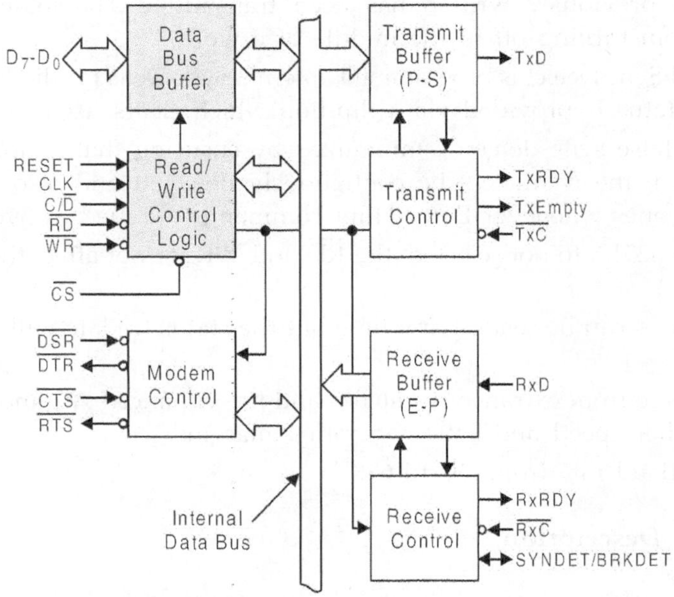

Fig. 6.39 8251A Block Diagram showing Data Bus Buffer and Read/Write Logic Functions

CLK (Clock)

The CLK input is used to generate internal device timing and is normally connected to the Phase 2 (TTL) output of the Clock Generator. No external inputs or outputs are referenced to CLK but the frequency of CLK must be greater than 30 times the Receiver or Transmitter data bit rates.

$\overline{WR}$ (Write)

A "low" on this input informs the 8251A that the CPU to writing data or control words to the 8251A.

$\overline{RD}$ (Read)

A "low" on this input informs the 8251A that the CPU is reading data or status information from the 8251A.

C/$\overline{D}$	$\overline{RD}$	$\overline{WR}$	$\overline{CS}$	
0	0	1	0	8251A Data → Data Bus
0	1	0	0	Data Bus → 8251A DATA
1	0	1	0	Status → DATA BUS
1	1	0	0	Data Bus → Control
X	1	1	0	Data Bus → 3-State
X	X	X	1	Data Bus → 3-State

$C/\overline{D}$ (Control/Data)

This input, in conjunction with the $\overline{WR}$ and $\overline{RD}$ inputs, informs the 8251A that the word on the Data Bus is either a data character, control word or status information.

1–Control/Status; 0 = Data.

$\overline{CS}$ (Chip Select)

A "low" on this input selects the 8251A. No reading or writing will occur unless the device is selected. When $\overline{CS}$ is high, the Data Bus is in the float state and $\overline{RD}$ and $\overline{WR}$ have no effect on the Chip.

Modem Control

The 8251A has a set of control inputs and outputs that can be used to simplify the interface to almost any modem. The modem control signals are general purpose in nature and can be used for functions other than modem control, if necessary.

$\overline{DSR}$ (Data Set Ready)

The $\overline{DSR}$ input signal is a general-purpose, 1-bit inverting input port. Its condition can be tested by the CPU using a Status Read operation. The $\overline{DSR}$ input is normally used to test modem conditions such as Data Set Ready.

$\overline{DTR}$ (Data Terminal Ready)

The $\overline{DTR}$ output signal is a general-purpose, 1-bit inverting output port. It can be set "low" by programming the appropriate bit in the Command Instruction word. The $\overline{DTR}$ output signal is normally used for modem control such as Data Terminal Ready.

$\overline{RTS}$ (Request to Send)

The $\overline{RTS}$ output signal is a general-purpose, 1-bit inverting output port it can be set "low" by programming the appropriate bit in the Command Instruction word. The $\overline{RTS}$ output signal is normally used for modem control such as Request to Send.

$\overline{CTS}$ (Clear to Send)

A "low" on this input enables the 8251A to transmit serial data if the Tx Enable bit in the Command byte is set to a "one". If either a Tx Enable off or $\overline{CTS}$ off condition occurs while the Tx is in operation, the Tx will transmit all the data in the USART, written prior to Tx Disable command before shutting down.

Transmitter Buffer

The Transmitter Buffer accepts parallel data from the Data Bus Buffer, converts it to a serial bit stream, inserts the appropriate characters or bits (based on the communication technique) and outputs a composite serial stream of data on the TxD output pin on the falling edge of $\overline{TxC}$. The transmitter will begin transmission upon being enabled if $\overline{CTS}$ = 0. The TxD line will be held in the marking state immediately upon a master Reset or when Tx Enable or $\overline{CTS}$ is off or the transmitter is empty.

Transmitter Control

The Transmitter Control manages all activities associated with the transmission of serial data. It accepts and issues signals both externally and internally to accomplish this function.

TxRDY (Transmitter Ready)

This output signals the CPU that the transmitter is ready to accept a data character. The TxRDY output pin can be used as an interrupt to the system, since it is masked by TxEnable; or, for Polled operation, the CPU can check TxRDY using a Status Head operation, TxRDY is automatically reset by the leading edge of $\overline{WR}$ when a data character is loaded from the CPU.

Note that when using the Polled operation, the TxRDY status bit is *not* masked by TxEnable, but will only indicate the Empty/Full Status or the Tx Data Input Register.

TxE (Transmitter Empty)

When the 8251A has no characters to send, the TxEMPTY output will go "high". It resets upon receiving a character from CPU if the transmitter is enabled. TxEMPTY remains high when the transmitter is disabled TxEMPTY can be used to indicate the end of a transmission mode, so that the CPU "knows" when to "turn the line around" in the half-duplex operational mode.

In the Synchronous mode, a "high" on this output indicates that a character has not been loaded and the SYNC character or characters are about to be or are being transmitted automatically as "fillers". TxEMPTY does not go low when the SYNC characters are being shifted out.

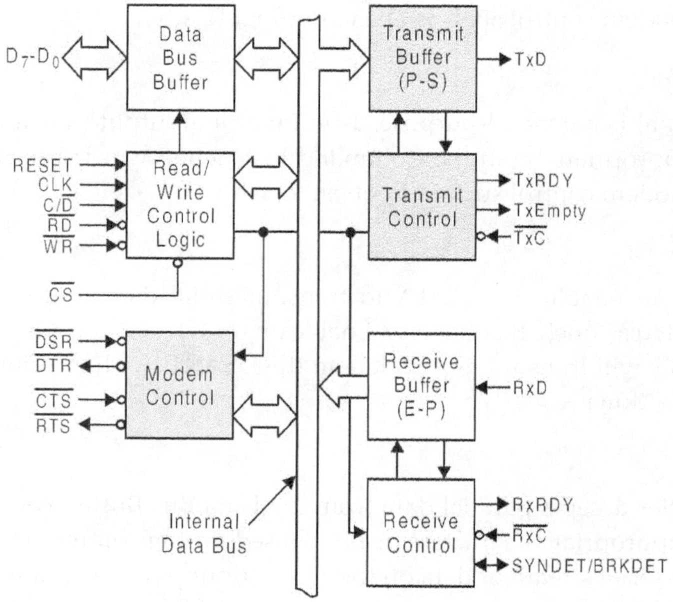

Fig. 6.40 8251A Block Diagram Showing Modem and Transmitter Buffer and Control Functions

$\overline{\text{TxC}}$ (Transmitter Clock)

The Transmitter Clock controls the rate at which the character is to be transmitted. In the Synchronous transmission mode, the Baud Rate (1x) is equal to the $\overline{\text{TxC}}$ frequency. In Asynchronous transmission mode, the baud rate is a fraction of the actual $\overline{\text{TxC}}$ frequency. A portion of the mode instruction selects this factor; it can be 1, 1/16 or 1/64, the $\overline{\text{TxC}}$.

For Example:

If Baud Rate equals 110 Baud,

$\overline{\text{TxC}}$ equals 110 Hz in the 1x mode.

$\overline{\text{TxC}}$ equals 1.72 kHz in the 16x mode.

$\overline{\text{TxC}}$ equals 7.04 kHz in the 64x mode.

The falling edge of $\overline{\text{TxC}}$ shifts the serial data out of the 8251A.

Receiver Buffer

The Receiver accepts serial data, converts this serial input to parallel format, checks for bits or characters that are unique to the communication technique and sends an "assembled" character to the CPU. Serial data is input to RxD pin, and is clocked in on the rising edge of $\overline{\text{RxC}}$.

Receiver Control

This functional block manages all receiver-related activities which consists of the following features.

The RxD initialization circuit prevents the 8251A from mistaking an unused input line for an active low data line to the "break condition". Before starting to receive serial characters on the RxD line, a valid "1" must first be detected after a chip master Reset. Once this has been determined, a search for a valid low (Start bit) is enabled. This feature is only active in the asynchronous mode, and is only done once for each master Reset.

The False Start bit detection circuit prevents false starts due to transient noise spike by first detecting the falling edge and then strobing the normal center of the Start bit (RxD = low).

Parity error detection sets the corresponding status bit.

The Framing Error status bit is set if the Stop bit is absent at the end of the data byte (asynchronous mode).

RxRDY (Receiver Ready)

This output indicates that the 8251A contains a character that is ready to be input to the CPU. Rx RDY can be connected to the interrupt structure of the CPU or, for polled operation, the CPU can check the condition of RxRDY using a Status Read operation.

RxEnable, when off, holds RxRDY in the Reset Condition. For Asynchronous mode, to set RxRDY. The Receiver must be enabled to sense a Start Bit and a complete character must be assembled and transferred to the Data Output Register. For Synchronous mode, to set RxRDY, the Receiver must be enabled and a character must finish assembly and be transferred to the Data Output Register.

Failure to read the received character from the Rx Data Output Register prior to the assembly of the next Rx Data character will set overrun condition error and the previous character will be written over end lost. If the Rx Data is being read by the CPU when the internal transfer is occurring, overrun error will be set and the old character will be lost.

$\overline{RxC}$ (Receiver Clock)

The Receiver Clock controls the rate at which the character is to be received. In Synchronous Mode, the Baud Rate (1x) is equal to the actual frequency of RxC. In Asynchronous Mode, the Baud Rate is a fraction of the actual $\overline{RxC}$ frequency. A portion of the mode instruction selects this factor 1, $1/_{16}$ or $1/_{64}$, the $\overline{RxC}$.

For Example:

Baud Rate equals 300 Baud, if

$\overline{RxC}$ equals 300 Hz in the 1x mode;

$\overline{RxC}$ equals 4800 Hz in the 16x mode,

$\overline{RxC}$ equals 19.2 kHz in the 64x mode.

Baud rate equals 2400 Baud, if

$\overline{RxC}$ equals 2400 Hz in the 1x mode;

$\overline{RxC}$ equals 38.4 kHz in the 16 mode:

$\overline{RxC}$ equals 153.6 kHz in the 64 mode.

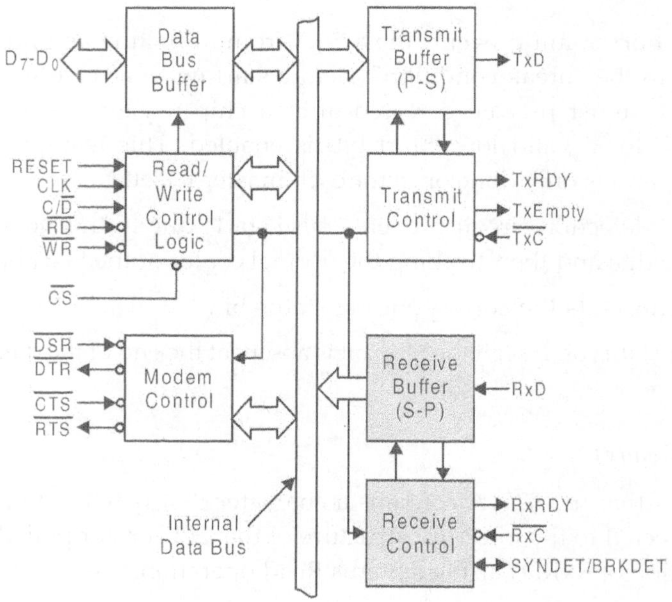

Fig. 6.41 8251A Block Diagram Showing Receiver Buffer and Control Functions

Data is sampled into the 8251A on the rising edge of $\overline{RxC}$.

Note: In most communication systems, the 8251A will be handling both the transmission and reception operations of a single link. Consequently, the Receive and Transmit Baud Rates will be the same, Both $\overline{TxC}$ and $\overline{RxC}$ will require identical frequencies for this operation and can be tied together and connected to a single frequency source (Baud Rate Generator) to simplify the interface.

SYNDET (SYNC Detect/BRKDET Break Detect)

This pin is used in Synchronous Mode for SYNDET and may be used as either input or output, programmable through the Control Word. It is reset to output mode low upon RESET. When used as an output (internal Sync mode), the SYNDET pin will go "high" to indicate that the 8251A has located the SYNC character in the Receive mode, if the 8251A is programmed to use double Sync characters (bi-sync), then SYNDET will go "high" in the middle of the last bit of the second Sync character. SYNDET is automatically reset upon a Status Read operation.

When used as an input (external SYNC detect mode}, a positive going signal will cause the 8251A to start assembling data characters on the rising edge of the next $\overline{RxC}$. Once in SYNC, the "high" input signal can be removed. When External SYNC Detect is programmed, Internal SYNC Detect is disabled.

BREAK (Async Mode Only)

This output will go high whenever the receiver remains low through two consecutive stop bit sequences (including the start bits, data bits, and parity bits). Break Detect may also be read as a Status bit. It is reset only upon a master chip Reset or Rx Data returning to a "one" state.

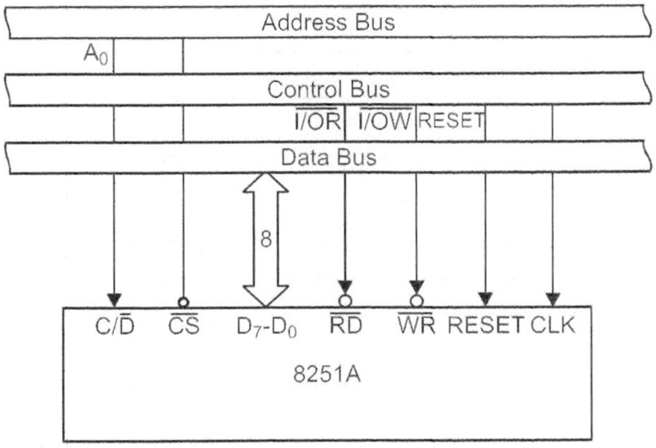

Fig. 6.42 8251A Interface to 8080 Standard System Bus

6.7.3 Operation Description

General

The complete functional definition of the 8251A is programmed by the system's software. A set of control words must be sent out by the CPU to initialize the 8251A to support the desired communications format. These control words will program the: BAUD RATE CHARACTER LENGTH NUMBER OF STOP BITS, SYNCHRONOUS or ASYNCHRONOUS OPERATION, EVEN/ODD/OFF

PARITY, etc. In the Synchronous Mode, options are also provided to select either internal or external character synchronization.

Once programmed, the 8251A is ready to perform its communication functions. The TxRDY output is raised "high" to signal the CPU that the 8251A is ready to receive a data character from the CPU. This output (TxRDY) is reset automatically when the CPU writes a character into the 8251A. On the other hand, the 8251A receives serial data from the MODEM or I/O device. Upon receiving an entire character, the RxRDY output is raised "high" to signal the CPU that the 8251A has a complete character ready for the CPU to fetch. RxRDY is reset automatically upon the CPU data, read operation.

The 8251A cannot begin transmission until the Tx Enable (Transmitter Enable) bit is set in the Command Instruction and it has received a Clear To Send ($\overline{CTS}$) Input. The TxD output will be held in the marking state upon Reset.

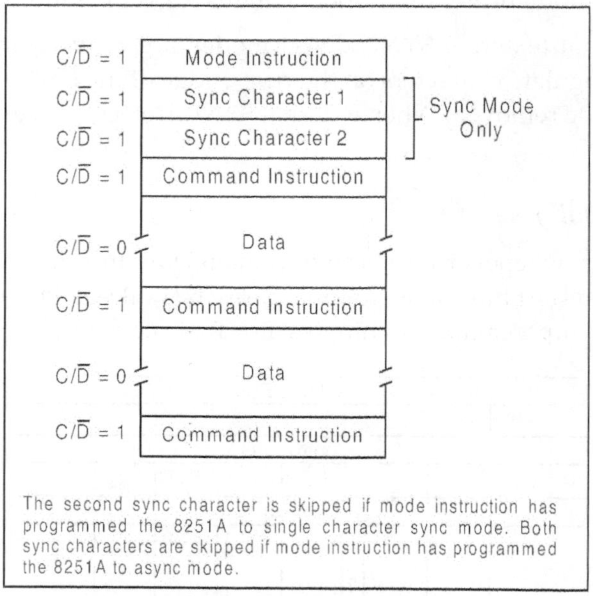

Fig. 6.43 Typical Data Block

Programming the 8251A

Prior to starting data transmission or reception, the 8251A must be loaded with a set of control words generated by the CPU. These control signals define the complete functional definition of the 8251A and must immediately follow a Reset operation (internal or external).

The control words are split into two formats:

1. Mode Instruction
2. Command Instruction

Mode Instruction. This instruction defines the general operational characteristics of the 8251A. It must follow a Reset operation (internal or external). Once the Mode Instruction has been written into the 8251A by the CPU, SYNC characters or Command Instructions may be written.

Command Instruction. This instruction defines a word that is used to control the actual operation of the 8251A.

Both the Mode and Command Instructions must conform to a specified sequence for proper device operation (see Fig. 6.43). The Mode Instruction must be written immediately following a Reset operation, prior to using the 8251A for data communication.

All control words written into the 8251A after the Mode Instruction will load the Command Instruction. Command Instructions can be written into the 8251A at any time in the data block during the operation of the 8251A. To return to the Mode Instruction format, the master Reset bit in the Command Instruction word can be set to initiate an internal. Reset operation which automatically places the 8251A back into the Mode Instruction format. Command Instructions must follow the Mode Instruction or Sync characters.

Asynchronous Mode (Transmission)

Whenever a data character is sent by the CPU the 8251A automatically adds a Start bit (low level) followed by the data bits (least significant bit first), and the programmed number of Stop bits to each character. Also, an even or odd Parity bit is inserted prior to the Stop bit(s), as defined by the Mode Instruction. The character is then transmitted as a serial data stream on the TxD output. The serial data is shifted out on the falling edge of $\overline{\text{TxC}}$ at a rate equal to 1, $1/_{16}$, or $1/_{64}$ that of the $\overline{\text{TxC}}$, as defined by the Mode instruction BREAK characters can be continuously sent to the TxD if commanded to do so.

When no data characters have been loaded into the 8251A, the TxD output remains "high" (marking) unless a Break (continuously low) has been programmed.

Asynchronous Mode (Receive)

The RxD line is normally high. A falling edge on this line triggers the beginning of a START bit. The validity of this START bit is checked by again strobing this bit at its nominal center (16X or 64X mode only). If a low is detected again, it is a valid START bit, and the bit counter will start counting. The bit counter thus locates the center of the data bits, the parity bit (if it exists) and the stop bits. If parity error occurs, the parity error flag is set. Data and parity bits are sampled on the RxD pin with the rising edge of the $\overline{\text{RxC}}$. If a low level is detected as the STOP bit the Framing Error flag will be set. The STOP bit signals the end of a character. Note that the *receiver* requires only *one* stop bit regardless of the number of step bits programmed. This character is then loaded into the parallel I/O buffer of the 8251A. The RxRDY pin is raised to signal the CPU that a character is ready to be fetched. If a previous character has not been fetched by the CPU, the present character replaces it in the I/O buffer, and the OVERRUN Error flag is raised (thus the previous character is lost). All of the error flags can be reset by an Error Reset instruction. The occurrence of any of these errors will not affect the operation of the 8251A.

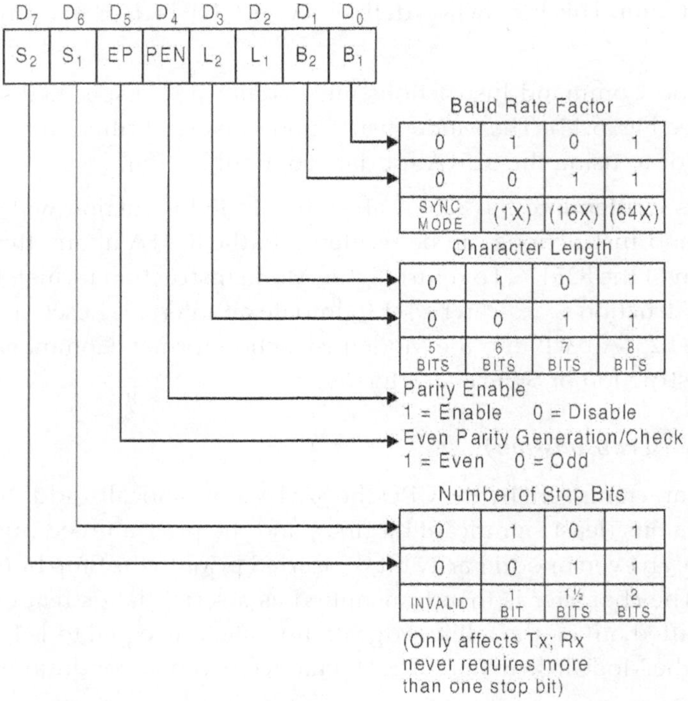

Fig. 6.44 Mode Instruction Format, Asynchronous Mode

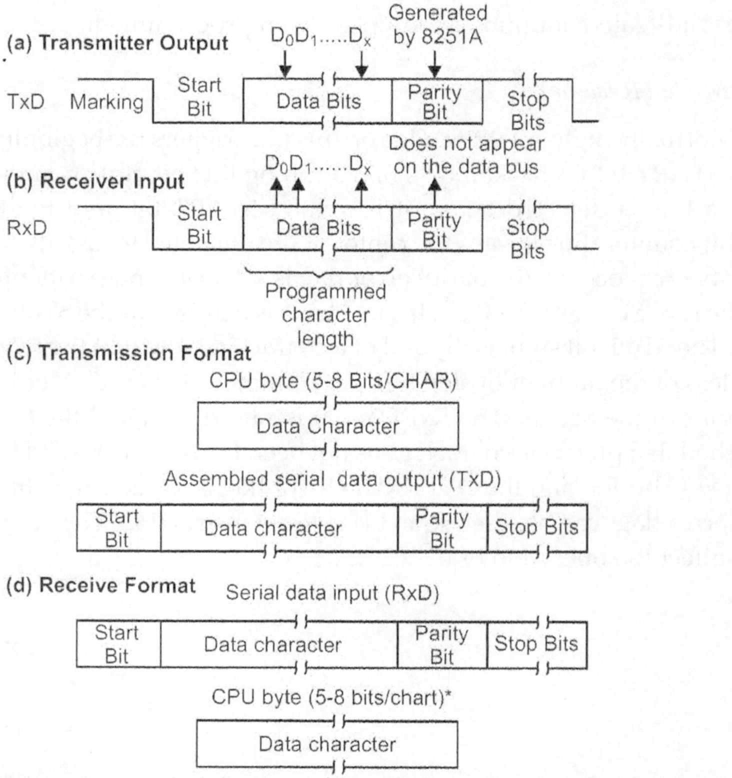

***Note:** If character length is defined as 5, 6 or 7 bits the unused bits are set to "zero".

Fig. 6.45 Asynchronous Mode

Synchronous Mode (Transmission)

The TxD output to continuously high until the CPU sends its first character to the 8251A which usually is a SYNC character. When the $\overline{CTS}$ line goes low, the first character is serially transmitted out. All characters are shifted out on the falling edge of $\overline{TxC}$. Data is shifted out at the same rate as the $\overline{TxC}$.

Once transmission has started, the data stream at the TxD output must continue at the $\overline{TxC}$ rate. If the CPU does not provide the 8251A with a data character before the 8251A Transmitter Buffers become empty, the SYNC characters (or character if in single SYNC character mode) will be automatically inserted in the TxD data stream. In this case, the TxEMPTY pin is raised high to signal that the 8251A is empty and SYNC characters are being sent out. TxEMPTY does not go low when the SYNC is being shifted out (see figure below). The TxEMPTY pin is internally reset by a date character being written into the 8251A.

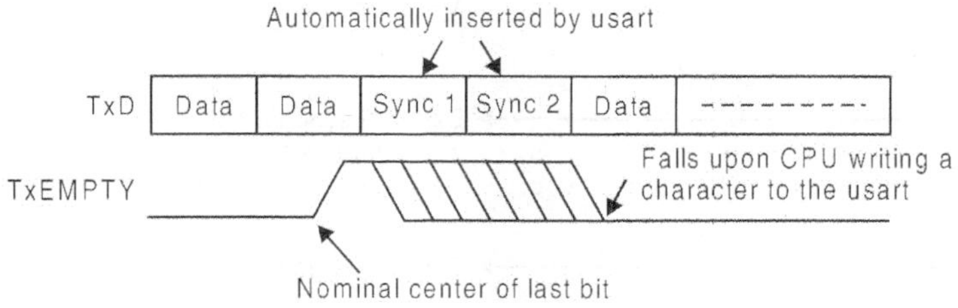

Fig. 6.46(a) Synchronous Mode (Transmission)

Synchronous Mode (Receive)

In this mode, character synchronization can be internally or externally achieved. If the SYNC mode has been programmed. ENTER HUNT command should be included in the first command instruction word written. Data on the RxD pin is then sampled on the rising edge of $\overline{RxC}$. The content of the Rx buffer is compared at every bit boundary with the first SYNC character until a match occurs. If the 8151A has been programmed for two SYNC characters, the subsequent received character is also compared; when both SYNC characters have been detected, the USART ends the HUNT mode and is in character synchronization. The SYNDET pin is then set high, and is reset automatically by a STATUS READ. If parity is programmed, SYNDET will not be set until the middle of the parity bit instead of the middle of the last data bit.

In the external SYNC mode, synchronization is achieved by applying a high level on the SYNDET pin, thus forcing the 8251A out of the HUNT mode. The high level can be removed after the one $\overline{RxC}$ cycle. An ENTER HUNT command has no effect in the asynchronous mode of operation.

Parity error and overrun error are both checked in the same way as in Asynchronous Rx mode. Parity is checked when not in Hunt, regardless of whether the Receiver is enabled or not.

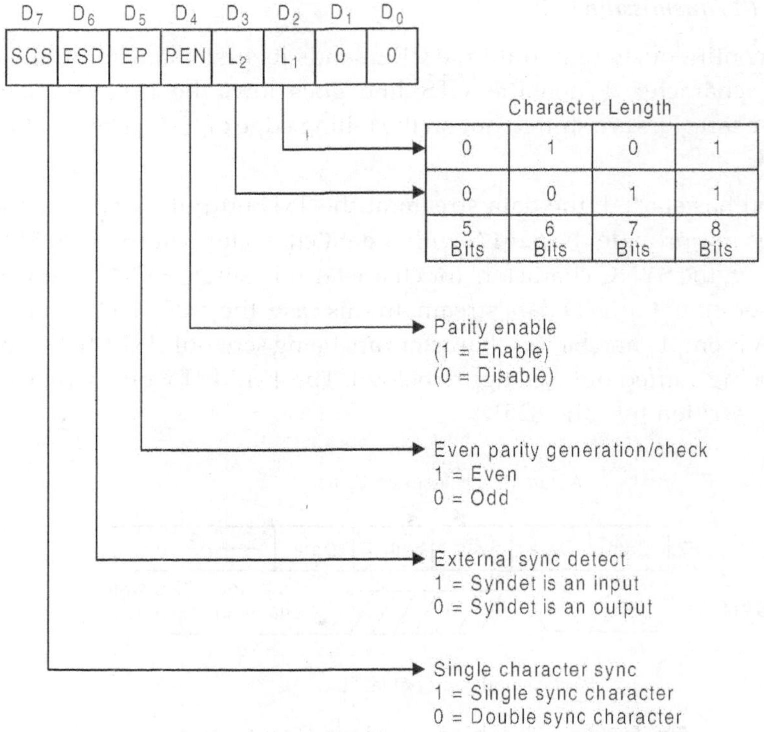

Note: In external sync mode, programming double character sync will affect only the Tx.

Fig. 6.46(b) Mode Instruction Format, Synchronous Mode

The CPU can command the receiver to enter the HUNT mode if synchronization is lost. This will also set all the used character bits in the buffer to a "one" thus preventing a possible false SYNDET caused by data that happens to be in the Rx Buffer at ENTER HUNT time. Note that the SYNDET F/F is reset at each Status Read regardless of whether internal or external SYNC has been programmed. This does not cause the 8251A to return to the HUNT mode. When in SYNC mode, but not in HUNT, Sync Detection is still functional, but only occurs at the "known" word boundaries. Thus, if one Status Read indicates SYNDET and a second Status Read also indicates SYNDET, then the programmed SYNDET characters have been received since the previous Status Read. (If double character sync has been programmed, then both sync characters have been contiguously received to gate a SYNDET indication). When external SYNDET mode is selected, internal Sync Detect is disabled, and the SYNDET F/F may be set at any bit boundary.

6.7.4 Command Instruction Definition

Once the functional definition of the 8251A has been programmed by the Mode Instruction and the Sync characters are loaded (if in Sync Mode) then the device is ready to be used for data communication. The Command Instruction controls the actual operation of the selected format. Function such as: Enable Transmit/Receive, Error Reset and Modem Controls are provided by the Command instruction.

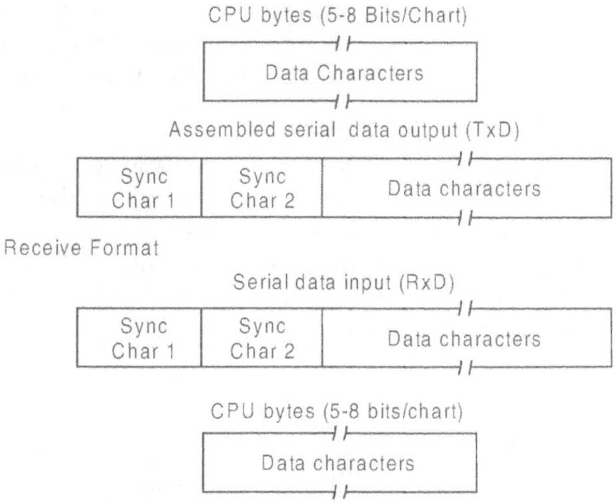

Fig. 6.47 Data Format, Synchronous Mode

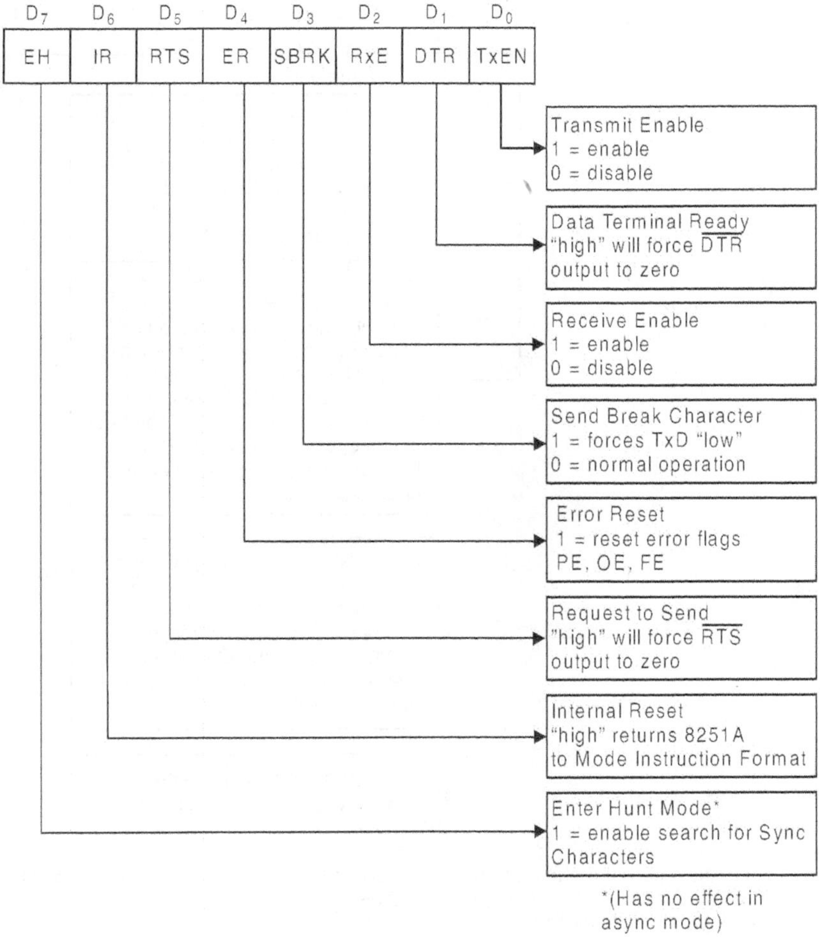

Note: Error Reset must be performed whenever RxEnable and Enter Hunt are programmed.

Fig. 6.48 Command Instruction Format.

Once the Mode Instruction has been written into the 8251A and Sync characters inserted, of necessary, then all further "control writes" ($C/\overline{D}$ = 1) will load a Command Instruction. A Reset Operation (internal or external) will return the 8251A to the Mode Instruction format.

Note: Internal Rest on Power-up:

When power is first applied, the 8251A may come up in the Mode, Sync character or Command format. To guarantee that the device is in the Command Instruction format before the Reset command is issued, it is safest to execute the worst-case initialization sequence (sync mode with two sync characters). Loading three 00Hs consecutively into the device with $C/\overline{D}$ = 1 configures sync operation and writes two dummy 00H sync characters. An Internal Reset command (40H) may then be issued to return the device to the "idle" state.

6.7.5 Status Read Definition

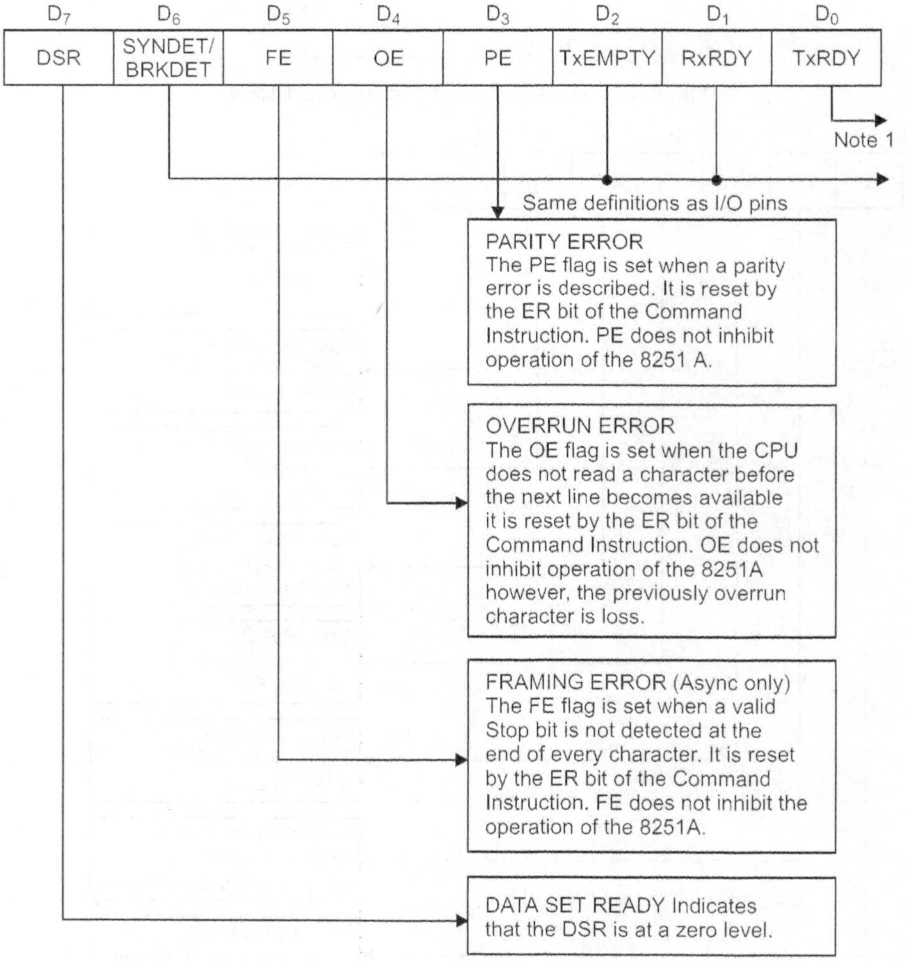

Note:

1. TxRDY status bit has different meaning from the TxRDY output pin. The former is not conditioned by $\overline{CTS}$ and TxEN; the latter is conditioned by both $\overline{CTS}$ and TxEN.
   i.e., TxRDY status bit = DB Buffer Empty
   TxRDY pin out = DB Buffer Empty • ($\overline{CTS}$ = 0) • (TxEN = 1)

Fig. 6.49 Status Read Format

In data communication system, it is often necessary to examine the "status" of the active device to ascertain if errors have occurred or other conditions that require the processor's attention. The 8251A has facilities that allow the programmer to "read" the status of the device at any time during the functional operation. (Status update is inhibited during status read).

A normal "read" command is issued by the CPU with $C/\overline{D}$ = 1 to accomplish this function.

Some of the bits in the Status Read Format have identical meanings to external output pins so that the 8251A can be used in a completely polled or interrupt-driven environment. TxRDY is an exception.

Note: that status update can have a maximum delay of 28 clock periods from the actual event affecting the status.

6.7.6 Applications of the 8251A

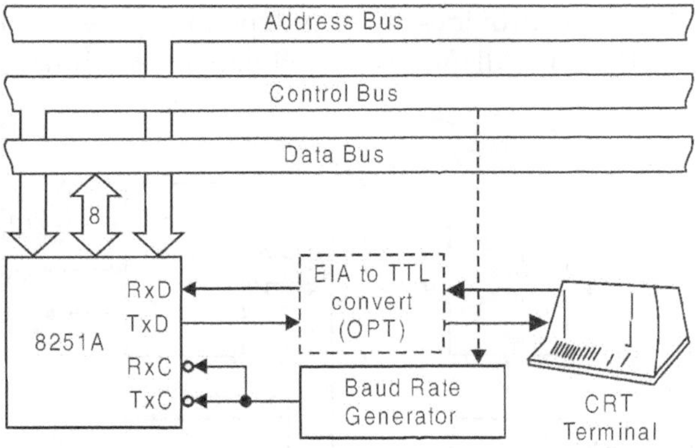

Fig. 6.50 Asynchronous Serial Interface to CRT Terminal

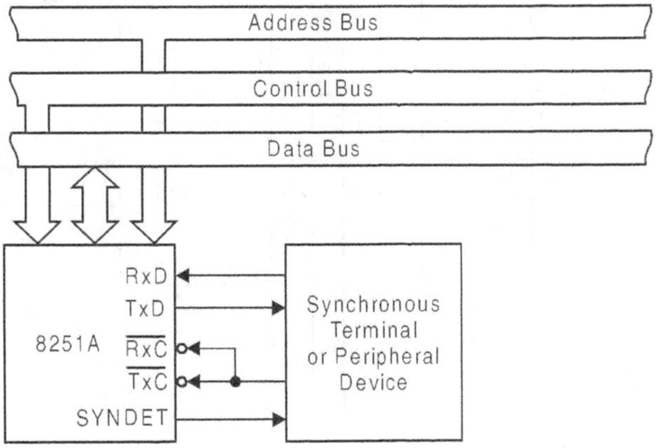

Fig. 6.51 Synchronous Interface to Terminal or Peripheral Device

6.8 8254

- Compatible with All Intel and Most Other Microprocessors
- Handles Inputs from DC to 10 MHz
  - 8 MHz 8254
  - 10 MHz 8254-2
- Status Read-Back Command
- Six Programmable Counter Modes
- Three Independent 16-Bit Counters
- Binary or BCD Counting
- Single + 5V Supply

The 8254 is a counter/timer device designed to solve the common timing control problems in microcomputer system design. It provides three independent 16-bit counters, each capable of handling clock inputs up to 10 MHz. All modes are software programmable. The 8254 is a superset of the 8253.

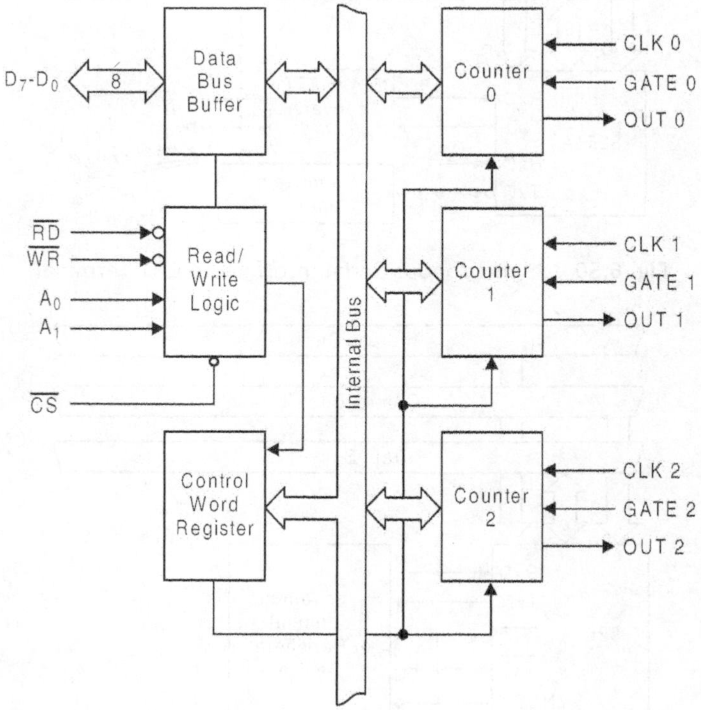

Fig. 6.52 8254 Block Diagram

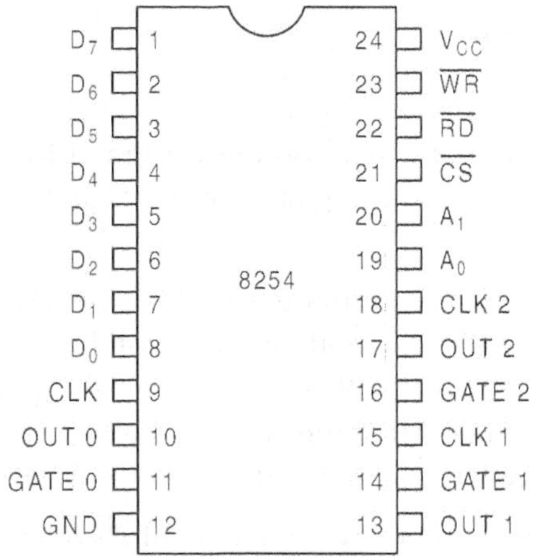

Fig. 6.53 Pin Configuration

6.8.1 PIN Description

Symbol	Pin No.	Type	Name and Function
D_7-D_0	1-8	I/O	*DATA:* Bi-directional three state data bus lines, connected to system data bus.
CLK 0	9	I	*CLOCK 0:* Clock input of Counter 0.
OUT 0	10	O	*OUTPUT 0:* Output of Counter 0.
GATE 0	11	I	*GATE 0:* Gate input of Counter 0.
GND	12		*GROUND:* Power supply connection.
V_{CC}	24		*POWER:* + 5V power supply connection.
$\overline{WR}$	23	I	*WRITE CONTROL:* This input is low during CPU write operations.
$\overline{RD}$	22	I	*READ CONTROL:* This input is low during CPU read operations.
$\overline{CS}$	21	I	*CHIP SELECT:* A low on this input enables the 8254 to respond to $\overline{RD}$ and $\overline{WR}$ signals. $\overline{RD}$ and $\overline{WR}$ are ignored otherwise.
A_1, A_0	20-19	I	*ADDRESS:* Used to select one of the three Counters or the Control Word Register for read or write operations. Normally connected to the system address bus.

A_1	*A_0*	*Selects*
0	0	Counter 0
0	1	Counter 1
1	0	Counter 2
1	1	Control Word Register

Symbol	Pin No.	Type	Name and Function
CLK 2	18	I	*CLOCK 2:* Clock input of Counter 2.
OUT 2	17	O	*OUT 2:* Output of Counter 2.
GATE 2	16	I	*GATE 2:* Gate input of Counter 2.
CLK 1	15	I	*CLOCK 1:* Clock input of Counter 1.
GATE 1	14	I	*GATE 1:* Gate input of Counter 1.
OUT 1	13	O	*OUT 1:* Output of Counter 1.

6.8.2 Functional Description

General

The 8254 is a programmable interval timer/counter designed for use with Intel microcomputer systems. It is a general purpose, multi-timing element that can be treated as an array of I/O ports in the system software.

The 8254 solves one of the most common problems in any microcomputer system, the generation of accurate time delays under software control. Instead of setting up timing loops in software, the programmer configures the 8254 to match his requirements and programs one of the counters for the desired delay. After the desired delay, the 8254 will interrupt the CPU. Software overhead is minimal and variable length delays can easily be accommodated.

Some of the other counter/timer functions common to microcomputers which can be implemented with the 8254 are:

- Real time clock
- Event-counter
- Digital one-shot
- Programmable rate generator
- Square wave generator
- Binary rate multiplier
- Complex waveform generator
- Complex motor controller

Block Diagram

Data Bus Buffer

This 3-state, bi-directional, 8-bit buffer is used to interface the 8254 to the system bus (see Fig. 6.54).

Read/Write Logic

The Read/Write Logic accepts inputs from the system bus and generates control signals for the other functional blocks of the 8254. A_1 and A_0 select one of the three counters or the Control Word Register to be read from/written into. A "low" on the $\overline{RD}$ input tells the 8254 that the CPU is reading one of the counters. A "low" on the $\overline{WR}$ input tells the 8254 that the CPU is writing either a Control Word or an initial count. Both $\overline{RD}$ and $\overline{WR}$ are qualified by $\overline{CS}$; RD and $\overline{WR}$ are ignored unless the 8254 has been selected by holding $\overline{CS}$ low.

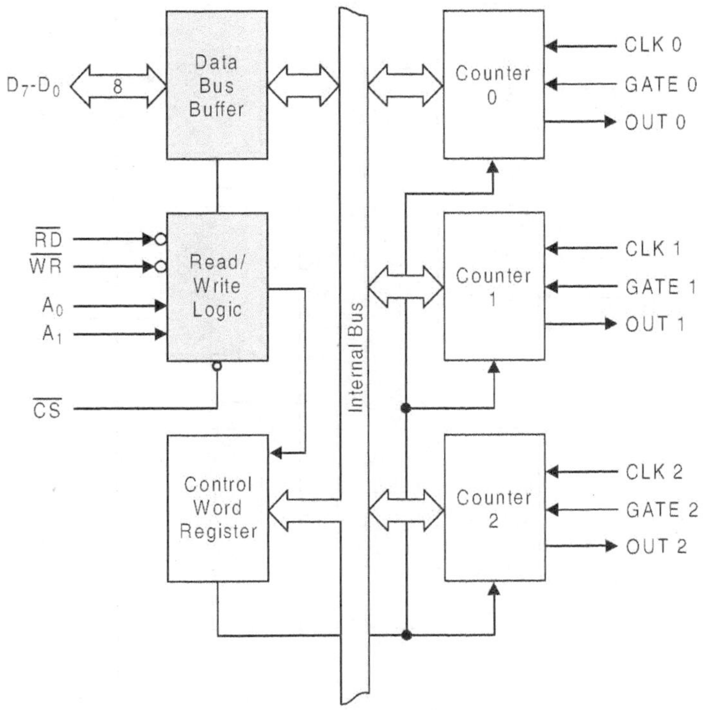

Fig. 6.54 Block Diagram Showing Data Bus Buffer and Read/Write Logic Functions

Control Word Register

The Control Word Register (see Fig. 6.55) is selected by the Read/Write Logic when A_1, A_0 = 11. If the CPU then does a write operation to the 8254, the data is stored in the Control Word Register and is interpreted as a Control Word used to define the operation of the Counters.

The Control Word Register can only be written to; status information is available with the Read-Back Command.

Counter 0, Counter 1, Counter 2

These three functional blocks are identical in operation, so only a single Counter will be described. The internal block diagram of a single counter is shown in Fig. 6.56.

The Counters are fully independent. Each Counter may operate in a different Mode.

The Control Word Register is shown in the Fig. 6.56; it is not part of the Counter itself, but its contents determine how the Counter operates.

The status register, shown in Fig. 6.56, when latched, contains the current contents of the Control Word Register and status of the output and null count flag.

The actual counter is labelled CE (for "Counting Element"). It is a 16-bit presettable synchronous down counter.

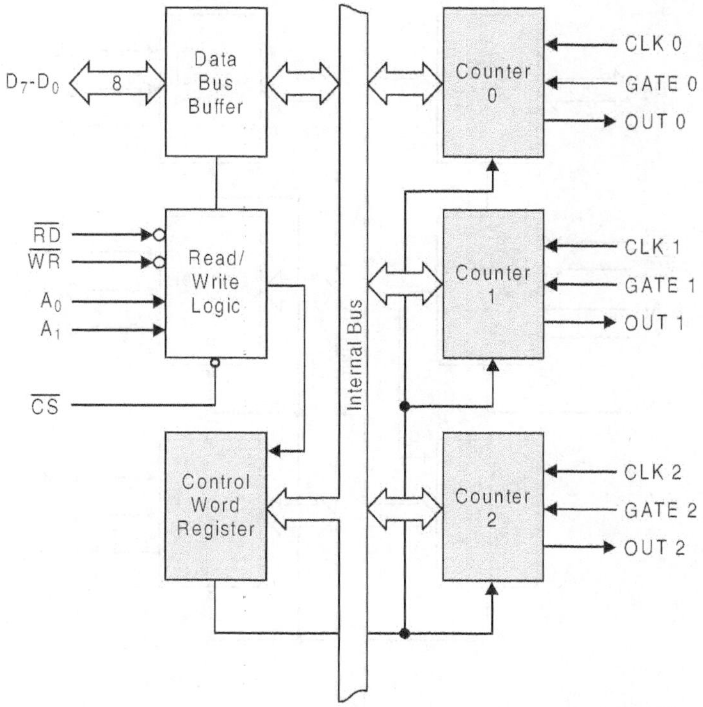

Fig. 6.55 Block Diagram Showing Control Word Register and Counter Functions.

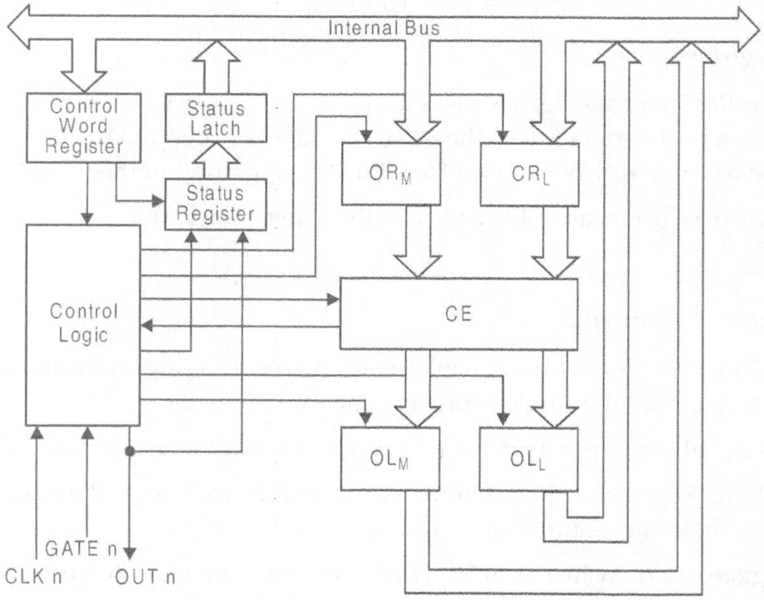

Fig. 6.56 Internal Block Diagram of a Counter.

OL_M and OL_L are two 8-bit latches. OL stands for "Output Latch"; the subscripts M and L stand for "Most significant byte" and "Least significant byte" respectively. Both are normally referred to as one unit and called just OL. These latches normally "follow" the CE, but if a suitable Counter Latch Command is sent to the 8254, the latches "latch" the present count until read by the CPU and then return to "following" the CE. One latch at a time is enabled by the counter's Control Logic to drive the internal bus. This is how the 16-bit Counter communicates over the 8-bit internal bus. Note that the CE itself cannot be read; whenever you read the count, it is the OL that is being read.

Similarly, there are two 8-bit registers called CR_M and CR_L (for "Count Register"). Both are normally referred to as one unit and called just CR. When a new count is written to the Counter, the count is stored in the CR and later transferred to the CE. The Control Logic allows one register at a time to be loaded from the internal bus. Both bytes are transferred to the CE simultaneously. CR_M and CR_L are cleared when the Counter is programmed. In this way, if the Counter has been programmed for one byte counts (either most significant byte only or least significant byte only) the other byte will be zero. Note that the CE cannot be written into; whenever a count is written, it is written into the CR.

The Control Logic is also shown in the diagram. CLK n, GATE n, and OUT n are all connected to the outside world through the Control Logic.

8254 System Interface

The 8254 is a component of the Intel Microcomputer Systems and interfaces in the same manner as all other peripherals of the family. It is treated by the system's software as an array of peripheral I/O ports; three are counters and the fourth is a control register for MODE programming.

Basically, the select inputs A_0, A_1 connect to the A_0, A_1 address bus signals of the CPU. The CS can be derived directly from the address bus using a linear select method. Or it can be connected to the output of a decoder, such as an Intel 8205 for larger systems.

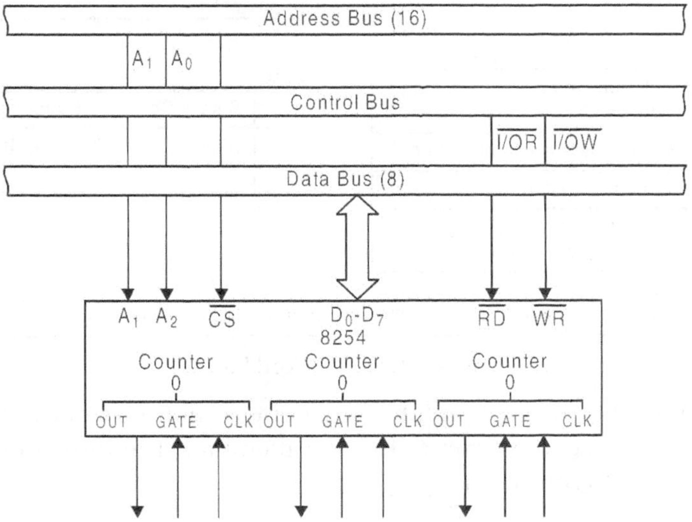

Fig. 6.57 8254 System Interface

6.8.3 Operational Description

General

After power-up, the state of the 8254 is undefined. The Mode, count value, and output of all Counters are undefined.

How each Counter operates is determined when it is programmed. Each Counter must be programmed before it can be used. Unused counters need not be programmed.

Programming the 8254

Counters are programmed by writing a Control Word and then an initial count.

The Control Words are written into the Control Word Register, which is selected when A_1, A_0 = 11. The Control Word itself specifies which Counter is being programmed.

Control Word Format

$A_1, A_0 = 11$ $\overline{CS} = 0$ $\overline{RD} = 1$ $\overline{WR} = 0$

D7	D6	D5	D4	D3	D2	D1	D0
SC1	SC0	RW1	RW0	M2	M1	M0	BCD

SC – Select Counter

SC1	SC0	
0	0	Select Counter 0
0	1	Select Counter 1
1	0	Select Counter 2
1	1	Read-Back Command (see Read Operations)

M – Mode

M2	M1	M0	
0	0	0	Mode 0
0	0	1	Mode 1
X	1	0	Mode 2
X	1	1	Mode 3
1	0	0	Mode 4
1	0	1	Mode 5

RW – Read/Write

RW1	RW0	
0	0	Counter Latch Command (see Read Operations)
0	1	Read/Write least significant byte only
1	0	Read/Write most significant byte only
1	1	Read/Write least significant byte first, then most significant byte

BCD

0	Binary Counter 16-bits
1	Binary Coded Decimal (BCD) Counter (4 Decades)

Note:
Don't care bits (X) should be 0 to insure compatibility with future Intel products.

Fig. 6.58 Control Word Format

By contrast, initial counts are written into the Counters, not the Control Word Register. The A_1A_0 inputs are used to select the Counter to be written into. The format of the initial count is determined by the Control Word used.

Write Operations

The programming procedure for the 8254 is very flexible. Only two conventions need to be remembered:

1. For each Counter, the Control Word must be written before the initial count is written.
2. The initial count must follow the count format specified in the Control Word (least significant byte only, most significant byte only, or least significant byte and then most significant byte).

Since the Control Word Register and the three Counters have separate addresses (selected by the A_1, A_0 inputs), and each Control Word specifies the Counter it applies to (SC0, SC1 bits), no special instruction sequence is required.

A new initial count may be written to a Counter at any time without affecting the Counter's programmed Mode in any way. Counting will be affected as described in the Mode definitions. The new count must follow the programmed count format.

If a Counter is programmed to read/write two-byte counts, the following precaution applies: A program must not transfer control between writing the first and second byte to another routine which also writes into that same Counter. Otherwise, the Counter will be loaded with an incorrect count.

		A_1	A_0			A_1	A_0
Control Word—Counter	0	1	1	Control Word—Counter	2	1	1
LSB of count—Counter	0	0	0	Control Word—Counter	1	1	1
MSB of count—Counter	0	0	0	Control Word—Counter	0	1	1
Control Word—Counter	1	1	1	LSB of count—Counter	2	1	0
LSB of count—Counter	1	0	1	MSB of count—Counter	2	1	0
MSB of count—Counter	1	0	1	LSB of count—Counter	1	0	1
Control Word—Counter	2	1	1	MSB of count—Counter	1	0	1
LSB of count—Counter	2	1	0	LSB of count—Counter	0	0	0
MSB of count—Counter	2	1	0	MSB of count—Counter	0	0	0
		A_1	A_0			A_1	A_0
Control Word—Counter	0	1	1	Control Word—Counter	1	1	1
Control Word—Counter	1	1	1	Control Word —Counter	0	1	1
Control Word—Counter	2	1	1	LSB of count—Counter	1	0	1
LSB of count—Counter	2	1	0	Control Word—Counter	2	1	1
LSB of count—Counter	1	0	1	LSB of count—Counter	0	0	0
LSB of count—Counter	0	0	0	MSB of count—Counter	1	0	1
MSB of count—Counter	0	0	0	LSB of count—Counter	2	1	0
MSB of count—Counter	1	0	1	MSB of count—Counter	0	0	0
MSB of count—Counter	2	1	0	MSB of count—Counter	2	1	0

Note: In all four examples, all Counters are programmed to read/write two-byte counts. These are only four of many possible programming sequences.

Fig. 6.59 A Few Possible Programming Sequences

Read Operations

It is often desirable to read the value of a Counter without disturbing the count in progress. This is easily done in the 8254.

There are three possible methods for reading the counters: a simple read operation, the Counter Latch Command, and the Read-Back Command. Each is explained below.

The first method is to perform a simple read operation. To read the Counter, which is selected with the A_1, A_0 inputs, the CLK input of the selected Counter must be inhibited by using either the GATE input or external logic. Otherwise, the count may be in the process of changing when it is read, giving an undefined result.

Counter Latch Command

The second method uses the "Counter Latch Command". Like a Control Word, this command is written to the Control Word Register, which is selected when $A_1A_0 = 11$. Also like a Control Word, the SC0, SC1 bits select one of the three Counters, but two other bits, D5 and D4, distinguish this command from a Control Word.

$A_1, A_0 = 11$; CS = 0; RD = 1; WR = 0

D7	D6	D5	D4	D3	D2	D1	D0
SC1	SC0	0	0	X	X	X	X

SS1, SS0 – specify counter to be latched

SC1	SC0	Counter
0	0	0
0	1	1
1	0	2
1	1	Read-Back Command

D5, D4 – 00 designates Counter Latch Command

X – don't care

Note:
Don't care bits (X) should be 0 to insure compatibility with future products.

Fig. 6.60 Counter Latching Command Format

The selected Counter's output latch (OL) latches the count at the time the Counter Latch Command is received. This count is held in the latch until it is read by the CPU (or until the Counter is reprogrammed). The count is then unlatched automatically and the OL returns to "following" the counting element (CE). This allows reading the contents of the Counters "on the fly" without affecting counting in progress. Multiple Counter Latch Commands may be used to latch more than one Counter. Each latched Counter's OL holds its count until it is read. Counter Latch Commands do not affect the programmed Mode of the Counter in any way.

If a Counter is latched and then, some time later, latched again before the count is read, the second Counter Latch Command is ignored. The count read will be the count at the time the first Counter Latch Command was issued.

With either method, the count must be read according to the programmed format; specifically, if the Counter is programmed for two byte counts, two bytes must be read. The two bytes do not have to be read one right after the other; read or write or programming operations of other Counters may be inserted between them.

Another feature of the 8254 is that reads and writes of the same Counter may be interleaved; for example, if the Counter is programmed for two byte counts, the following sequence is valid.

1. Read least significant byte.
2. Write new least significant byte.

3. Read most significant byte.
4. Write new most significant byte.

If a Counter is programmed to read/write two-byte counts, the following precaution applies: A program must not transfer control between reading the first and second byte to another routine which also reads from that same Counter. Otherwise, an incorrect count will be read.

Read-Back Command

The third method uses the Read-Back Command. This command allows the user to check the count value, programmed Mode, and current states of the OUT pin and Null Count flag of the selected counter(s).

The command is written into the Control Word Register and has the format shown in Fig. 6.61. The command applies to the counters selected by setting their corresponding bits D3, D2, D1 = 1.

A0, A1 = 11 $\overline{\text{CS}}$ = 0 $\overline{\text{RD}}$ = 1 $\overline{\text{WR}}$ = 0

D7	D6	D5	D4	D3	D2	D1	D0
1	1	$\overline{\text{COUNT}}$	$\overline{\text{STATUS}}$	CNT 2	CNT 1	CNT 0	0

D5: 0 = Latch count of selected counters(s)
D4: 0 = Latch status of selected counters(s)
D3: 1 = Select Counter 2
D2: 1 = Select Counter 1
D1: 1 = Select Counter 0
D0: Reserved for future expansion; Must be 0

Fig. 6.61 Read-Back Command Format

The read-back command may be used to latch multiple counter output latches (OL) by setting the $\overline{\text{COUNT}}$ bit 05 = 0 and selecting the desired counter(s). This single command is functionally equivalent to several counter latch commands, one for each counter latched. Each counter's latched count is held until it is read (or the counter is reprogrammed). The counter is automatically unlatched when read, but other counters remain latched until they are read. If multiple count read-back commands are issued to the same counter without reading the count, all but the first are ignored; *i.e.*, the count which will be read is the count at the time the first read-back command was issued.

The read-back command may also be used to latch status information of selected counter(s) by setting $\overline{\text{STATUS}}$ bit D4 = 0. Status must be latched to be read; status of a counter is accessed by a read from that counter.

The counter status format is shown in Fig. 6.62. Bits D5 through D0 contain the counter's programmed Mode exactly as written in the last Mode Control Word. OUTPUT bit D7 contains the current state of the OUT pin. This allows the user to monitor the counter's output via software, possibly eliminating some hardware from a system.

D7	D6	D5	D4	D3	D2	D1	D0
Output	Null Count	RW1	RW0	M2	M1	M0	BCD

D7 1 = OUT Pin is 1
0 = OUT Pin is 0

D6 1 = Null Count
0 = Count available for reading

D_5-D_0 Counter programmed mode (see Fig. 6.58)

Fig. 6.62 Status Byte

NULL COUNT bit D6 indicates when the last count written to the counter register (CR) has been loaded into the counting element (CE). The exact time this happens depends on the Mode of the counter and is described in the Mode Definitions, but until the count is loaded into the counting element (CE), it can't be read from the counter. If the count is latched or read before this time, the count value will not reflect the new count just written. The operation of Null Count is shown in Fig. 6.63.

This Action	Causes
A. Write to the control word register; (1)	Null Count = 1
B. Write to the count register (CR); (2)	Null Count = 1
C. New Count is loaded into CE (CR → CE);	Null Count = 0

Note:

1. Only the counter specified by the control word will have its Null Count set to 1. Null count bits of other counters are unaffected.
2. If the counter is programmed for two-byte counts (least significant byte then most significant byte) Null Count goes to 1 when the second byte is written.

Fig. 6.63 Null Count Operation

If multiple status latch operations of the counter(s) are performed without reading the status, all but the first are ignored; *i.e.*, the status that will be read is the status of the counter at the time the first status read-back command was issued.

Both count and status of the selected counter(s) may be latched simultaneously by setting both $\overline{\text{COUNT}}$ and $\overline{\text{STATUS}}$ bits D5.D4 = 0. This is functionally the same as issuing two separate read-back commands at once, and the above discussions apply here also. Specifically, if multiple count and/or status read-back commands are issued to the same counter(s) without any intervening reads, all but the first are ignored. This is illustrated in Fig. 6.64.

If both count and status of a counter are latched, the first read operation of that counter will return latched status, regardless of which was latched first. The next one or two reads (depending on whether the counter is programmed for one or two type counts) return latched count Subsequent reads return unlatched count.

Command								Description	Result
D_7	D_6	D_5	D_4	D_3	D_2	D_1	D_0		
1	1	0	0	0	0	1	0	Read back count and status of Counter 0	Count and status latched for Counter 0
1	1	1	0	0	1	0	0	Read back status of Counter 1	Status latched for Counter 1
1	1	1	0	1	1	0	0	Read back status of Counters 2, 1	Status latched for Counter 2, but not Counter 1
1	1	0	1	1	0	0	0	Read back count of Counter 2	Count latched for Counter 2
1	1	0	0	0	1	0	0	Read back count and status of Counter 1	Count latched for Counter 1, but not status
1	1	1	0	0	0	1	0	Read back status of Counter 1	Command ignored, status already latched for Counter 1

Fig. 6.64 Read-Back Command Example

$\overline{CS}$	$\overline{RD}$	$\overline{WR}$	A_1	A_0	
0	1	0	0	0	Write into Counter 0
0	1	0	0	1	Write into Counter 1
0	1	0	1	0	Write into Counter 2
0	1	0	1	1	Write Control Word
0	0	1	0	0	Read from Counter 0
0	0	1	0	1	Read from Counter 1
0	0	1	1	0	Read from Counter 2
0	0	1	1	0	1 No-Operation (3-State)
1	X	X	X	1	X No-Operation (3-State)
0	1	1	X	0	X No-Operation (3-State)

Fig. 6.65 Read/Write Operations Summary

6.8.4 Mode Definitions

The following are defined for use in describing the operation of the 8254.

CLK Pulse: a rising edge, then a falling edge, in that order, of a Counter's CLK input.

Trigger: a rising edge of a Counter's GATE input.

Counter loading: the transfer of a count from the CR to the CE

Mode 0: Interrupt on Terminal Count

Mode 0 is typically used for event counting. After the Control Word is written, OUT is initially low, and will remain low until the Counter reaches zero. OUT then goes high and remains high until a new count or a new Mode 0 Control Word is written into the Counter.

GATE = 1 enables counting; GATE = 0 disables counting. GATE has no effect on OUT.

After the Control Word and initial count are written to a Counter, the initial count will be loaded on the next CLK pulse. This CLK pulse does not decrement the count, so for an initial count of N, OUT does not go high until N + 1 CLK pulses after the initial count is written.

If a new count is written to the Counter, it will be loaded on the next CLK pulse and counting will continue from the new count. If a two-byte count is written, the following happens:

1. Writing the first byte disables counting. OUT is set low immediately (no dock pulse required)
2. Writing the second byte allows the new count to be loaded on the next CLK pulse.

This allows the counting sequence to be synchronized by software. Again, OUT does not go high until N + 1 CLK pulses after the new count of N is written.

If an initial count is written while GATE = 0, it will still be loaded on the next CLK pulse. When GATE goes high, OUT will go high N CLK pulses later; no CLK pulse is needed to load the Counter as this has already been done.

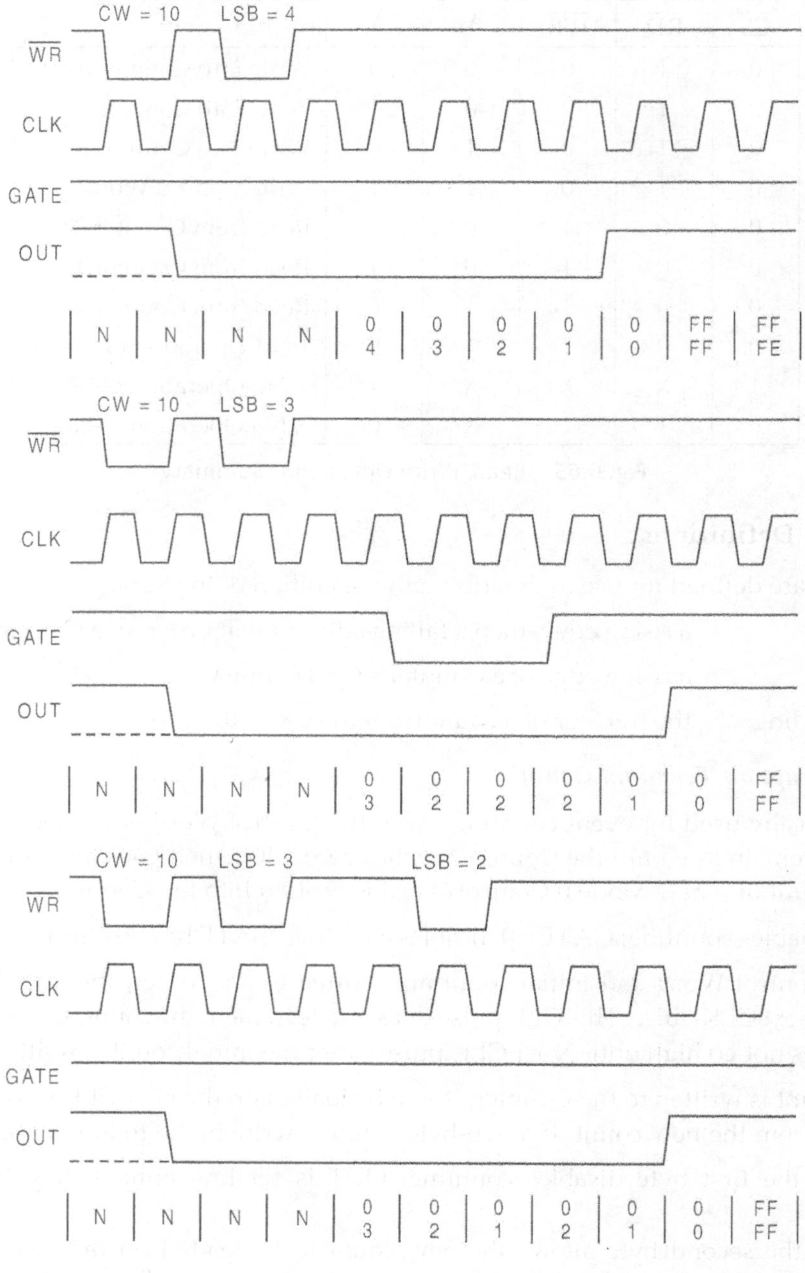

Note:

The following conventions apply to all mode timing diagrams:

1. Counters are programmed for binary (not BCD) counting and for reading/writing least significant byte (LSB) only.
2. The counter is always selected (CS always low).
3. CW stands for "Control Word", CW = 10 means a control word of 10 HEX is written to the counter.
4. LSB stands for "Least Significant Byte" of count.
5. Numbers below diagrams are count values. The lower number is the least significant byte. The upper number is the most significant byte. Since the counter is programmed to read/write LSB only, the most significant byte cannot be read.
   N Stands for an undefined count.
   Vertical lines show transitions between count values.

Fig. 6.66 Mode 0

Mode 1: Hardware Retriggerable One-shot

OUT will be initially high. OUT will go low on the CLK pulse following a trigger to begin the one-shot pulse, and will remain low until the Counter reaches zero.

OUT will then go high and remain high until the CLK pulse after the next trigger.

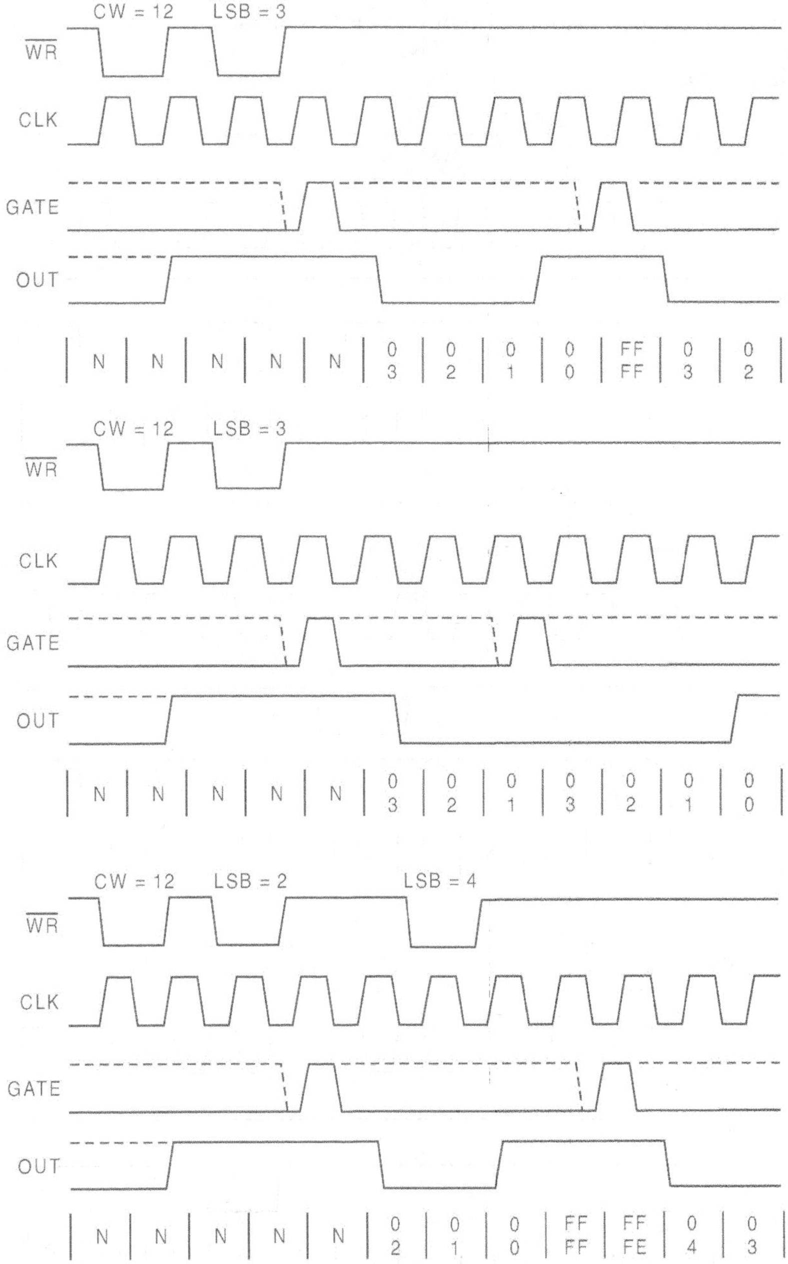

Fig. 6.67 Mode 1

After writing the Control Word and initial count, the Counter is armed. A trigger results in loading the Counter and setting OUT low on the next CLK pulse, thus starting the one-shot pulse.

An initial count of N will result in a one-shot pulse N CLK cycles in duration. The one-shot is retriggerable, hence OUT will remain low for N CLK pulses after any trigger. The one-shot pulse can be repeated without rewriting the same count into the counter. GATE has no effect on OUT.

If a new count is written to the Counter during a one-shot pulse, the current one-shot is not affected unless the counter is retriggered. In that case, the Counter is loaded with the new count and the one-shot pulse continues until the new count expires.

Mode 2: Rate Generator

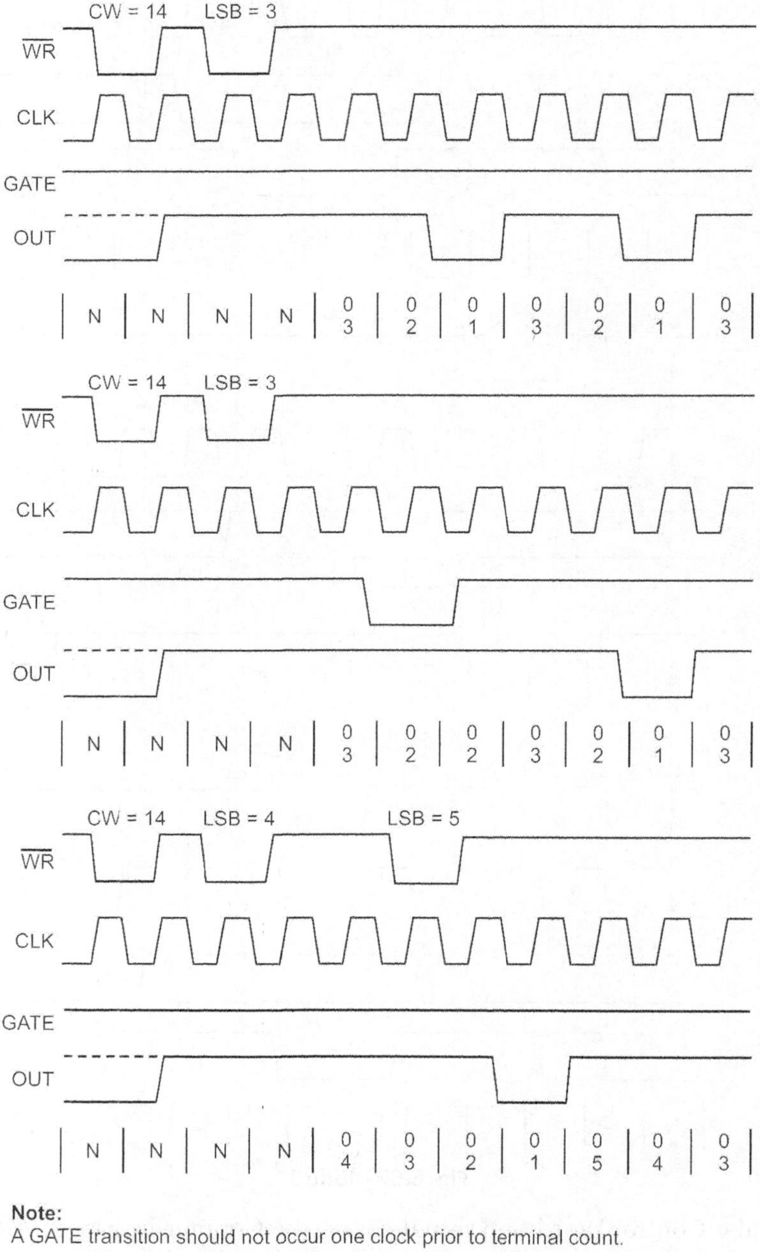

Fig. 6.68 Mode 2

This Mode functions like a divide-by-N counter. It is typically used to generate a Real Time Clock interrupt. OUT will initially be high. When the initial count has decremented to 1, OUT goes low for one CLK pulse. OUT then goes high again, the Counter reloads the initial count and the process is repeated. Mode 2 is periodic; the same sequence is repeated indefinitely. For an initial count of N, the sequence repeats every N CLK cycles.

GATE = 1 enables counting; GATE = 0 disables counting. If GATE goes low during an output pulse, OUT is set high immediately. A trigger reloads the Counter with the initial count on the next CLK pulse; OUT goes low N CLK pulses after the trigger. Thus the GATE input can be used to synchronize the Counter.

After writing a Control Word and initial count, the Counter will be loaded on the next CLK pulse. OUT goes low N CLK Pulses after the initial count is written. This allows the Counter to be synchronized by software also.

Writing a new count while counting does not affect the current counting sequence. If a trigger is received after writing a new count but before the end of the current period, the Counter will be loaded with the new count on the next CLK pulse and counting will continue from the new count. Otherwise, the new count will be loaded at the end of the current counting cycle. In mode 2, a COUNT of 1 is illegal.

Mode 3: Square Wave Mode

Mode 3 is typically used for Baud rate generation. Mode 3 is similar to Mode 2 except for the duty cycle of OUT. OUT will initially be high. When half the initial count has expired, OUT goes low for the remainder of the count. Mode 3 is periodic; the sequence above is repeated indefinitely. An initial count of N results in a square wave with a period of N CLK cycles.

GATE = 1 enables counting; GATE = 0 disables counting. If GATE goes low while OUT is low, OUT is set high immediately; no CLK pulse is required. A trigger reloads the Counter with the initial count on the next CLK pulse. Thus the GATE input can be used to synchronize the Counter.

After writing a Control Word and initial count, the Counter will be loaded on the next CLK pulse. This allows the Counter to be synchronized by software also writing a new count while counting does not affect the current counting sequence. If a trigger is received after writing a new count but before the end of the current half-cycle of the square wave, the Counter will be loaded with the new count on the next CLK pulse and counting will continue from the new count. Otherwise, the new count will be at the end of the current half-cycle.

Mode 3 is implemented as follows:

Even counts: OUT is initially high. The initial count is loaded on one CLK pulse and then is decremented by two on succeeding CLK pulses. When the count expires OUT changes value and the Counter is reloaded with the initial count. The above process is repeated indefinitely.

Odd counts: OUT is initially high. The initial count minus one (an even number) is loaded on one CLK pulse and then is decremented by two on succeeding CLK pulses. One CLK pulse after the count expires, OUT goes low and the Counter is reloaded with the initial count minus one. Succeeding CLK pulses decrement the count by two. When the count expires, OUT goes high again and the Counter is reloaded with the initial count minus one. The above process is repeated indefinitely. So for odd counts, OUT will be high for (N + 1)/2 counts and low for (N – 1)/2 counts.

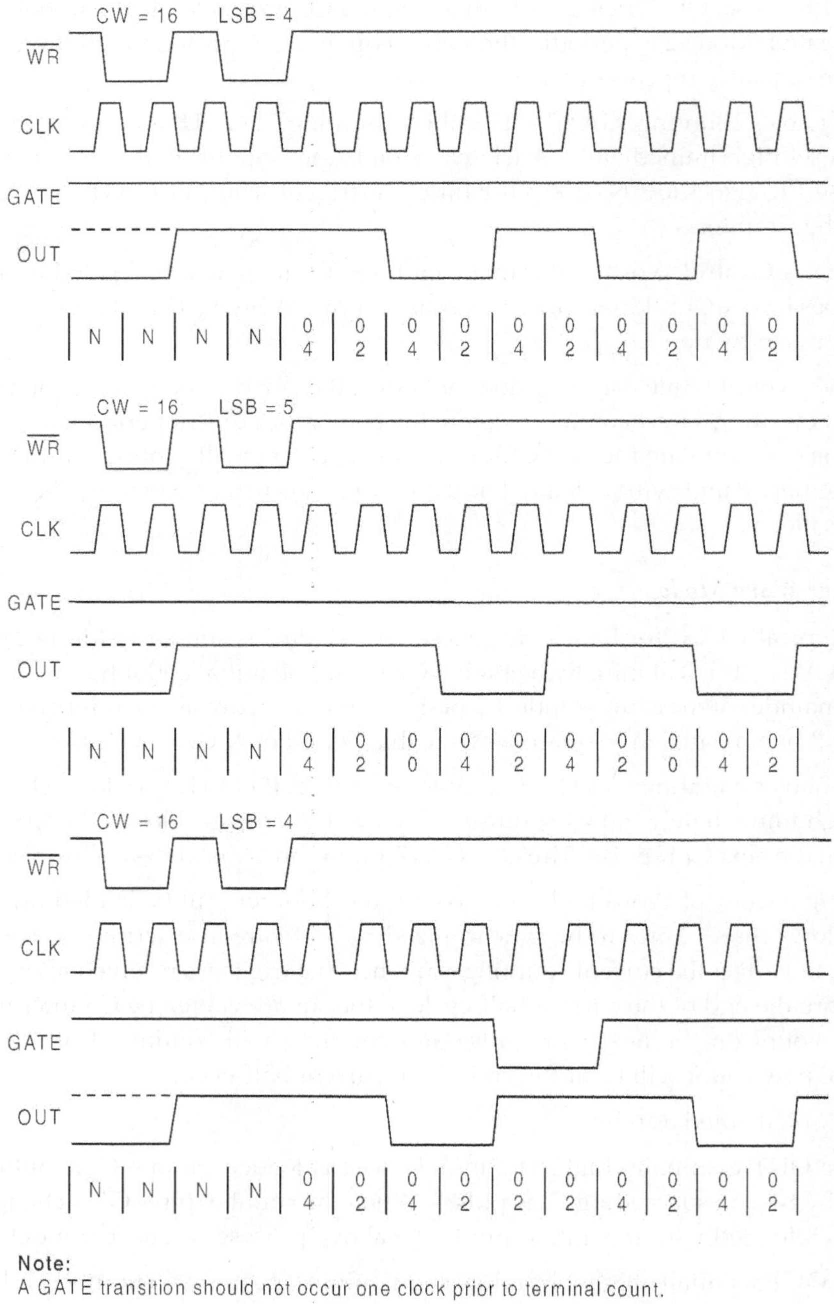

Fig. 6.69 Mode 3

Mode 4: Software Triggered Strobe

OUT will be initially high. When the initial count expires, OUT will go low for one CLK pulse and then go high again. The counting sequence is "triggered" by writing the initial count.

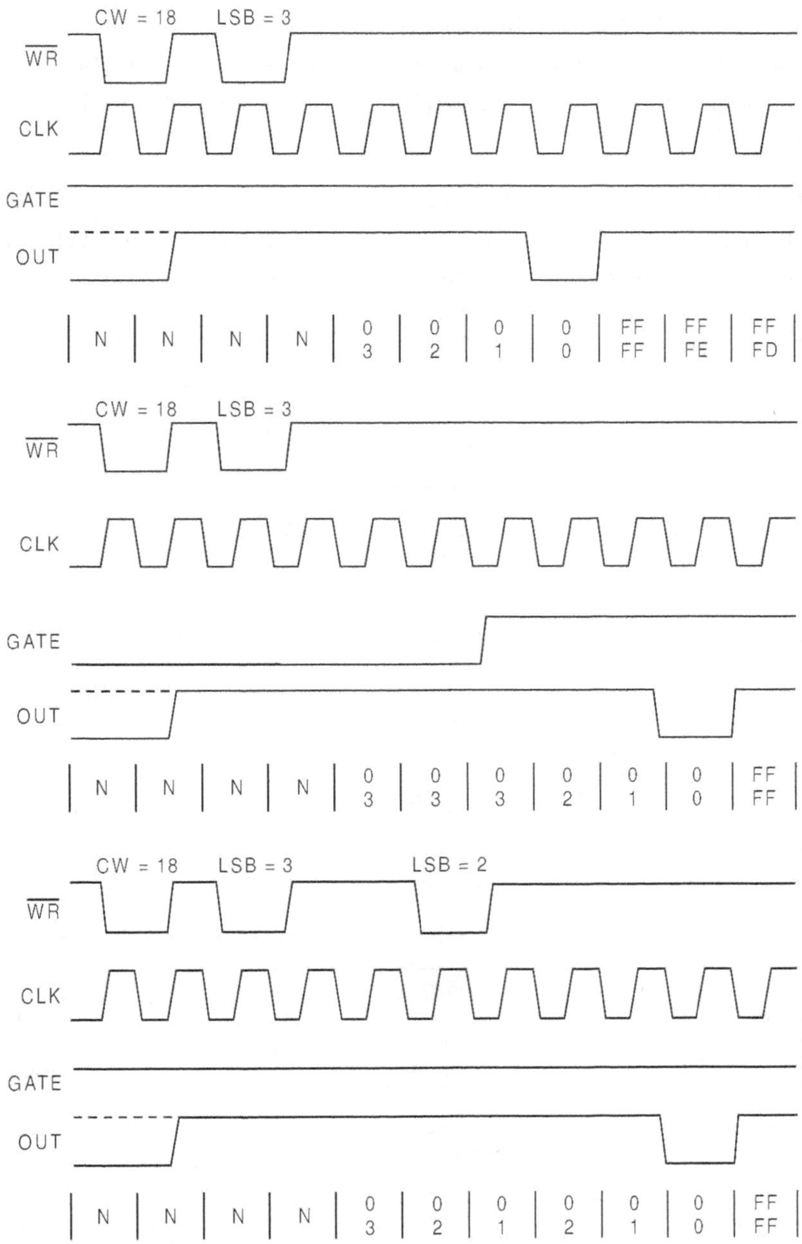

Fig. 6.70 Mode 4

GATE = 1 enables counting; GATE = 0 disables counting. GATE has no effect on OUT.

After writing a Control Word and initial count, the Counter will be loaded on the next CLK pulse. This CLK pulse does not decrement the count, so for an initial count of N, OUT does not strobe low until N + 1 CLK pulses after the initial count is written.

If a new count is written during counting, it will be loaded on the next CLK pulse and counting will continue from the new count If a two-byte count is written, the following happens:

1. Writing the first byte has no effect on counting.
2. Writing the second byte allows the new count to be loaded on the next CLK pulse.

This allows the sequence to be "retriggered" by software. OUT strobes low N + 1 CLK pulses after the new count of N is written.

Mode 5: Hardware Triggered Strobe (Retriggerable)

OUT will initially be high. Counting is triggered by a rising edge of GATE. When the initial count has expired, OUT will go low for one CLK pulse and then go high again.

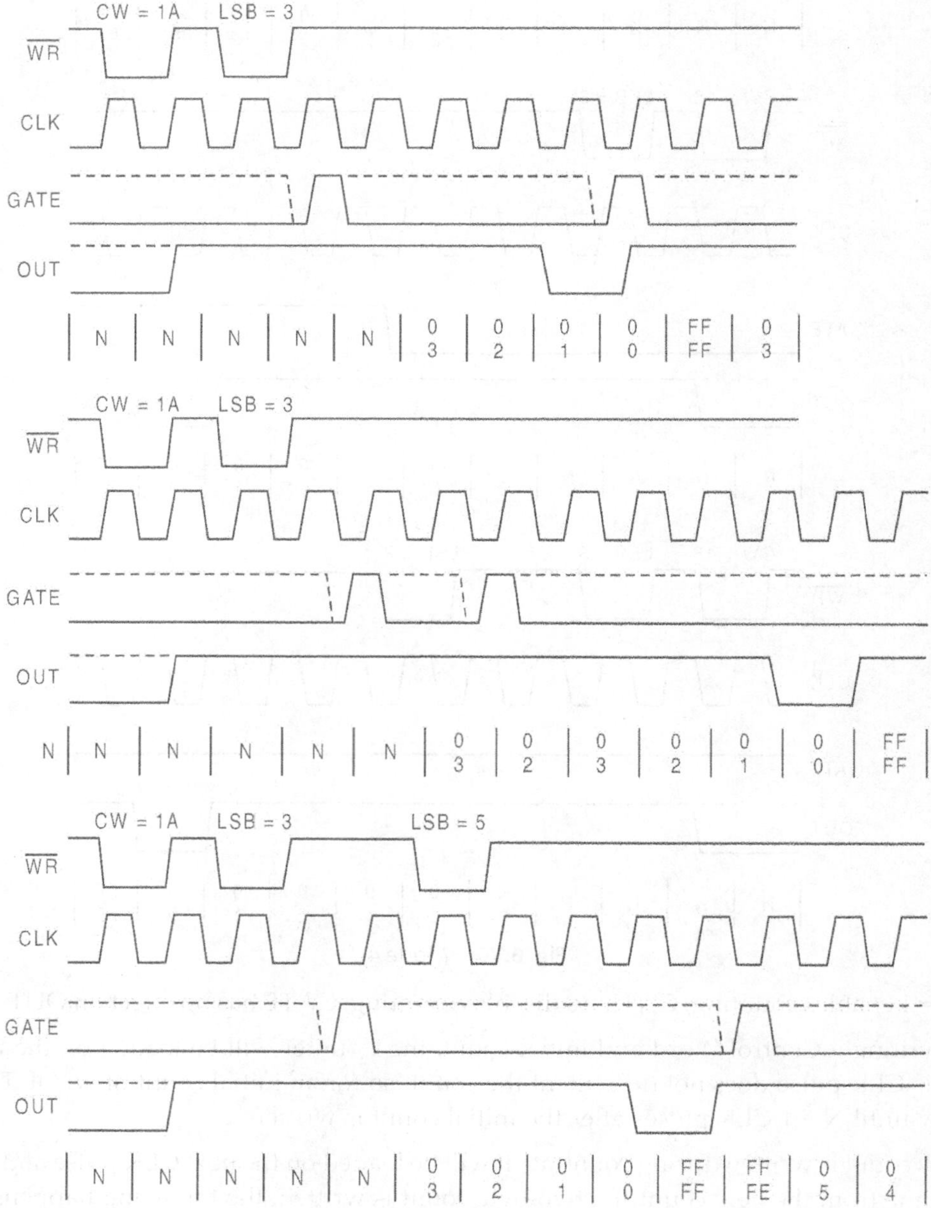

Fig. 6.71 Mode 5

After writing the Control Word and initial count, the counter will not be loaded until the CLK pulse after a trigger. This CLK pulse does not decrement the count, so for an initial count of N, OUT does not strobe low until N + 1 CLK pulses after a trigger.

A trigger results in the Counter being loaded with the initial count on the next CLK pulse. The counting sequence is retriggerable. OUT will not strobe low for N + 1 CLK pulses after any trigger. GATE has no effect on OUT.

If a new count is written during counting, the current counting sequence will not be affected. If a trigger occurs after the new count is written but before the current count expires, the Counter will be loaded with the new count on the next CLK pulse and counting will continue from there.

Signal Status Modes	Low or Going Low	Rising	High
0	Disables Counting	— —	Enables Counting
1	— —	1. Initiates Counting 2. Resets Output after Next Clock	— —
2	1. Disables Counting 2. Sets Output Immediately High	Initiates Counting	Enables Counting
3	1. Disables Counting 2. Sets Output Immediately High	Initiates Counting	Enables Counting
4		— —	Enables Counting
5	— —		— —

Fig. 6.72 Gate Pin Operations Summary

Mode	Min Count	Max Count
0	1	0
1	1	0
2	2	0
3	2	0
4	1	0
5	1	0

Note: 0 is equivalent to 2^{16} for binary counting and 10^4 for BCD counting.

Fig. 6.73 Minimum and Maximum Initial Counts

Operation Common to All Modes

Programming. When a Control Word is written to a Counter, all Control Logic is immediately reset and OUT goes to a known initial state; no CLK pulses are required for this.

Gate. The GATE input is always sampled on the rising edge of CLK. In Modes 0, 2, 3, and 4 the GATE input is level sensitive, and the logic level is sampled on the rising edge of CLK. In Modes 1, 2, 3, and 5 the GATE input is rising-edge sensitive. In these Modes, a rising edge of GATE (trigger) sets an edge-sensitive flip-flop in the Counter. This flip-flop is then sampled on the next rising edge of CLK; the flip-flop is reset immediately after it is sampled. In this way, a trigger will be detected no matter when it occurs—a high logic level does not have to be maintained until the next rising edge of CLK. Note that in Modes 2 and 3, the GATE input is both edge- and level-sensitive. In Modes 2 and 3, if a CLK source other than the system dock is used, GATE should be pulsed immediately following $\overline{WR}$ of a new count value.

Counter. New counts are loaded and Counters are decremented on the falling edge of CLK.

The largest possible initial count is 0; this is equivalent to 2^{16} for binary counting and 10^4 for BCD counting.

The Counter does not stop when it reaches zero. In Modes 0, 1, 4, and 5, the Counter "wraps around" to the highest count, either FFFF hex for binary counting or 9999 for BCD counting, and continues counting. Modes 2 and 3 are periodic; the Counter reloads itself with the initial count and continues counting from there.

6.9 8259A PROGRAMMABLE INTERRUPT CONTROLLER

- 8086, 8088 Compatible
- MCS-80, MCS-85 Compatible
- Eight-Level Priority Controller
- Expandable to 64 Levels
- Programmable Interrupt Modes
- Individual Request Mask Capability
- Single + 5V Supply (No Clocks)
- Available in 28-Pin DIP and 28-Lead PLCC Package
- Available in
  - Standard Temperature Range
  - Extended Temperature Range

The 8259A Programmable Interrupt Controller handles up to eight vectored priority interrupts for the CPU. It is cascadable for up to 64 vectored priority interrupts without additional circuitry. It is packaged in a 28-pin DIP, uses NMOS technology and requires a single + 5V supply. Circuitry is static, requiring no clock input.

The 8259A is designed to minimize the software and real time overhead in handing multi-level priority interrupts. It has several modes, permitting optimization for a variety of system requirements.

The 8259A is fully upward compatible with the Intel 8259. Software originally written for the 8259 will operate the 8259A in all 8259 equivalent modes (MCS-80/85. Non-Buffered, Edge Triggered).

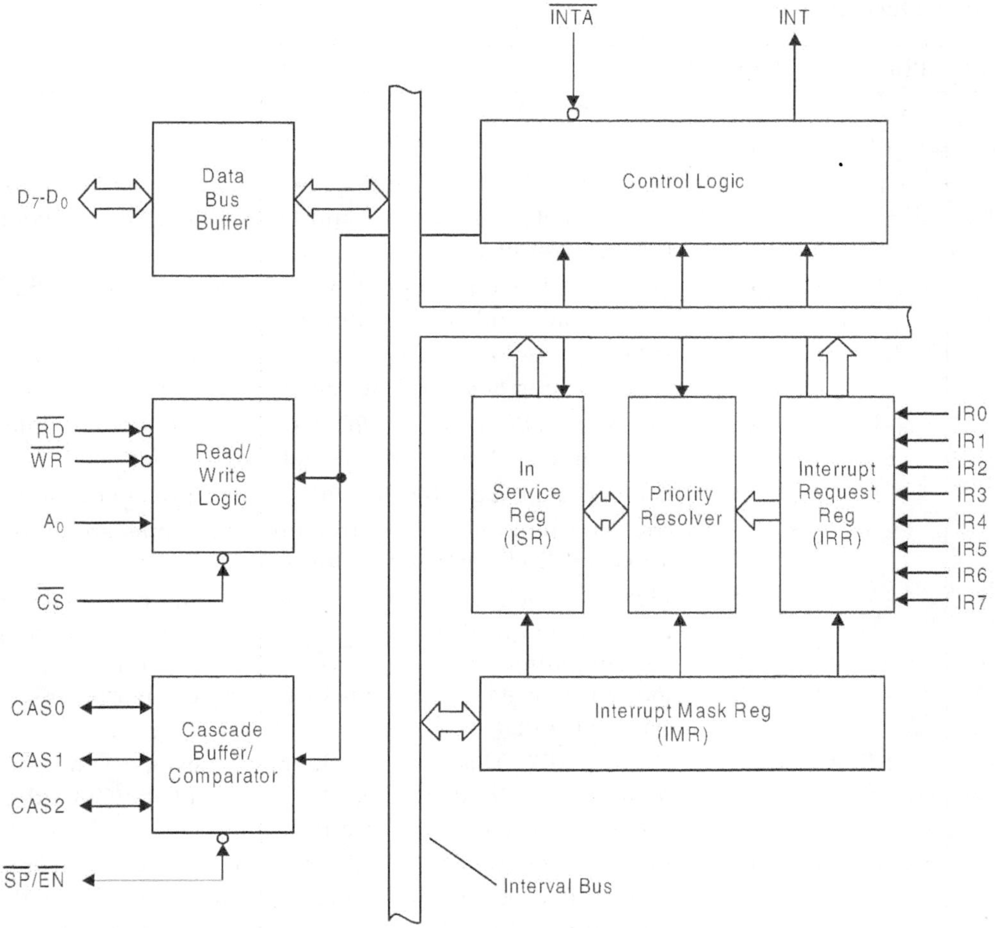

Fig. 6.74 Block diagram

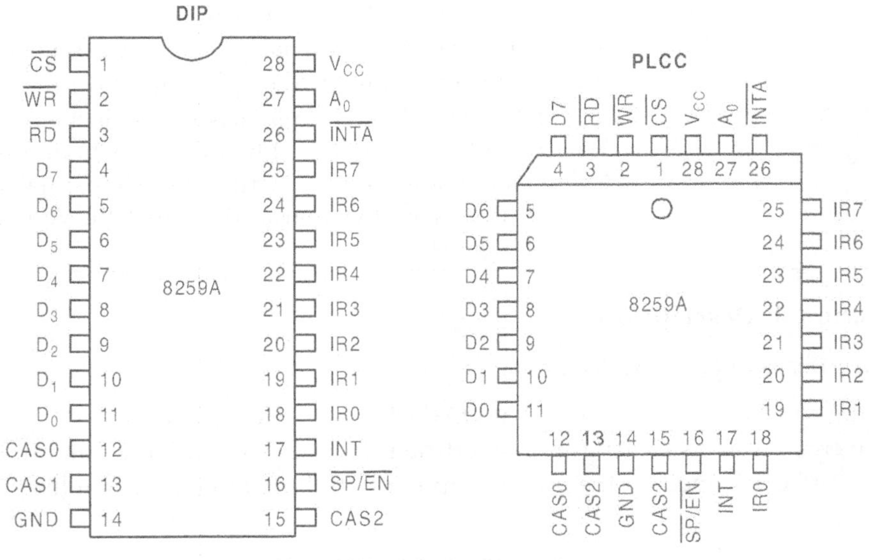

Fig. 6.75 Pin Configuration

6.9.1 Pin Description

Symbol	Pin No.	Type	Name and Function
V_{CC}	28	I	*SUPPLY:* + 5V Supply.
GND	14	I	*GROUND*
$\overline{CS}$	1	I	*CHIP SELECT:* A low on this pin enables $\overline{RD}$ and $\overline{WR}$ communication between the CPU and the 8259A. INTA functions are independent of CS.
$\overline{WR}$	2	I	*WRITE:* A low on this pin when CS is low enables the 8259A to accept command words from the CPU.
$\overline{RD}$	3	I	*READ:* A low on this pin when CS is low enables the 8259A to release status onto the data bus for the CPU.
D_7-D_0	4-11	I/O	*BIDIRECTIONAL DATA BUS:* Control, status and interrupt-vector information is transferred via this bus.
CAS_0-CAS_2	12, 13, 15	I/O	*CASCADE LINES:* The CAS lines form a private 8259A bus to control a multiple 8259A structure. These pins are outputs for a master 8259A and inputs for a slave 8259A.
$\overline{SP}/\overline{EN}$	16	I/O	*SLAVE PROGRAM/ENABLE BUFFER:* This is a dual function pin. When in the Buffered Mode it can be used as an output to control buffer transceivers (EN). When not in the buffered mode it is used as an input to designate a master (SP = 1) or slave (SP = 0).
INT	17	o	*INTERRUPT:* This pin goes high whenever a valid interrupt request is asserted. It is used to interrupt the CPU, thus it is connected to the CPU's interrupt pin.
IR_0-IR_7	18-25	I	*INTERRUPT REQUESTS:* Asynchronous inputs. An interrupt request is executed by raising an IR input (low to high), and holding it high until it is acknowledged (Edge Triggered Mode), or just by a high level on an IR input (Level Triggered Mode).
$\overline{INTA}$	26	I	*INTERRUPT ACKNOWLEDGE:* This pin is used to enable 8259A interrupt-vector data onto the data bus by a sequence of interrupt acknowledge pulses issued by the CPU.
A_0	27	I	*AO ADDRESS LINE:* This pin acts in conjunction with the $\overline{CS}$, $\overline{WR}$, and $\overline{RD}$ pins. It is used by the 8259A to decipher various Command Words the CPU writes and status the CPU wishes to read. It is typically connected to the CPU AO address line (A1 for 8086, 8088).

6.9.2 Functional Description

Interrupts in Microcomputer Systems

Microcomputer system design requires that I/O devices such as keyboards, displays, sensors and other components receive servicing in a an efficient manner so that large amounts of the total system tasks can be assumed by the microcomputer with little or no effect on throughput.

The most common method of servicing such devices is the *Polled* approach. This is where the processor must test each device in sequence and in effect "ask" each one if it needs servicing. It is easy to see that a large portion of the main program is looping through this continuous polling cycle and that such a method would have a serious detrimental effect on system throughput, thus limiting the tasks that could be assumed by the microcomputer and reducing the cost effectiveness of using such devices.

A more desirable method would be one that would allow the microprocessor to be executing its main program and only stop to service peripheral devices when it is told to do so by the device itself. In effect, the method would provide an external asynchronous input that would inform the processor that it should complete whatever instruction that is currently being executed and fetch a new routine that will service the requesting device. Once this servicing is complete, however, the processor would resume exactly where it left off.

This method is called *Interrupt*. It is easy to see that system throughput would drastically increase, and thus more tasks could be assumed by the microcomputer to further enhance its cost effectiveness.

The Programmable Interrupt Controller (PIC) functions as an overall manager in an Interrupt-Driven system environment. It accepts requests from the peripheral equipment, determines which of the incoming requests is of the highest importance (priority), ascertains whether the incoming request has a higher priority value than the level currently being serviced, and issues an interrupt to the CPU based on this determination.

Each peripheral device or structure usually has a special program or "routine" that is associated with its specific functional or operational requirements; this is referred to as a "service routine". The PIC, after issuing an Interrupt to the CPU, must somehow input information into the CPU that can "point" the Program Counter to the service routine associated with the requesting device. This "pointer" is an address in a vectoring table and will often be referred to, in this document, as vectoring data.

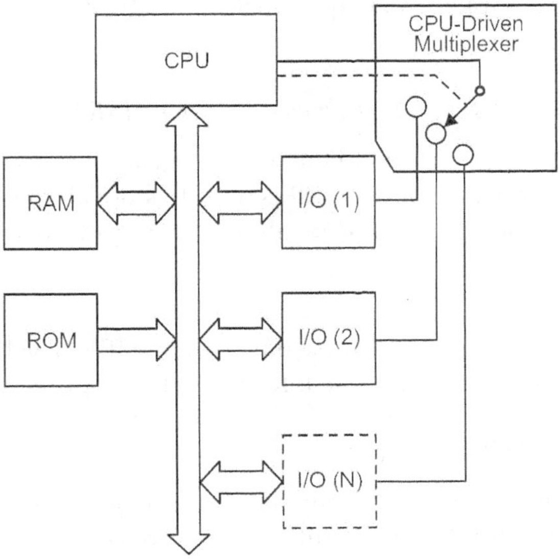

Fig. 6.76 (*a*) Polled Method

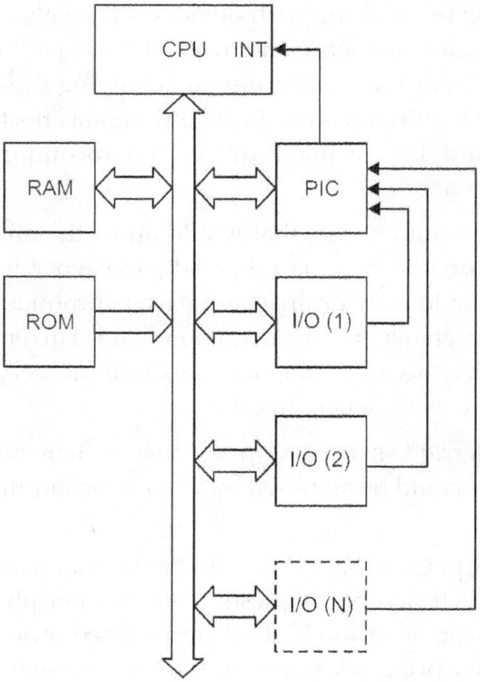

Fig. 6.76 (b) Interrupt Method

The 8259A is a device specifically designed for use in real time, interrupt driven microcomputer systems. It manages eight levels or requests and has built-in features for expandability to other 8259A's (up to 64 levels). It is programmed by the system's software as an I/O peripheral. A selection of priority modes is available to the programmer so that the manner in which the requests are processed by the 8259A can be configured to match his system requirements. The priority modes can be changed or reconfigured dynamically at any time during the main program. This means that the complete interrupt structure can be defined as required, based on the total system environment.

Interrupt Request Register (IRR) and In-service Register (ISR)

The interrupts at the IR input lines are handled by two registers in cascade, the Interrupt Request Register (IRR) and the In-Service (ISR). The IRR is used to store all the interrupt levels which are requesting service; and the ISR is used to store all the interrupt levels which are being serviced.

Priority Resolver

This logic block determines the priorities of the bits set in the IRR. The highest priority is selected and strobed into the corresponding bit of the ISR during $\overline{\text{INTA}}$ pulse.

Interrupt Mask Register (IMR)

The IMR stores the bits which mask the interrupt lines to be masked. The IMR operates on the IRR. Masking of a higher priority input will not affect the interrupt request lines of lower quality.

INT (Interrupt)

This output goes directly to the CPU interrupt input. The V_{OH} level on this line is designed to be fully compatible with the 8080A, 8085A and 8086 input levels.

INTA (Interrupt Acknowledge)

$\overline{\text{INTA}}$ pulses will cause the 8259A to release vectoring information onto the data bus.

Data Bus Buffer

This 3-state, bidirectional 8-bit buffer is used to interface the 8259A to the system Data Bus. Control words and status information are transferred through the Data Bus Buffer.

Read/Write Control Logic

The function of this block is to accept OUTput commands from the CPU. It contains the Initialization Command Word (ICW) registers and Operation Command Word (OCW) registers which store the various control formats for device operation. This function block also allows the status of the 8259A to be transferred onto the Data Bus.

$\overline{CS}$ (Chip Select)

A LOW on this input enables the 8259A. No reading or writing of the chip will occur unless the device is selected.

$\overline{WR}$ (Write)

A LOW on this input enables the CPU to write control words (ICWs and OCWs) to the 8259A.

$\overline{RD}$ (Read)

A LOW on this input enables the 8259A to send the status of the Interrupt Request Register (IRR), In Service Register (ISR), the Interrupt Mask Register (IMR), or the Interrupt level onto the Data Bus.

A_0

This input signal is used in conjunction with $\overline{\text{WR}}$ and $\overline{\text{RD}}$ signals to write commands into the various command registers, as well as reading the various status registers of the chip. This line can be tied directly to one of the address lines.

The Cascade Buffer/Comparator

This function block stores and compares the IDs of all 8259A's used in the system. The associated three I/O pins (CAS0-2) are outputs when the 8259A is used as a master and are inputs when the 8259A is used as a slave. As a master, the 8259A sends the ID of the interrupting slave device onto the CAS0-2 lines. The slave thus selected will send its preprogrammed subroutine address onto the Data Bus during the next one or two consecutive $\overline{\text{INTA}}$ pulses.

Interrupt Sequence

The powerful features of the 8259A in a microcomputer system are its programmability and the interrupt routine addressing capability. This allows direct or indirect jumping to the specific interrupt routine requested without any polling of the interrupting devices. The normal sequence of events during an interrupt depends on the type of CPU being used.

The events occur as follows

1. One or more of the INTERRUPT REQUEST lines (IR7-0) are raised high, setting the corresponding IRR bit(s).
2. The 8259A evaluates these requests, and sends an INT to the CPU, if appropriate.
3. The CPU acknowledges the INT and responds with an $\overline{\text{INTA}}$ pulse.
4. Upon receiving an $\overline{\text{INTA}}$ from the CPU group, the highest priority ISR bit is set, and the corresponding IRR bit is reset. The 8259A will also release a CALL instruction code (11001101) onto the 8-bit Data Bus through its D7-0 pins.
5. This CALL instruction will initiate two more $\overline{\text{INTA}}$ pulses to be sent to the 8259A from the CPU group.
6. These two $\overline{\text{INTA}}$ pulses allow the 8259A to release its preprogrammed subroutine address onto the Data Bus. The lower 8-bit address is released at the first $\overline{\text{INTA}}$ pulse and the higher 8-bit address is released at the second $\overline{\text{INTA}}$ pulse.
7. This completes the 3-byte CALL instruction released by the 8259A. In the AEOI mode the ISR bit is reset at the end of the third $\overline{\text{INTA}}$ pulse. Otherwise, the ISR bit remains set until an appropriate EOI command is issued at the end of the interrupt sequence.

The events occurring in an 8086 system are the same until step 4.

5. Upon receiving an $\overline{\text{INTA}}$ from the CPU group, the highest priority ISR bit is set and the corresponding IRR bit is reset. The 8259A does not drive the Data Bus during this cycle.
6. The 8086 will initiate a second $\overline{\text{INTA}}$ pulse. During this pulse, the 8259A releases an 8-bit pointer onto the Data Bus where it is read by the CPU.
7. This completes the interrupt cycle. In the AEOI mode, the ISR bit is reset at the end of the second $\overline{\text{INTA}}$ pulse. Otherwise, the ISR bit remains set until an appropriate EOI command is issued at the end of the interrupt subroutine.

If no interrupt request is present at step 4 of either sequence (*i.e.*, the request was too short in duration) the 8259A will issue an interrupt level 7. Both the vectoring bytes and the CAS lines will look like an interrupt level 7 was requested.

When the 8259A PIC receives an interrupt, INT becomes active and an interrupt acknowledge cycle is started. If a higher priority interrupt occurs between the two INTA pulses, the INT line goes inactive immediately after the second INTA pulse. After an unspecified amount of time the INT line is activated again to signify the higher priority interrupt waiting for service. This inactive time is not specified and can vary between parts.

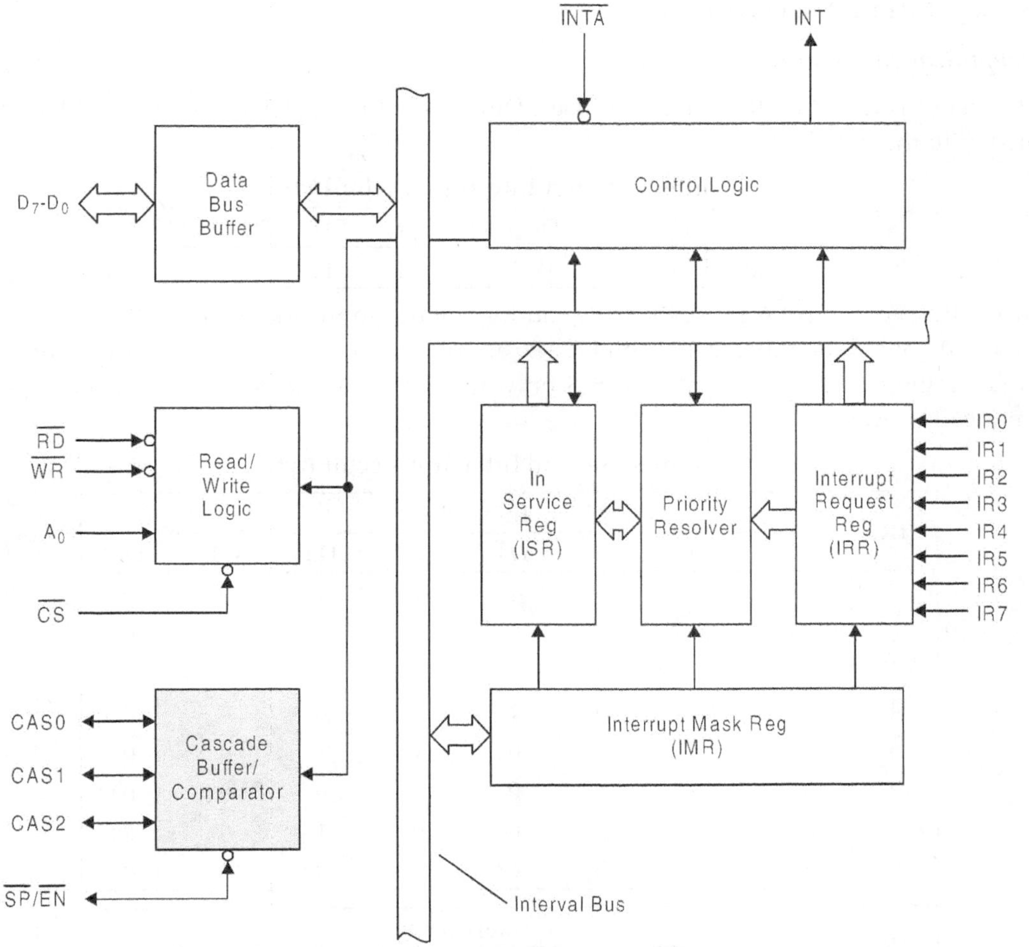

Fig. 6.77 8259A Block Diagram

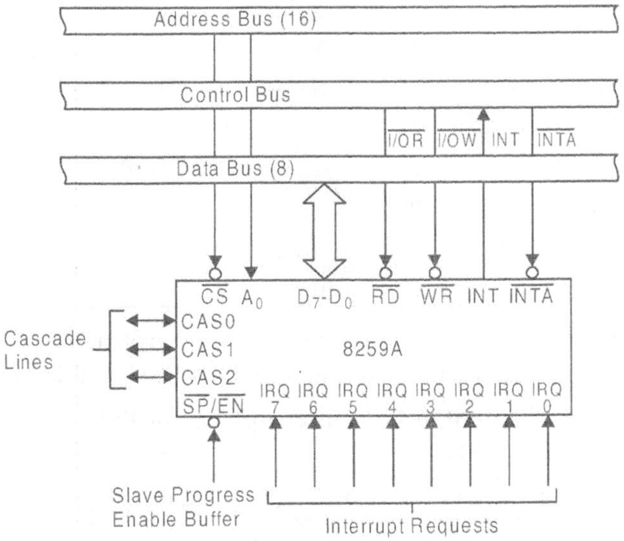

Fig. 6.78 8259A Interface to Standard System Bus

6.9.3 Operational Description

Interrupt Sequence Outputs

This sequence is timed by three $\overline{INTA}$ pulses. During the first $\overline{INTA}$ pulse, the CALL opcode is enabled onto the data bus.

Content of First Interrupt Vector Byte

	D7	*D6*	*D5*	*D4*	*D3*	*D2*	*D1*	*D0*
Call Code	1	1	0	0	1	1	0	1

During the second $\overline{INTA}$ pulse the lower address of the appropriate service routine is enabled onto the data bus. When Interval = 4 bits A_5-A_7 are programmed, while A_0-A_4 are automatically inserted by the 8259A. When Interval = 8 only A_6 and A_7 are programmed, while A_0-A_5 are automatically inserted.

Content of Second Interrupt Vector Byte

IR	Interval = 4							
	D7	**D6**	**D5**	**D4**	**D3**	**D2**	**D1**	**D0**
7	A7	A6	A5	1	1	1	0	0
6	A7	A6	A5	1	1	0	0	0
5	A7	A6	A5	1	0	1	0	0
4	A7	A6	A5	1	0	0	0	0
3	A7	A6	A5	0	1	1	0	0
2	A7	A6	A5	0	1	0	0	0
1	A7	A6	A5	0	0	1	0	0
0	A7	A6	A5	0	0	0	0	0

IR	Interval = 8							
	D7	**D6**	**D5**	**D4**	**D3**	**D2**	**D1**	**D0**
7	A7	A6	1	1	1	0	0	0
6	A7	A6	1	1	0	0	0	0
5	A7	A6	1	0	1	0	0	0
4	A7	A6	1	0	0	0	0	0
3	A7	A6	0	1	1	0	0	0
2	A7	A6	0	1	0	0	0	0
1	A7	A6	0	0	1	0	0	0
0	A7	A6	0	0	0	0	0	0

During the third $\overline{INTA}$ pulse the higher address of the appropriate service routine, which was programmed as byte 2 of the initialization sequence (A_8-A_{15}), is enabled onto the bus.

Content of First Interrupt Vector Byte

D7	*D6*	*D5*	*D4*	*D3*	*D2*	*D1*	*D0*
A15	A14	A13	A12	A11	A10	A9	A8

8086, 8088

8086 mode is similar except that only two Interrupt Acknowledge cycles are issued by the processor and no CALL opcode is sent to the processor. The first interrupt acknowledge cycle is similar in that the 8259A uses it to internally freeze the state of the interrupts for priority resolution and as a master it issues the interrupt code on the cascade lines at the end of the $\overline{INTA}$ pulse. On this first cycle, it does not issue any data to the processor and leaves its data bus buffers disabled. On the second interrupt acknowledge cycle in 8086 mode, the master (or slave if so programmed) will send a byte of data to the processor with the acknowledged interrupt code composed as follows (note the state of the ADI) mode control is ignored and A_5-A_{11} are unused in 8086 mode):

Content of Interrupt Vector Byte for 8086 System Mode

	D7	D6	D5	D4	D3	D2	D1	D0
IR7	T7	T6	T5	T4	T3	1	1	1
IR6	T7	T6	T5	T4	T3	1	1	0
IR5	T7	T6	T5	T4	T3	1	0	1
IR4	T7	T6	T5	T4	T3	1	0	0
IR3	T7	T6	T5	T4	T3	0	1	1
IR2	T7	T6	T5	T4	T3	0	1	0
IR1	T7	T6	T5	T4	T3	0	0	1
IRO	T7	T6	T5	T4	T3	0	0	0

6.9.4 Programming the 8259A

The 8259A accepts two types of command words generated by the CPU:

1. *Initialization Command Words (ICWs):* Before normal operation can begin, each 8259A in the system must be brought to a starting point—by a sequence of 2 to 4 bytes timed by $\overline{WR}$ pulses.
2. *Operation Command Words (OCWs):* These are the command words which command the 8259A to operate in various interrupt modes. These modes are:
   (*a*) Fully nested mode
   (*b*) Rotating priority mode
   (*c*) Special mask mode
   (*d*) Polled mode

The OCWs can be written into the 8259A anytime after initialization.

Initialization Command Words (ICWS)

General. Whenever a command is issued with A0 = 0 and D4 = 1, this is interpreted as Initialization Command Word 1 (ICW1). ICW1 starts the initialization sequence during which the following automatically occur.

(*a*) The edge sense circuit is reset, which means that following initialization, an interrupt request (IR) input must make a low-to-high transistion to generate an interrupt.
(*b*) The Interrupt Mask Register is cleared.
(*c*) IR7 input is assigned priority 7.
(*d*) The slave mode address is set to 7.
(*e*) Special Mask Mode is cleared and Status Read is set to IRR.
(*f*) If IC4 = 0, then all functions selected in ICW4 are set to zero.

Initialization Command Words 1 and 2 (ICW1, ICW2)

A_5-A_{15}: Page starting address of service routines. In an MCS 80/85 system, the 8 request levels will generate CALLs to 8 locations equally spaced in memory. These can be programmed to be spaced at intervals of 4 or 8 memory locations, thus the 8 routines will occupy a page of 32 or 64 bytes, respectively.

The address format is 2 bytes long (A_0-A_{15}). When the routine interval is 4, A_0-A_4 are automatically inserted by the 8259A, while A_5-A_{15} are programmed externally. When the routine interval is 8, A_0-A_5 are automatically inserted by the 8259A, white A_6-A_{15} are programmed externally.

The 8-byte interval will maintain compatibility with current software, while the 4-byte interval is best for a compact jump table.

In an 8086 system, A_{15}-A_{11} are inserted in the five most significant bits of the vectoring byte and the 8259A sets the three least significant bits according to the interrupt level. A_{10}-A_5 are ignored and ADI (Address interval) has no effect.

LTIM: If LTIM = 1, then the 8259A will operate in the level interrupt mode. Edge detect logic on the interrupt inputs will be disabled.

ADI: CALL address interval. ADI = 1 then interval = 4; ADI = 0 then interval = 8.

SNGL: Single. Means that this is the only 8259A in the system. If SNGL = 1 no ICW3 will be issued.

IC4: If this bit is set—ICW4 has to be read. If ICW4 is not needed, set IC4 = 0.

Initialization Command Word 3 (ICW3)

This word is read only when there is more than one 8259A in the system and cascading is used, in which case SNGL = 0. It will load the 8-bit slave register. The functions of this register are:

(*a*) In the master mode (either when SP = 1, or in buffered mode when M/S = 1 in ICW4) a "1" is set for each slave in the system. The master then win release byte 1 of the call sequence (for MCS-80/85 system) and will enable the corresponding slave to release bytes 2 and 3 (for 8086 only byte 2) through the cascade lines.

(*b*) In the slave mode (either when $\overline{SP}$ = 0, or if BUF = 1 and M/S = 0 in ICW4) bits 2–0 identify the slave. The slave compares its cascade input with these bits and, if they are equal, bytes 2 and 3 of the call sequence (or just byte 2 for 8086) are released by it on the Data Bus.

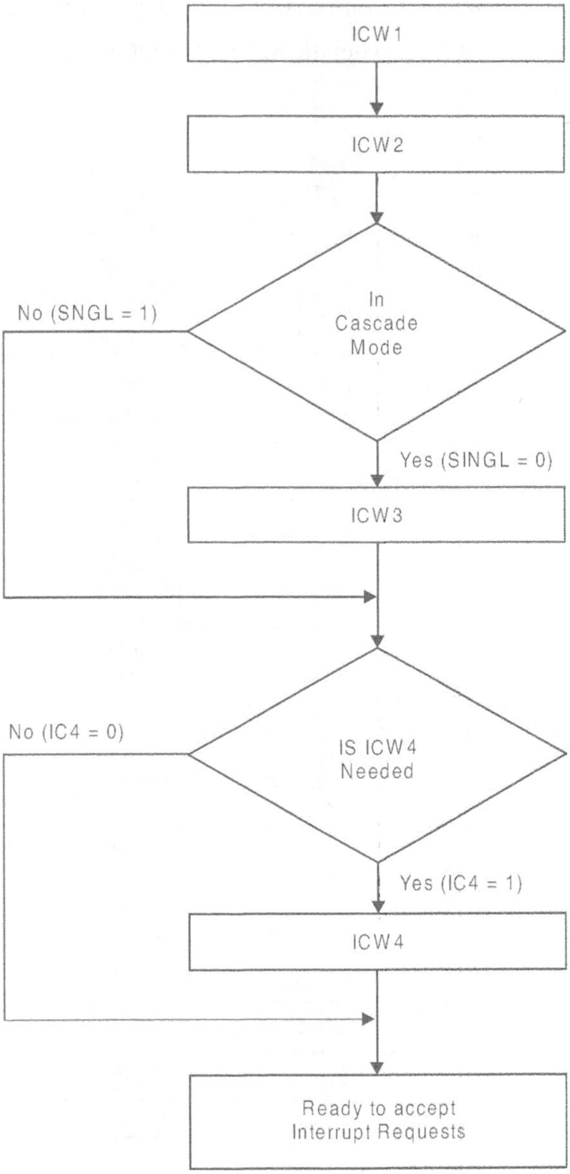

Fig. 6.79 Initialization Sequence

Initialization Command Word 4 (ICW4)

SFNM: If SFNM = 1 the special fully nested mode is programmed.

BUF: If BUF = 1 the buffered mode is programmed. In buffered mode $\overline{SP}/\overline{EN}$ becomes an enable output and the master/slave determination is by M/S.

M/S: If buffered mode is selected: M/S = 1 means the 8259A is programmed to be a master, M/S = 0 means the 8259A is programmed to be a slave. If BUF = 0, M/S has no function.

AEOI: If AEOI = 1 the automatic end of interrupt mode is programmed.

μPM: Microprocessor mode: μPM = 0 sets the 8259A for MCS-80, 85 system operation. μPM = 1 sets the 8259A for 8086 system operation.

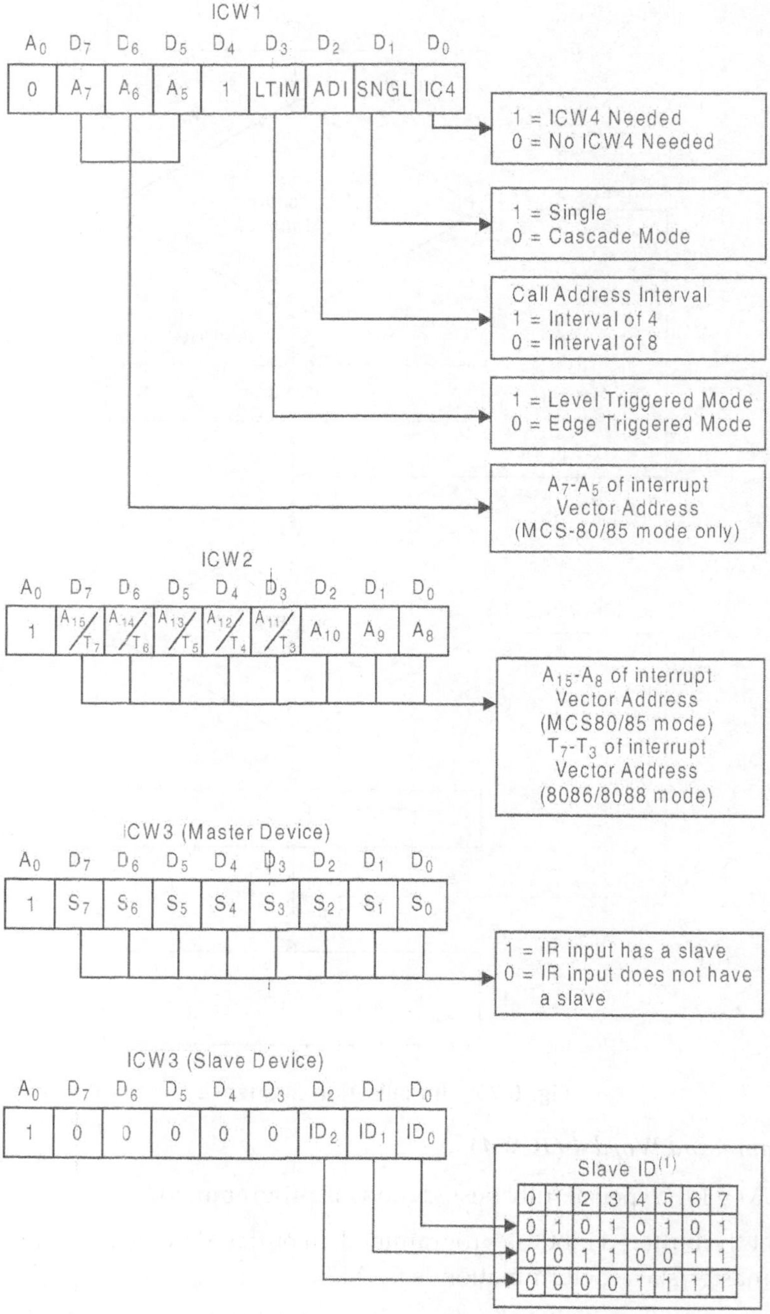

Fig 6.80 Initialization Command Word Format

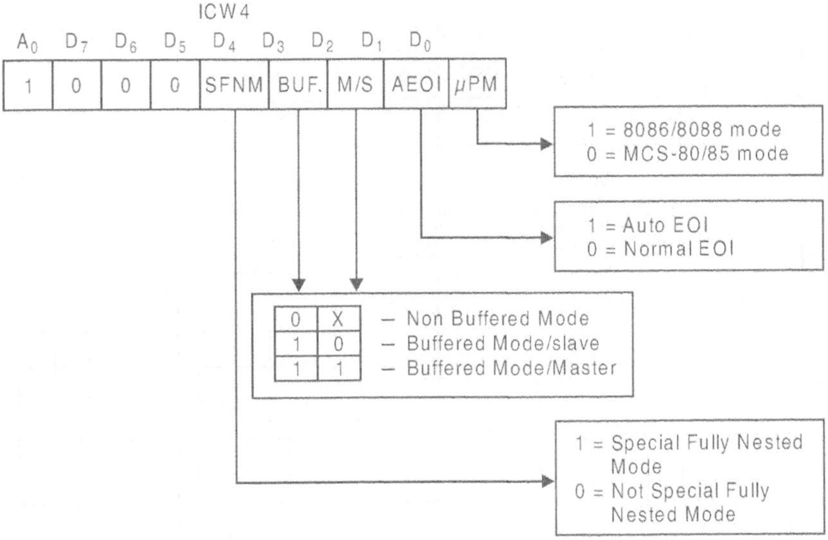

Note: Slave ID is equal to the corresponding master IR input.

Fig 6.80 Initialization Command Word Format (Continued)

Operation Command Words (OCWS)

After the Initialization Command Words (ICWs) are programmed into the 8259A, the chip is ready to accept interrupt requests at its input lines. However, during the 8259A operation, a selection of algorithms can command the 8259A to operate in various modes through the Operation Command Words (OCWs).

Operation Control Words (OCWs)

OCW1

A0	*D7*	*D6*	*D5*	*D4*	*D3*	*D2*	*D1*	*D0*
1	M7	M6	M5	M4	M3	M2	M1	M0

OCW2

A0	*D7*	*D6*	*D5*	*D4*	*D3*	*D2*	*D1*	*D0*
0	R	SL	EOI	0	0	L2	L1	L0

OCW3

A0	*D7*	*D6*	*D5*	*D4*	*D3*	*D2*	*D1*	*D0*
0	0	ESMM	SMM	0	1	P	PR	RIS

Operation Control Word 1 (OCW1)

OCW1 sets and clears the mask bits in the interrupt Mask Register (IMP). M_7-M_0 represent the eight mask bits. M = 1 indicates the channel is masked (inhibited), M = 0 indicates the channel is enabled.

Operation Control Word 2 (OCW2)

R, SL, EOI—These three bits control the Rotate and End of Interrupt modes and combinations of the two. A chart of these combinations can be found on the Operation Command Word Formal.

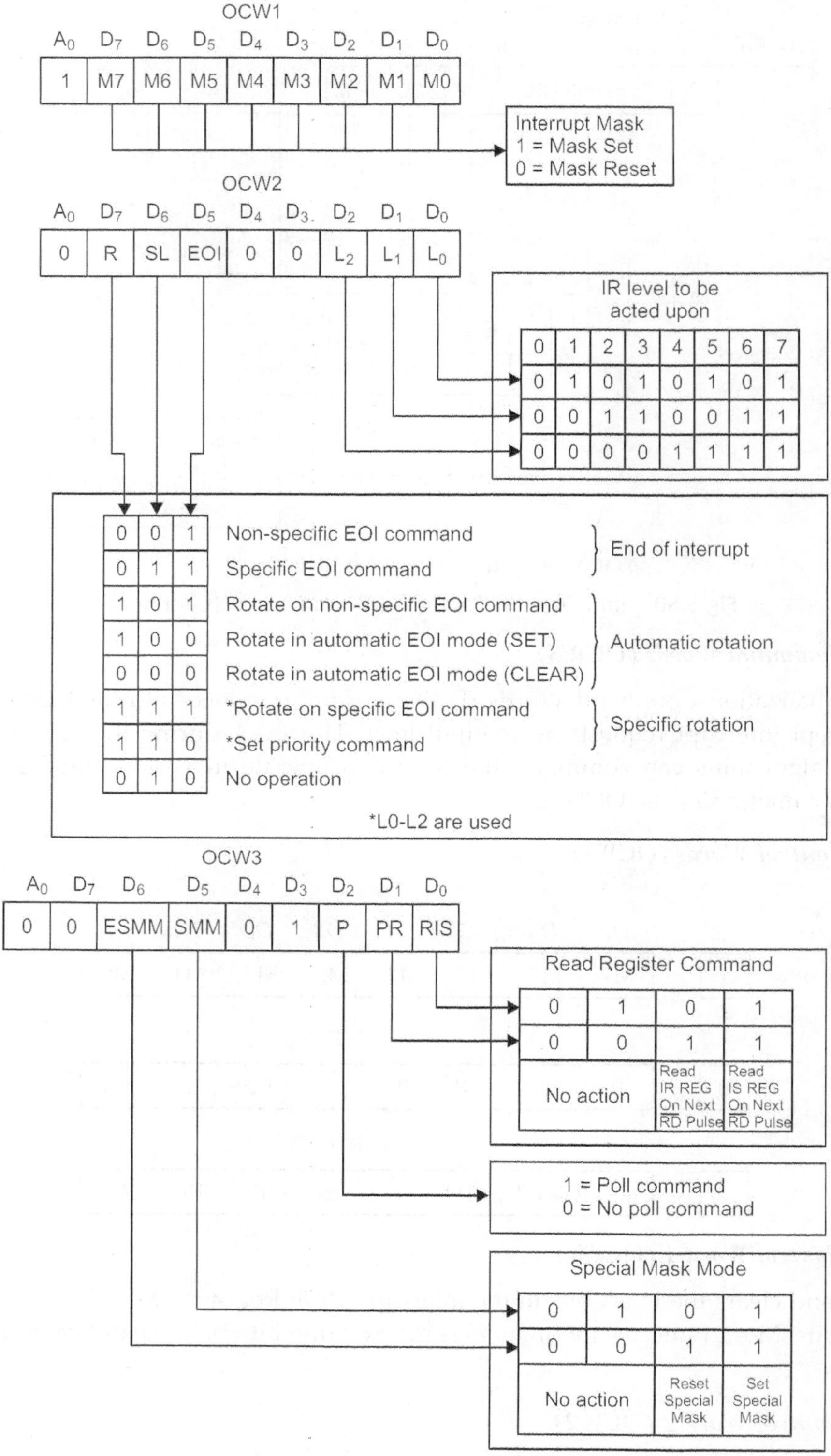

Fig. 6.81 Operation Command Word Format

$L2$, L_1, L_0—These bits determine the interrupt level acted upon when the SL bit is active.

Operation Control Word 3 (OCW3)

ESMM—Enable Special Mask Mode. When this bit is set to 1 it enables the SMM bit to set or reset the Special Mask Mode. When ESMM = 0 the SMM bit becomes a "don't care".

SMM—Special Mask Mode. If ESMM = 1 and SMM = 1 the 8259A will enter Special Mask Mode. If ESMM = 1 and SMM = 0 the 8259A will revert to normal mask mode. When ESMM = 0, SMM has no effect.

6.9.5 Operational Mode

Fully Nested Mode

This mode is entered after initialization unless another mode is programmed. The interrupt requests are ordered in priority from 0 through 7 (0 highest). When an interrupt is acknowledged the highest priority request is determined and its vector placed on the bus. Additionally, a bit of the Interrupt Service register (IS0-7) is set. This bit remains set until the microprocessor issues an End of Interrupt (EOI) command immediately before returning from the service routine, or if AEOI (Automatic End of Interrupt) bit is set, until the training edge of the last INTA. While the IS bit is set, all further interrupts of the same or lower priority are inhibited, while higher levels will generate an interrupt (which will be acknowledged only if the microprocessor internal Interrupt enable flip-flop has been re-enabled through software).

After the initialization sequence, IR0 has the highest priority and IR7 the lowest. Priorities can be changed, as will be explained, in the rotating priority mode.

End of Interrupt (EOI)

The In Service (IS) bit can be reset either automatically following the trailing edge of the last in sequence $\overline{INTA}$ pulse (when AEOI bit in ICW1 is set) or by a command word that must be issued to the 8259A before returning from a service routine (EOI command). An EOI command must be issued twice if in the Cascade mode, once for the master and once for the corresponding slave.

There are two forms of EOI command: Specific and Non-Specific. When the 8259A is operated in modes which perserve the fully nested structure, it can determine which IS bit to reset on EOI. When a Non-specific EOI command is issued the 8259A will automatically reset the highest IS bit of those that are set, since in the fully nested mode the highest IS level was necessarily the last level acknowledged and serviced. A non-specific EOI can be issued with OCW2 (EOI = 1, SL = 0, R = 0).

When a mode is used which may disturb the fully nested structure, the 8259A may no longer be able to determine the last level acknowledged. In this case a Specific End of Interrupt must be issued which includes as part of the command the IS level to be reset. A specific EOI can be issued with OCW2 (EOI = 1, SL = 1. R = 0, and L0-L2 is the binary level of the IS bit to be reset).

It should be noted that an IS bit that is masked by an IMR bit will not be cleared by a non-specific EOI if the 8259A is in the Special Mask Mode.

Automatic End of Interrupt (AEOI) Mode

If AEOI = 1 in ICW4, then the 8259A will operate in AEOI mode continuously until reprogrammed by ICW4. In this mode the 8259A will automatically perform a non-specific EOI operation at the trailing edge of the last interrupt acknowledge pulse (third pulse in MCS-80/85, second in 8086). Note that from a system standpoint, this mode should be used only when a nested multilevel interrupt structure is not required within a single 8259A. The AEOI mode can only be used in a master 8259A and not a slave.

Automatic Rotation (Equal Priority Devices)

In some applications, there are a number of interrupting devices of equal priority. In this mode a device, after being serviced, receives the lowest priority, so a device requesting an interrupt will have to wait, in the worst case until each of 7 other devices are serviced at most *once*. For example, if the priority and "in service" status is:

Before Rotate (IR4 the highest priority requiring service).

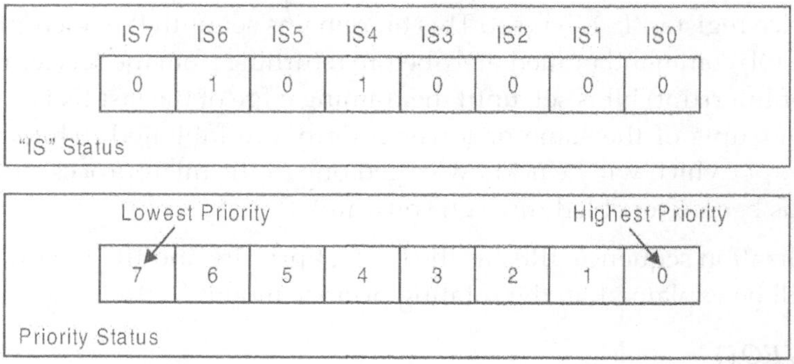

The following registers can be read via OCW3

Interrupt Request Register (IRR): 8-bit register which contains the levels requesting an interrupt to be acknowledged. The highest request level is reset from the IRR when an interrupt is acknowledged. (Not affected by IMR.)

In-Service Register (ISR): 8-bit register which contains the priority levels that are being serviced. The ISR is updated when an End of Interrupt Command is issued.

Interrupt Mask Register: 8-bit register which contains the interrupt request lines which are masked.

The IRR can be read when, prior to the RD pulse, a Read Register Command is issued with OCW3 (RR = 1, RIS = 0).

The ISR can be read, when, prior to the RD pulse, a Read Register Command is issued with OCW3 (RR = 1, RIS = 1).

There is no need to write an OCW3 before every status read operation, as long as the status read corresponds with the previous one; *i.e.*, the 8259A "remembers" whether the IRR or ISR has been previously selected by the OCW3. This is not true when poll is used.

After initialization the 8259A is set to IRR.

For reading the IMR. no OCW3 is needed. The output data bus will contain the IMR whenever RD is active and A0 = 1 (OCW1).

Polling overrides status read when P = 1, RR = 1 in OCW3.

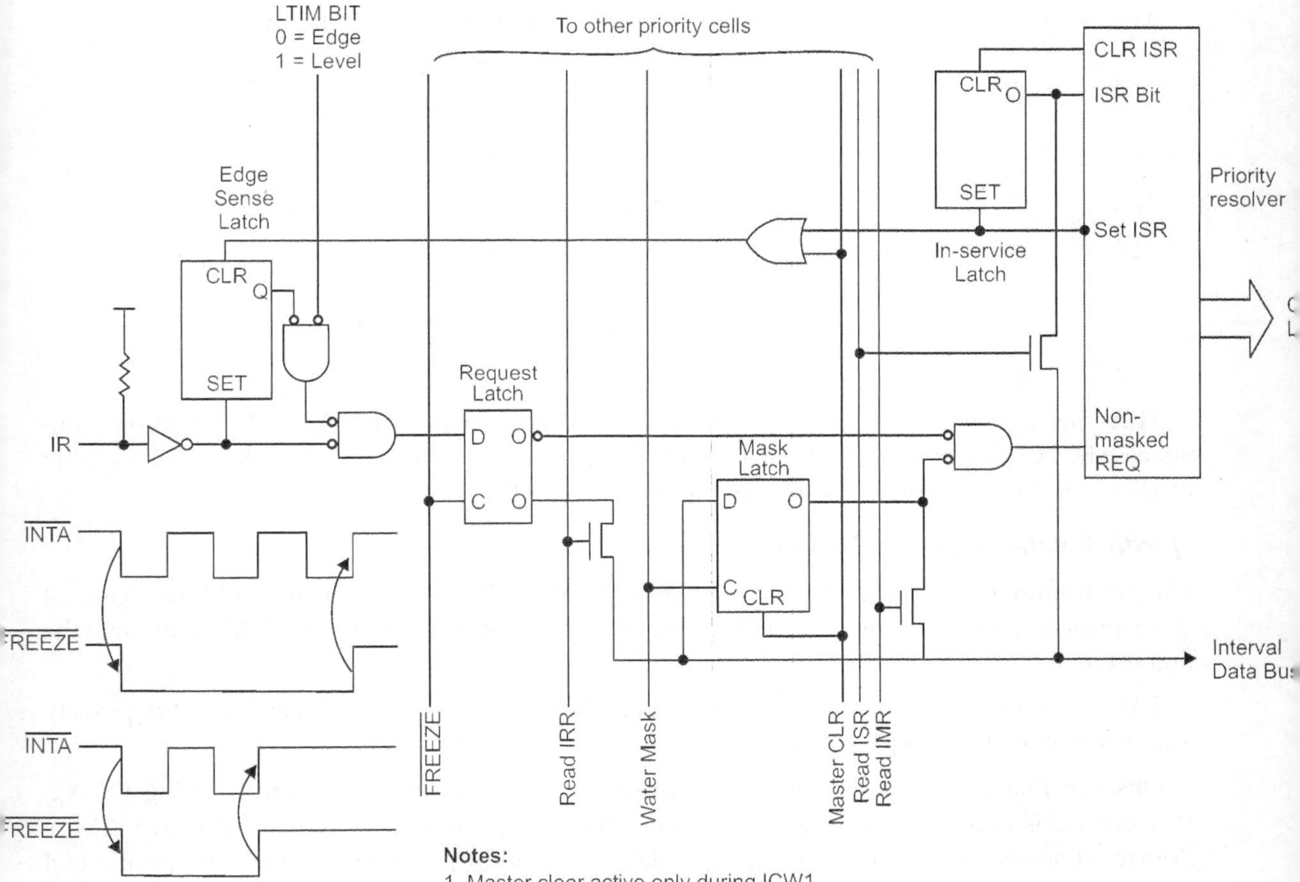

Notes:

1. Master clear active only during ICW1.
2. $\overline{\text{FREEZE}}$ is active during $\overline{\text{INTA}}$ and poll sequence only.
3. Truth Table for a D-Latch

C	D	Q	Operation
1	Di	DI	Follow
0	X	Qn-1	Hold

Fig. 6.82 Priority Cell—Simplified Logic Diagram.

Edge and Level Triggered Modes

This mode is programmed using bit 3 in ICW1.

If LTIM = '0', an interrupt request will be recognized by a low to high transition on an IR input. The IR input can remain high without generating another interrupt.

After Rotate (IR4 was serviced, all other priorities rotated correspondingly).

"IS" Status

IS7	IS6	IS5	IS4	IS3	IS2	IS1	IS0
0	1	0	0	0	0	0	0

Priority Status (Highest Priority → 0; Lowest Priority → 7)

2	1	0	7	6	5	4	3

There are two ways to accomplish Automatic Rotation using OCW2, the Rotation on Non-Specific EOI Command (R = 1, SL = 0, EOI = 1) and the Rotate in Automatic EOI Mode which is set by (R = 1, SL = 0, EOI = 0) and cleared by (R = 0, SL = 0, EOI = 0).

Specific Rotation (Specific Priority)

The programmer can change priorities by programming the bottom priority and thus fixing all other priorities; *i.e.*, if IR5 is programmed as the bottom priority device, then IR6 will have the highest one.

The Set Priority command is issued in OCW2 where: R = 1, SL = 1, L0-L2 is the binary priority level code of the bottom priority device.

Observe that in this mode internal status is updated by software control during OCW2. However, it is independent of the End of Interrupt (EOI) command (also executed by OCW2). Priority changes can be executed during an EOI command by using the Rotate on Specific EOI command in OCW2 (R = 1, SL = 1, EOI = 1 and L0-L2 = IR level to receive bottom priority).

Interrupt Masks

Each Interrupt Request input can be masked individually by the Interrupt Mask Register (IMR) programmed through OCW1. Each bit in the IMR masks one interrupt channel if it is set (1). Bit 0 masks IR0, Bit 1 masks IR1 and so forth. Masking an IR channel does not affect the other channels operation.

Special Mask Mode

Some applications may require an interrupt service routine to dynamically alter the system priority structure during its execution under software control. For example the routine may wish to inhibit lower priority requests for a portion of its execution but enable some of them for another portion.

The difficulty here is that if an Interrupt Request is acknowledged and an End of Interrupt command did not reset its IS bit (*i.e.*, while executing a service routine), the 8259A would have inhibited all lower priority requests with no easy way for the routine to enable them.

That is where the Special Mask Mode comes in. In the special Mask Mode, when a mask bit is set in OCW1, it inhibits further interrupts at that level *and enables* interrupts from *all other* levels (lower as well as higher) that are not masked.

Thus, any interrupts may be selectively enabled by loading the mask register.

The special Mask Mode is set by OWC3 where: SSMM = 1, SMM = 1, and cleared where SSMM = 1, SMM = 0.

Poll Command

In Poll mode, the INT output functions as it normally does. The microprocessor should ignore this output. This can be accomplished either by not connecting the INT output or by masking interrupts within the microprocessor, thereby disabling its interrupt input. Service to devices is achieved by software using a Poll command.

The Poll command is issued by setting P = "1" in OCW3. The 8259A treats the next $\overline{RD}$ pulse to the 8259A (*i.e.*, $\overline{RD}$ = 0, $\overline{CS}$ = 0) as an interrupt acknowledge, sets the appropriate IS bit if there is a request, and reads the priority level. Interrupt is frozen from $\overline{WR}$ to $\overline{RD}$.

The word enabled onto the data bus during RD is:

D7	*D6*	*D5*	*D4*	*D3*	*D2*	*D1*	*D0*
I	—	—	—	—	W2	W1	W0

W0-W2: Binary code of the highest priority level requesting service.

I: Equal to "1" if there is an interrupt.

This mode is useful if there is a routine command common to several levels so that the $\overline{INTA}$ sequence is not needed (saves ROM space). Another application is to use the poll mode to expand the number of priority levels to more than 64.

Reading the 8259A Status

The input status of several internal registers can be read to update the user information on the system.

If LTIM = '1', an interrupt request will be recognized by a 'high' level on IR Input, and there is no need for an edge detection. The interrupt request must be removed before the EOI command is issued or the CPU interrupts is enabled to prevent a second interrupt from occurring.

In both the edge and level triggered modes, the IR inputs must remain high until after the falling edge of the first $\overline{INTA}$. If the IR input goes low before this time a DEFAULT IR7 will occur when the CPU acknowledges the interrupt. This can be a useful safeguard for detecting interrupts caused by spurious noise glitches on the IR inputs. To implement this feature the IR7 routine is used for "dean up" simply executing a return instruction, thus ignoring the interrupt. If IR7 is needed for other purposes a default IR7 can still be detected by reading the ISR. A normal IR7 interrupt will set the corresponding ISR bit, a default IR7 won't. If a default IR7 routine occurs during a normal IR7 routine, however, the ISR will remain set. In this case, it is necessary to keep track of whether or not the IR7 routine was previously entered. If another IR7 occurs it is a default.

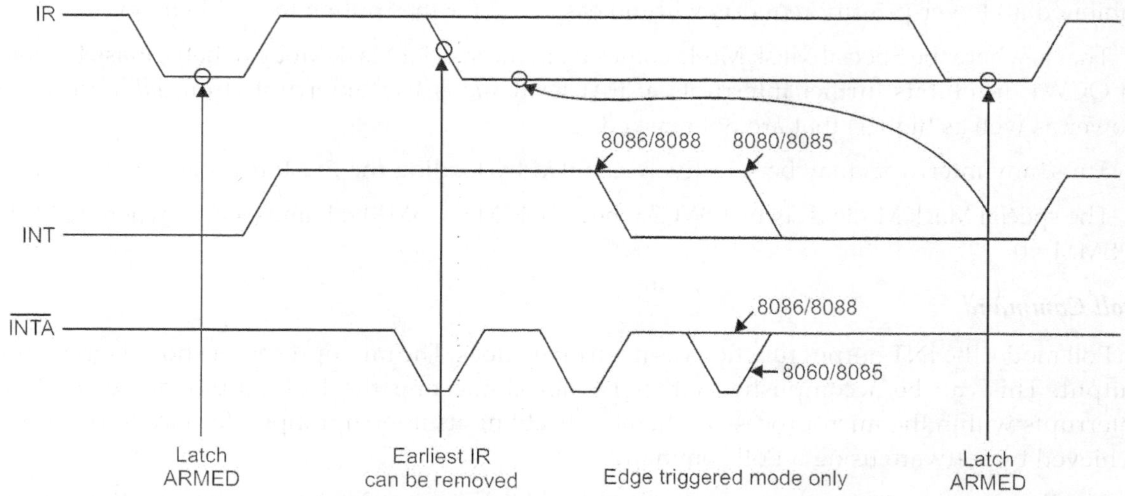

Fig. 6.83 IR Triggering Timing Requirements

The Special Fully Nest Mode

This mode will be used in the case of a big system where cascading is used, and the priority has to be conserved within each slave. In this case the fully nested mode will be programmed to the master (using ICW4). This mode is similar to the normal nested mode with the following exceptions:

(*a*) When an interrupt request from a certain slave is in service this slave is not locked out from the master's priority logic and further interrupt requests from higher priority IR's within the slave will be recognized by the master and win initiate interrupts to the processor. (In the normal nested mode a slave is masked out when its request is in service and no higher requests from the same slave can be serviced.)

(*b*) When exiting the Interrupt Service routine the software has to check whether the interrupt serviced was the only one from that slave. This is done by sending a non-specific End of Interrupt (EOI) command to the slave and then reading its In-Service register and checking for zero. If it is empty, a non-specific EOI can be sent to the master too. If not. no EOI should be sent.

Buffered Mode

When the 8259A is used in a large system where bus driving buffers are required on the data bus and the cascading mode is used, there exists the problem of enabling buffers.

The buffered mode will structure the 8259A to send an enable signal on $\overline{SP}/\overline{EN}$ to enable the buffers. In this mode, whenever the 8259A's data bus outputs are enabled, the $\overline{SP}/\overline{EN}$ output becomes active.

This modification forces the use of software programming to determine whether the 8259A is a master or a slave. Bit 3 in ICW4 programs the buffered mode, and bit 2 in ICW4 determines whether it is a master or a slave.

Cascade Mode

The 8259A can be easily interconnected in a system of one master with up to eight slaves to handle up to 64 priority levels.

The master controls the slaves through the 3 line cascade bus. The cascade bus acts like chip selects to the slaves during the $\overline{\text{INTA}}$ sequence.

In a cascade configuration, the slave interrupt outputs are connected to the master interrupt request inputs. When a slave request line is activated and afterwards acknowledged, the master will enable the corresponding slave to release the device routine address during bytes 2 and 3 of $\overline{\text{INTA}}$ (Byte 2 only for 8086/8088).

The cascade bus lines are normally low and will contain the slave address code from the trailing edge of the first $\overline{\text{INTA}}$ pulse to the trailing edge of the third pulse. Each 8259A in the system must follow a separate initialization sequence and can be programmed to work in a different mode. An EOI command must be issued twice: once for the master and once for the corresponding slave. An address decoder is required to activate the Chip Select (CS) input of each 8259A.

The cascade lines of the Master 8259A are activated only for slave inputs, non-slave inputs leave the cascade line inactive (low).

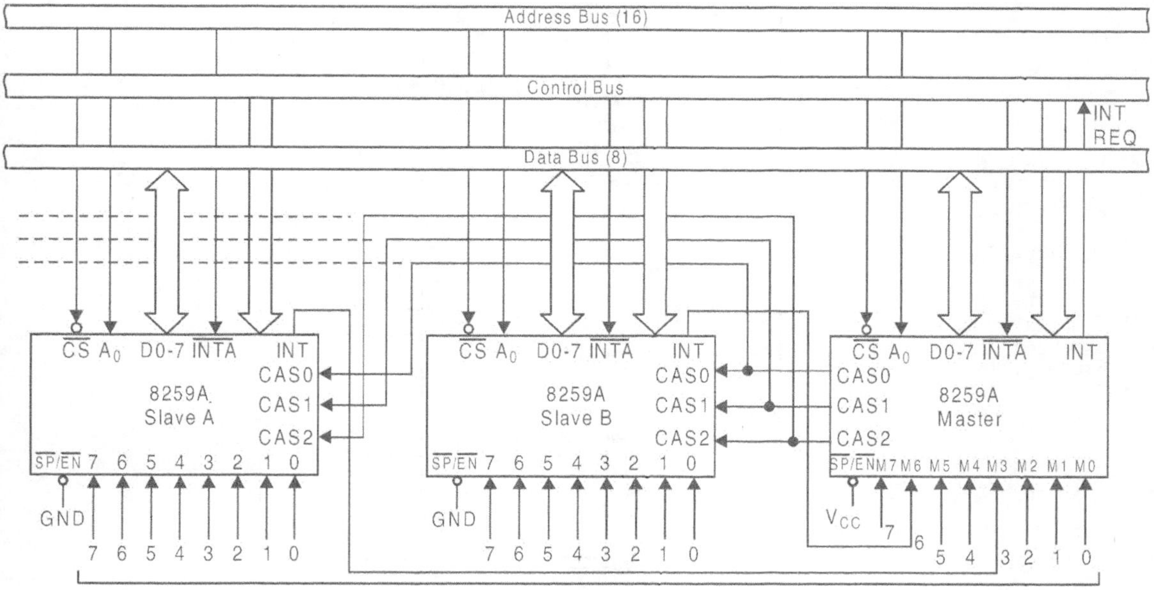

Fig. 6.84 Cascading the 8259A

6.10 8279/8279-5 PROGRAMMABLE KEYBOARD/DISPLAY INTERFACE

- Simultaneous Keyboard Display Operations
- Scanned Keyboard Mode
- Scanned Sensor Mode
- Strobed Input Entry Mode
- 8-Character Keyboard FIFO
- 2-Key Lockout or N-Key Rollover with Contact Debounce
- Dual 8- or 16-Numerical Display
- Single 16-Character Display
- Right or Left Entry 16-Byte Display RAM
- Mode Programmable from CPU
- Programmable Scan Timing
- Interrupt Output on Key Entry

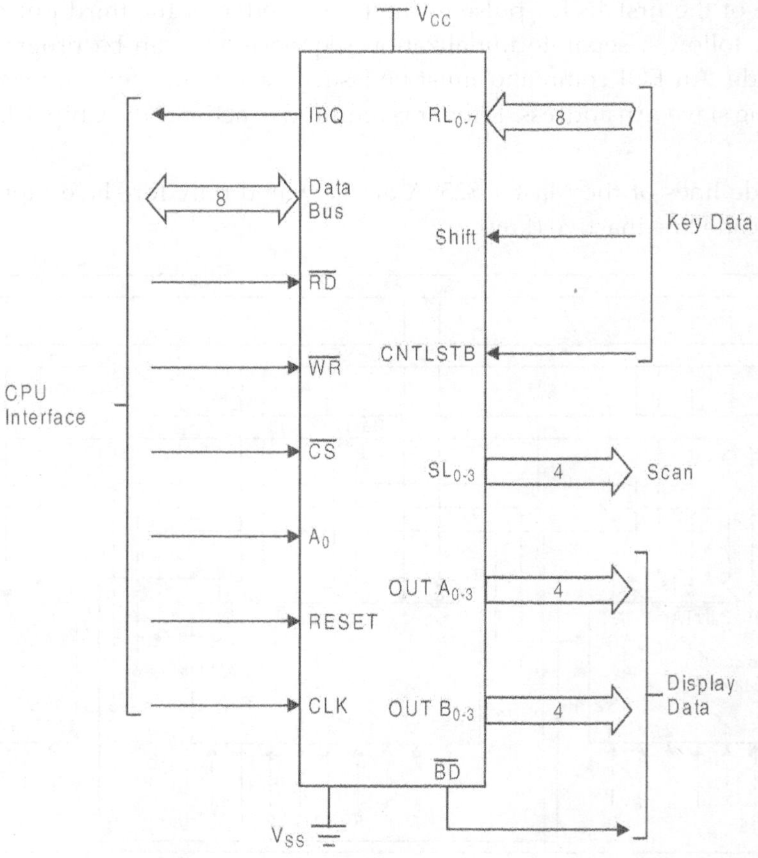

Fig. 6.85 Logic Symbol

The 8279 is a general purpose programmable keyboard and display I/O interface device designed for use with microprocessors. The keyboard portion can provide a scanned interface to a 64-contact key matrix. The keyboard portion will also interface to an array of sensors or a strobed interface keyboard, such as the hall effect and ferrite variety. Key depressions can be 2-key lock-out or N-key rollover. Keyboard entries are debounced and strobed in an 8-character FIFO. If more than 8 characters are entered, overrun status is set. Key entries set the interrupt output line to the CPU.

The display portion provides a scanned display interface for LED, incandescent, and other popular display technologies. Both numeric and alphanumeric segment displays may be used as well as simple indicators. The 8279 has 16x8 display RAM which can be organized into dual 16x4. The RAM can be loaded or interrogated by the CPU. Both right entry, calculator and left entry typewriter display formats are possible. Both read and write of the display RAM can be done with auto-increment of the display RAM address.

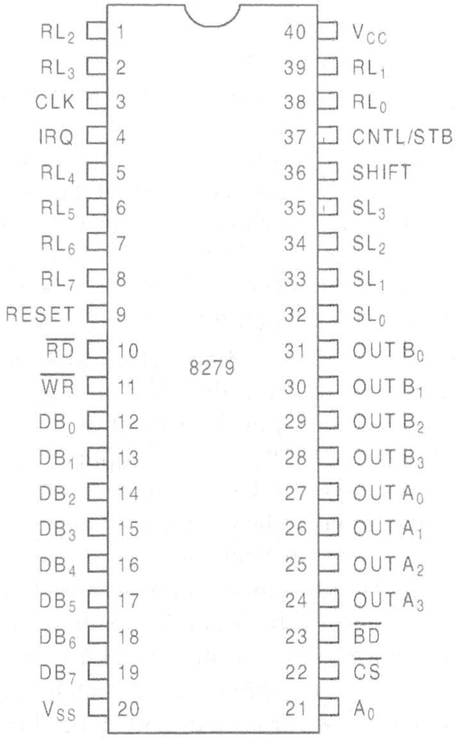

Fig. 6.86 Pin Configuration

6.10.1 Pin Description

The 8279 is packaged in a 40 pin DIP. The following is a functional description of each pin.

Symbol	Pin No.	Name and Function
DB_0-DB_7	19-12	*BI-DIRECTIONAL DATA BUS:* All data and commands between the CPU and the 8279 are transmitted on these lines.
CLK	3	*CLOCK:* Clock from system used to generate internal timing.
RESET	9	*RESET:* A high signal on this pin resets the 8279. After being reset the 8279 is placed in the following mode: 1. 16 8-bit character display—left entry. 2. Encoded scan keyboard—2 key lockout. Along with this the program clock prescaler is set to 31 .
CS	22	*CHIP SELECT:* A low on this pin enables the interface functions to receive or transmit.
A_0	21	*BUFFER ADDRESS:* A high on this line indicates the signals in or out are interpreted as a command or status. A low indicates that they are data.
$\overline{RD}$, $\overline{WR}$	10-11	*INPUT/OUTPUT READ AND WRITE:* These signals enable the data buffers to either send data to the external bus or receive it from the external bus.
IRQ	4	*INTERRUPT REQUEST:* In a keyboard mode, the interrupt line is high when there is data in the FIFO/Sensor RAM. The interrupt line goes low with each FIFO/Sensor RAM read and returns high if there is still information in the RAM. In a sensor mode, the interrupt line goes high whenever a change in a sensor is detected.
V_{SS}, V_{CC}	20, 40	*GROUND AND POWER SUPPLY PINS.*
SL_0-SL_3	32-35	*SCAN LINES:* Scan lines which are used to scan the key switch or sensor matrix and the display digits. These lines can be either encoded (1 of 16) or decoded (1 of 4).
RL_0-RL_7	38, 39, 1,2, 5-8	*RETURN LINE:* Return line inputs which are connected to the scan lines through the keys or sensor switches. They have active internal pullups to keep them high until a switch closure pulls one low. They also serve as an 8-bit input in the Strobed Input mode.
SHIFT	36	*SHIFT:* The shift input status is stored along with the key position on key closure in the Scanned Keyboard modes. It has an active internal pullup to keep it high until a switch closure pulls it low.
CNTL/STB	37	*CONTROL/STROBED INPUT MODE:* For keyboard modes, this line is used as a control input and stored like status on a key closure. The line is also the strobe line that enters the data into the FIFO in the Strobed Input mode. It has an active internal pullup to keep it high until a switch closure pulls it low.
OUT A_0- OUT A_3 OUT B_0- OUT B_3	27-24 31-28	*OUTPUTS:* These two ports are the outputs for the 16 × 4 display refresh registers. The data from these outputs is synchronized to the scan lines (SL_0-SL_3) for multiplexed digit displays. The two 4 bit ports may be blanked independently. These two ports may also be considered as one 8-bit port.
BD	23	*BLANK DISPLAY:* This output is used to blank the display during digit switching or by a display blanking command.

6.10.2 Functional Description

Since data input and display are an integral part of many microprocessor designs, the system designer needs an interface that can control these functions without placing a large load on the CPU. The 8279 provides this function for 8-bit microprocessors.

The 8279 has two sections: keyboard and display. The keyboard section can interface to regular typewriter style keyboards or random toggle or thumb switches. The display section drives alphanumeric displays or a bank of indicator lights. Thus the CPU is relieved from scanning the keyboard or refreshing the display.

The 8279 is designed to directly connect to the microprocessor bus. The CPU can program all operating modes for the 8279. These modes include:

Input Modes

- Scanned Keyboard—with encoded (8 × 8 key keyboard) or decoded (4 × 8 key keyboard) scan lines. A key depression generates a 6-bit encoding of key position. Position and shift and control status are stored in the FIFO. Keys are automatically debounced with 2-key lockout or N-key rollover.
- Scanned Sensor Matrix—with encoded (8 × 8 matrix switches) or decoded (4 × 8 matrix switches) scan lines. Key status (open or closed) stored in RAM addressable by CPU.
- Strobed Input—Data on return lines during control line strobe is transferred to FIFO.

Output Modes

- 8 or 16 character multiplexed displays that can be organized as dual 4-bit or single 8-bit ($B_0 = D_0$, $A_3 = D_7$).
- Right entry or left entry display formats.

Other features of the 8279 include:

- Mode programming from the CPU.
- Clock Prescaler
- Interrupt output to signal CPU when there is keyboard or sensor data available.
- An 8 byte FIFO to store keyboard information.
- 16 byte internal Display RAM for display refresh. This RAM can also be read by the CPU.

6.10.3 Operational Description

The following is a description of the major elements of the 8279 Programmable Keyboard/Display interface device. Refer to the block diagram in Fig. 6.87.

I/O Control and Data Buffers

The I/O control section uses the $\overline{CS}$, A_0, $\overline{RD}$ and $\overline{WR}$ lines to control data flow to and from the various internal registers and buffers. All data flow to and from the 8279 is enabled by $\overline{CS}$. The character of the information, given or desired by the CPU, is identified by A_0. A logic one means the information is a command or status. A logic zero means the information is data, $\overline{RD}$ and $\overline{WR}$ determine the direction of data flow through the Data Buffers. The Data Buffers are bi-directional buffers that connect the internal bus to the external bus. When the chip is not selected ($\overline{CS}$ = 1), the devices are in a high impedance state. The drivers input during $\overline{WS} \bullet \overline{CS}$ and output during $\overline{RD} \bullet \overline{CS}$.

Control and Timing Registers and Timing Control

These registers store the keyboard and display modes and other operating conditions programmed by the CPU. The modes are programmed by presenting the proper command on the data lines with A_0 = 1 and then sending a $\overline{WR}$. The command is latched on the rising edge of $\overline{WR}$. The command is then decoded and the appropriate function is set. The timing control contains the basic timing counter chain. The first counter is a ÷ N prescaler that can be programmed to yield an internal frequency of 100 kHz which gives a 5.1 ms keyboard scan time and a 10.3 ms debounce time. The other counters divide down the basic internal frequency to provide the proper key scan, row scan, keyboard matrix scan, and display scan times.

Scan Counter

The scan counter has two modes. In the encoded mode, the counter provides a binary count that must be externally decoded to provide the scan lines for the keyboard and display. In the decoded mode, the scan counter decodes the least significant 2 bits and provides a decoded 1 of 4 scan. Note that when the keyboard is in decoded scan, so is the display. This means that only the first 4 characters in the Display RAM are displayed.

In the encoded mode, the scan lines are active high outputs. In the decoded mode, the scan lines are active low outputs.

Return Buffers and Keyboard Debounce and Control

The 8 return lines are buffered and latched by the Return Buffers. In the keyboard mode, these lines are scanned, looking for key closures in that row. If the debounce circuit detects a closed switch, it waits about 10 ms to check if the switch remains closed. If it does, the address of the switch in the matrix plus the status of SHIFT and CONTROL are transferred to the FIFO. In the scanned Sensor Matrix modes, the contents of the return lines is directly transferred to the corresponding row of the Sensor RAM (FIFO) each key scan time. In Strobed Input mode, the contents of the return lines are transferred to the FIFO on the rising edge of the CNTL/STB line pulse.

FIFO/Sensor RAM and Status

This block is a dual function 8 × 8 RAM. In Keyboard or Strobed Input modes, it is a FIFO. Each new entry is written into successive RAM positions and each is then read in order of entry. FIFO status keeps track of the number of characters in the FIFO and whether it is full or empty. Too many reads or writes will be recognized as an error. The status can be read by an $\overline{RD}$ with $\overline{CS}$ low and A_0 high. The status logic also provides an IRQ signal when the FIFO is not empty. In Scanned Sensor Matrix mode, the memory is a Sensor RAM. Each row of the Sensor RAM is loaded with the status of the corresponding row of sensor in the sensor matrix. In this mode, IRQ is high if a change in a sensor is detected.

Display Address Registers and Display RAM

The Display Address Registers hold the address of the word currently being written or read by the CPU and the two 4-bit nibbles being displayed. The read/write addresses are programmed by CPU command. They also can be set to auto increment after each read or write. The Display RAM

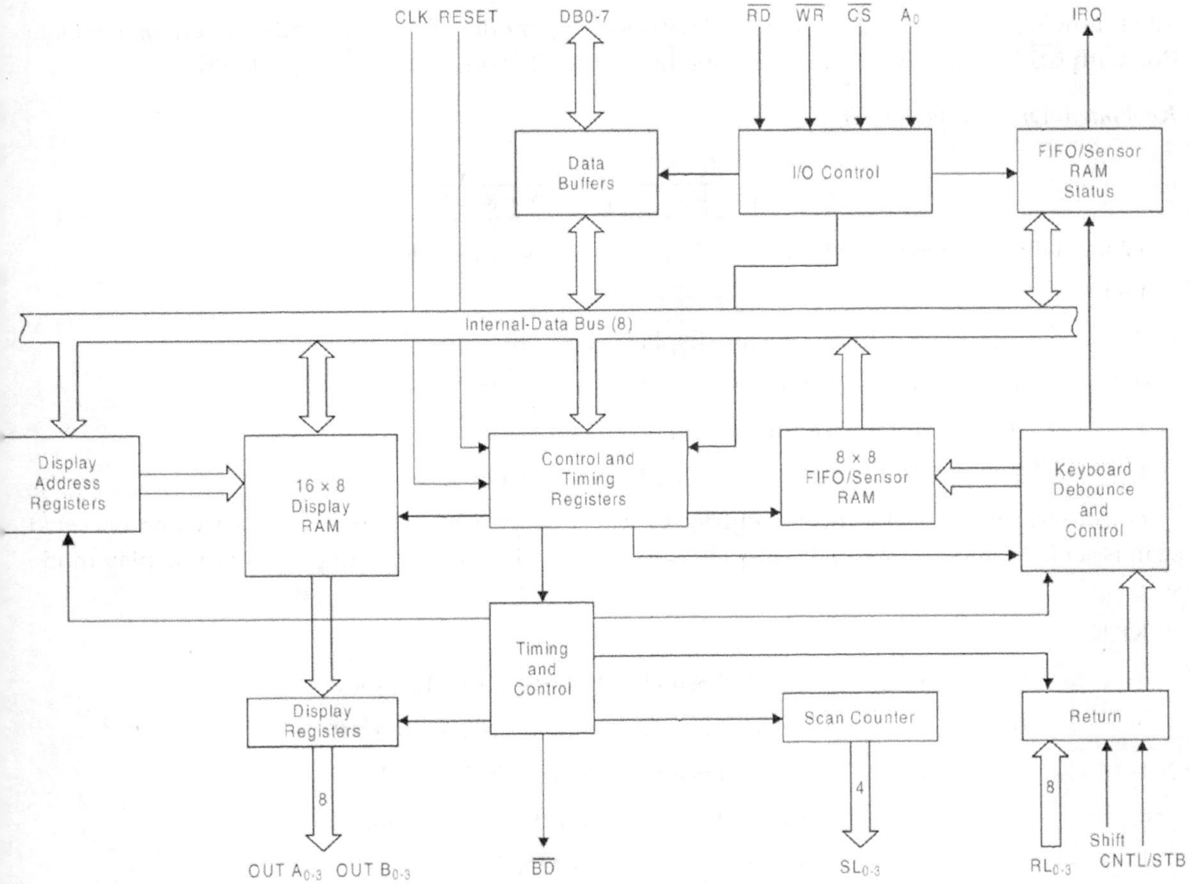

Fig. 6.87 Internal Block Diagram.

can be directly read by the CPU after the correct mode and address is set. The addresses for the A and B nibbles are automatically updated by the 8279 to match data entry by the CPU. The A and B nibbles can be entered independently or as one word, according to the mode that is set by the CPU. Data entry to the display can be set to either left or right entry.

6.10.4 Software Operation

8279 Commands

The following commands program the 8279 operating modes. The commands are sent on the Data Bus with $\overline{CS}$ low and A_0 high and are loaded to the 8279 on the rising edge of $\overline{WR}$.

Keyboard/Display Mode Set

	MSB							LSB
Code:	0	0	0	D	D	K	K	K

Where DD is the Display Mode and KKK is the Keyboard Mode.

DD

0	0	8 8-bit character display—Left entry
0	1	16 8-bit character display—Left entry*
1	0	8 8-bit character display—Right entry
1	1	16 8-bit character display—Right entry

For description of right and left entry, see Interface Considerations. Note that when decoded scan is set in keyboard mode, the display is reduced to 4 characters independent of display mode set.

KKK

0	0	0	Encoded Scan Keyboard—2 Key Lockout*
0	0	1	Decoded Scan Keyboard—2-Key Lockout
1	1	0	Encoded Scan Keyboard—N-Key Rollover
0	1	1	Decoded Scan Keyboard—N-Key Rollover
1	0	0	Encoded Scan Sensor Matrix
1	0	1	Decoded Scan Sensor Matrix
1	1	0	Strobed Input, Encoded Display Scan
1	1	1	Strobed Input, Decoded Display Scan

*Default after reset

Program Clock

Code:	0	0	1	P	P	P	P	P

All timing and multiplexing signals for the 8279 are generated by an internal prescaler. This prescaler divides the external clock (pin 3) by a programmable integer. Bits PPPPP determine the value of this integer which ranges from 2 to 31. Choosing a divisor that yields 100 kHz will give the specified scan and debounce times. For instance, if Pin 3 of the 8279 is being clocked by a 2 MHz signal, PPPPP should be set to 10100 to divide the clock by 20 to yield the proper 100 kHz operating frequency.

Read FIFO/Sensor RAM

Code:

0	1	0	AI	X	A	A	A

X = Don't Care

The CPU sets the 8279 for a read of the FIFO/Sensor RAM by first writing this command. In the Scan Keyboard Mode, the Auto-Increment flag (AI) and the RAM address bits (AAA) are irrelevant. The 8279 will automatically drive the data bus for each subsequent read ($A_0 = 0$) in the same sequence in which the data first entered the FIFO. All subsequent reads will be from the FIFO until another command is issued.

In the Sensor Matrix Mode, the RAM address bits AAA select one of the 8 rows of the Sensor RAM. If the AI flag is set (AI = 1), each successive read will be from the subsequent row of the sensor RAM.

Read Display RAM

Code:

0	1	1	AI	A	A	A	A

The CPU sets up the 8279 for a read of the Display RAM by first writing this command. The address bits AAAA select one of the 16 rows of the Display RAM. If the AI flag is set (A1 = 1), this row address will be incremented after each following read or *write* to the Display RAM. Since the same counter is used for both reading and writing, this command sets the next read or *write* address and the sense of the Auto-Increment mode for both operations.

Write Display RAM

Code:

1	0	0	AI	A	A	A	A

The CPU sets up the 8279 for a write to the Display RAM by first writing this command. After writing the command with $A_0 = 1$, all subsequent writes with $A_0 = 0$ will be to the Display RAM. The addressing and Auto-Increment functions are identical to those for the Read Display RAM. However, this command does not affect the source of subsequent Data Reads; the CPU will read from whichever RAM (Display of FIFO/Sensor) which was last specified. If, indeed, the Display RAM was last specified, the Write Display RAM will, nevertheless, change the next Read location.

Display Write Inhibit/Blanking

Code:

				A	B	A	B
1	0	1	X	IW	IW	BL	BL

The IW Bits can be used to mask nibble A and nibble B in applications requiring separate 4-bit display ports. By setting the IW flag (IW = 1) for one of the ports, the port becomes marked so that entries to the Display RAM from the CPU do not affect that port. Thus, if each nibble is input to a BCD decoder, the CPU may write a digit to the Display RAM without affecting the other digit being displayed. It is important to note that bit B_0 corresponds to bit D_0 on the CPU bus, and that bit A_3 corresponds to bit D_7.

If the user wishes to blank the display, the BL flags are available for each nibble. The last Clear command issued determines the code to be used as a "blank." This code defaults to all zeros after a reset. Note that both BL flags must be set to blank a display formatted with a single 8-bit port.

Clear

Code:

1	1	0	C_D	C_D	C_D	C_F	C_A

The C_D bits are available in this command to clear all rows of the Display RAM to a selectable blanking code as follows:

C_D	C_D	C_D	
	0	X	All Zeros (X = Don't Care)
	1	0	AB = Hex 20 (0010 0000)
	1	1	All Ones

Enable clear display when = 1 (or by C_A = 1)

During the time the Display RAM is being cleared (~160 μs), it may not be written to. The most significant bit of the FIFO status word is set during this time. When the Display RAM becomes available again, it automatically resets.

If the C_F bit is asserted ($C_F = 1$), the FIFO status is cleared and the interrupt output line is reset. Also, the Sensor RAM pointer is set to row 0.

C_A, the Clear All bit, has the combined effect of C_D and C_F; it uses the C_D clearing code on the Display RAM and also clears FIFO status. Furthermore, it resynchronizes the internal timing chain.

End Interrupt/Error Mode Set

Code:

1	1	1	E	X	X	X	X

X = Don't Care

For the sensor matrix modes this command lowers the IRQ line and enables further writing into RAM. (The IRQ line would have been raised upon the detection of a change in a sensor value. This would have also inhibited further writing into the RAM until reset).

For the N-key rollover mode—if the E bit is programmed to "1" the chip will operate in the special Error mode.

Status Word

The status word contains the FIFO status, error, and display unavailable signals. This word is read by the CPU when A_0 is high and $\overline{CS}$ and $\overline{RD}$ are low.

Data Read

Data is read when A_0, $\overline{CS}$ and $\overline{WR}$ are all low. The source of the data is specified by the Read FIFO or Read Display commands. The trailing edge of $\overline{RD}$ will cause the address of the RAM being read to be incremented if the Auto-Increment flag is set. FIFO reads always increment (if no error occurs) independent of AI.

Data Write

Data that is written with A_0, $\overline{CS}$ and $\overline{WR}$ low is always written to the Display RAM. The address is specified by the latest Read Display or Write Display command. Auto-Incrementing on the rising edge of $\overline{WR}$ occurs if AI is set by the latest display command.

6.10.5 Interface Considerations

Scanned Keyboard Mode, 2-Key Lockout

There are three possible combinations of conditions that can occur during debounce scanning. When a key is depressed, the debounce logic is set. Other depressed keys are looked for during the next two scans. If none are encountered, it is a single key depression and the key position is entered into the FIFO along with the status of CNTL and SHIFT lines. If the FIFO was empty, IRQ will be set to signal the CPU that there is an entry in the FIFO. If the FIFO was full, the key will not be entered and the error flag will be set. If another closed switch is encountered, no entry to the FIFO can occur. If all other keys are released before this one, then it will be entered to the FIFO. If this key is released before any other, it will be entirely ignored. A key is entered to the FIFO only once per depression, no matter how many keys were pressed along with it or in what order they were released. If two keys are depressed within the debounce cycle, it is a simultaneous depression. Neither key will be recognized until one key remains depressed alone. The last key will be treated as a single key depression.

Scanned Keyboard Mode, N-Key Rollover

With N-key Rollover each key depression is treated independently from all others. When a key is depressed, the debounce circuit waits 2 keyboard scans and then checks to see if the key is still down. If it is , the key is entered into the FIFO. Any number of keys can be depressed and another can be recognized and entered into the FIFO. If a simultaneous depression occurs, the keys are recognized and entered according to the order the keyboard scan found them.

Scanned Keyboard—Special Error Modes

For N-key rollover mode the user can program a special error mode. This is done by the "End Interrupt/Error Mode Set" command. The debounce cycle and key-validity check are as in normal N-key mode. If during a *single debounce cycle*, two keys are found depressed, this is considered a simultaneous multiple depression, and sets an error flag. This flag will prevent any further writing into the FIFO and will set interrupt (if not yet set). The error flag could be read in this mode by reading the FIFO STATUS word. (See "FIFO STATUS" for further details.) The error flag is reset by sending the normal CLEAR command with CF = 1.

Sensor Matrix Mode

In Sensor Matrix mode, the debounce logic is inhibited. The status of the sensor switch is inputted directly to the Sensor RAM. In this way the Sensor RAM keeps an image of the state of the switches in the sensor matrix. Although debouncing is not provided, this mode has the advantage that the CPU knows how long the sensor was closed and when it was released. A keyboard mode can only indicate a validated closure. To make the software easier, the designer should functionally group the sensors by row since this is the format in which the CPU will read them.

The IRQ line goes high if any sensor value change is detected at the end of a sensor matrix scan. The IRQ line is cleared by the first data read operation if the Auto-Increment flag is set to zero, or by the End Interrupt command if the Auto-Increment flag is set to one.

Note: Multiple changes in the matrix Addressed by (SL_{0-3} = 0) may cause multiple interrupts. (SL_0 = 0 in the Decoded Mode.) Reset may cause the 8279 to see multiple changes.

Data Format

In the Scanned Keyboard mode, the character entered into the FIFO corresponds to the position of the switch in the keyboard plus the status of the CNTL and SHIFT lines (non-inverted). CNTL is the MSB of the character and SHIFT is the next most significant bit. The next three bits are from the scan counter and indicate the row the key was found in. The last three bits are from the column counter and indicate to which return line the key was connected.

MSB			LSB
CNTL	SHIFT	SCAN	RETURN

Scanned Keyboard Data Format

In Sensor Matrix mode, the data on the return lines is entered directly in the row of the Sensor RAM that corresponds to the row in the matrix being scanned. Therefore, each switch position maps directly to a Sensor RAM position. The SHIFT and CNTL inputs are ignored in this mode. Note that switches are not necessarily the only thing that can be connected to the return lines in this mode. Any logic that can be triggered by the scan lines can enter data to the return line inputs. Eight multiplexed input ports could be tied to the return lines and scanned by the 8279.

MSB							LSB
RL_7	RL_6	RL_5	RL_4	RL_3	RL_2	RL_1	RL_0

In Strobed Input mode, the data is also entered to the FIFO from the return lines. The data is entered by the rising edge of a CNTL/STB line pulse. Data can come from another encoded keyboard or simple switch matrix. The return lines can also be used as a general purpose strobed input.

Display

Left Entry. Left Entry mode is the simplest display format in that each display position directly corresponds to a byte (or nibble) in the Display RAM. Address 0 in the RAM is the left-most display character and address 15 (or address 7 in 8 character display) is the right most display character. Entering characters from position zero causes the display to fill from the left. The 17th (9th) character is entered back in the left most position and filling again proceeds from there.

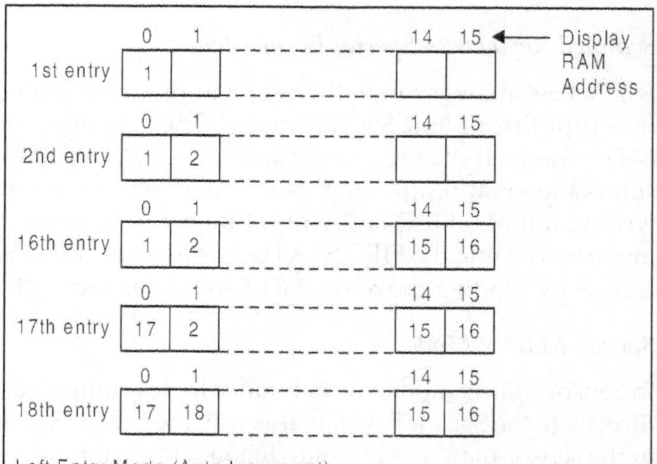

Left Entry Mode (Auto Increment)

Right Entry. Right entry is the method used by most electronic calculators. The first entry is placed in the right most display character. The next entry is also placed in the right most character after the display is shifted left one character. The left most character is shifted off the end and is lost.

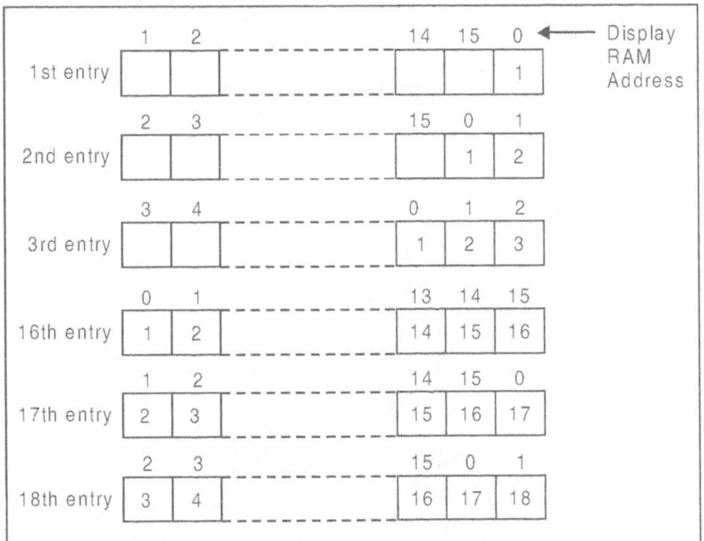

Right Entry Mode (Auto Increment)

Note that now the display position and register address do not correspond. Consequently, entering a character to an arbitrary position in the Auto Increment mode may have unexpected results. Entry starting at Display RAM address 0 with sequential entry is recommended.

Auto Increment

In the Left Entry mode, Auto Incrementing causes the address where the CPU will next write to be incremented by one and the character appears in the next location. With non-Auto Incrementing the entry is both to the same RAM address and display position. Entry to an arbitrary address in the Auto Increment mode has no undesirable side effects and the result is predictable:

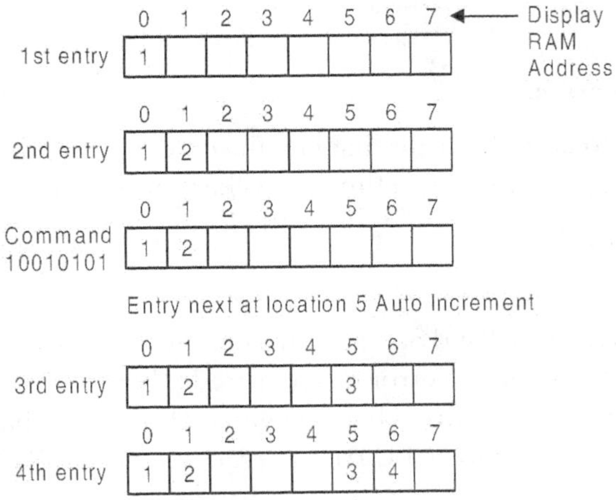

Left Entry Mode
(Auto Increment)

In the Right Entry mode, Auto Incrementing and non Incrementing have the same effect as in the Left Entry except if the address sequence is interrupted.

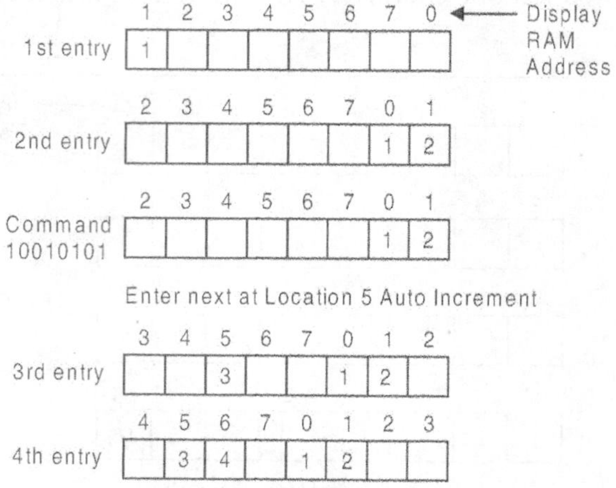

Right Entry Mode
(Auto Increment)

Starting at an arbitrary location operates as shown below.

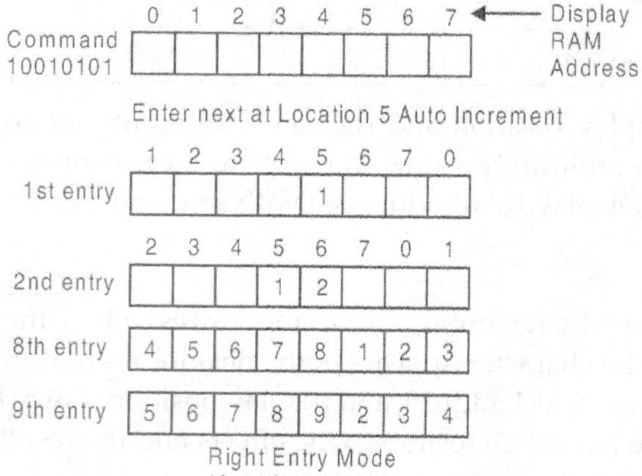

Right Entry Mode
(Auto Increment)
Enter appears to be from the initial entry point

8/16 Character Display Formats

If the display mode is set to an 8 character display, the on duty-cycle is double what it would be for a 16 character display (*e.g.*, 5.1 ms scan time for 8 characters *vs.* 10.3 ms for 16 characters with 100 kHz internal frequency).

G. FIFO Status

FIFO status is used in the Keyboard and Strobed Input modes to indicate the number of characters in the FIFO and to indicate whether an error has occurred. There are two types of errors possible: overrun and underrun. Overrun occurs when the entry of another character into a full FIFO is attempted. Underrun occurs when the CPU tries to read an empty FIFO.

The FIFO status word also has a bit to indicate that the Display RAM was unavailable because a Clear Display or Clear All command had not completed its clearing operation.

In a Sensor Matrix mode, a bit is set in the FIFO status word to indicate that at least one sensor closure indication is contained in the Sensor RAM.

In Special Error Mode, the S/E bit is showing the error flag and serves as an indication to whether a simultaneous multiple closure error has occurred.

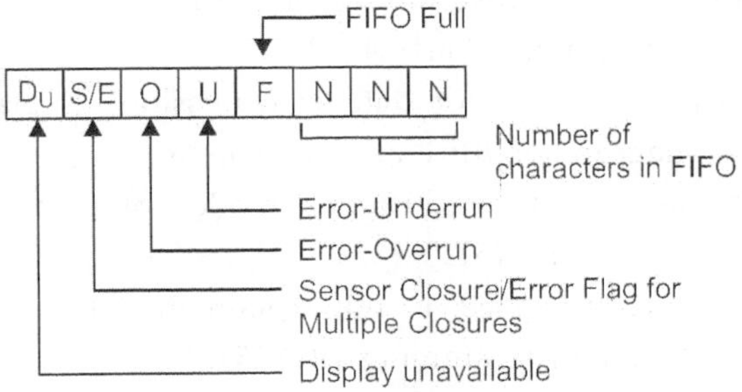

Fig. 6.88 FIFO Status Word

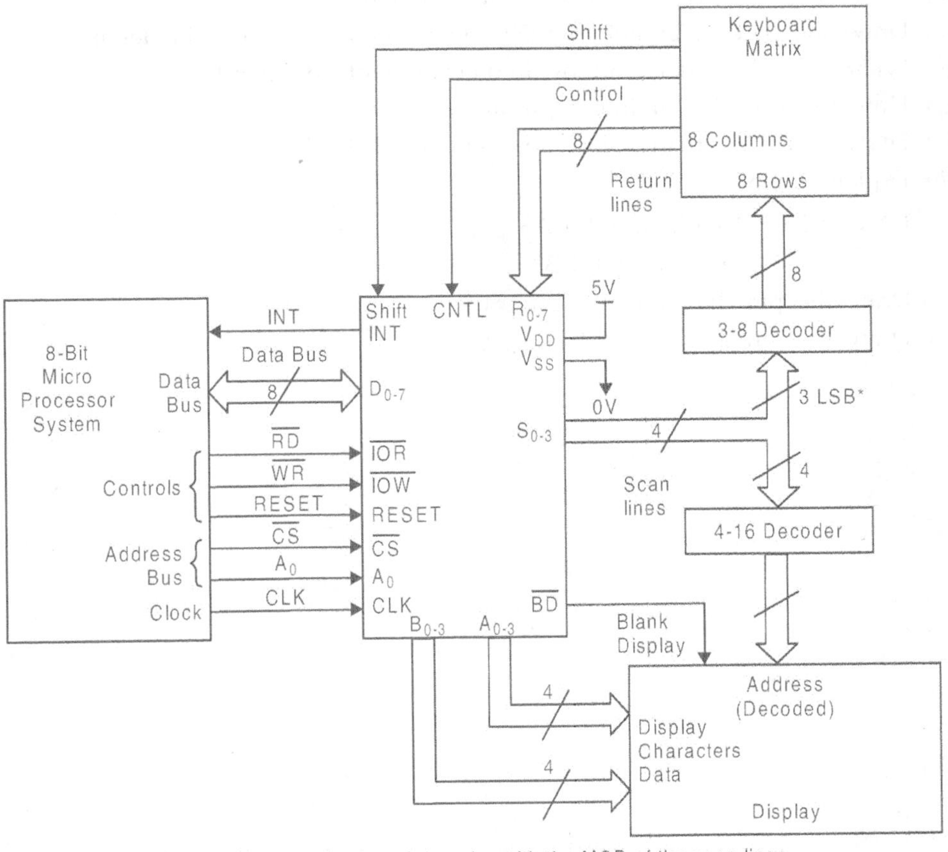

*Do not drive the keyboard decoder with the MSB of the scan lines.

Fig. 6.89 System Block Diagram

REVIEW QUESTIONS

1. What do you understand by Direct Memory Access? What are the advantages of DMA?
2. (*a*) Explain the Functional Block Diagram of 8257 in detail.
   (*b*) Explain the DMA operation in detail.
3. (*a*) Explain in Functional Block Diagram and Pin Diagram of 8255.
   (*b*) Explain the different operating modes of 8255.
   (*c*) Explain one 'or' two applications of 8255.
4. (*a*) What are the various modes of data transmission? Explain.
   (*b*), Draw the block diagram of 8151 and explain it.
   (*c*) Explain asynchronous and synchronous modes of transmission by 8251.
   (*d*) Draw the status word of 8251.
5. (*a*) Draw the block diagram of 8254 and explain in detail.
   (*b*) Write a short note on programming of 8254.
   (*c*) Explain—Read, Write and Read-Back Command for 8254.
   (*d*) What are the various Modes of operation for 8254?
6. (*a*) Draw the Block diagram of 8259 and explain each block in detail.
   (*b*) What do you understand by initialization of command words?
   (*c*) How the priorties of interrupts are resolved?
7. (*a*) Draw and explain the Block diagram of 8279.
   (*b*) Explain for 8279—

   CNTL/STB, $\overline{\text{BD}}$, OUT A0-OUT A3

   OUT B0-OUT B3
8. (*a*) Draw the pin diagram of ADC 0808.
   (*b*) Draw the pin diagram of DAC 0808.

INDEX